NUMERICAL MATHEMATICS AND SCIENTIFIC COMPUTATION

Modern Fortran Explained

Incorporating Fortran 2023

Michael Metcalf

Formerly of CERN, Geneva, Switzerland

John Reid

JKR Associates, Oxford, UK

Malcolm Cohen

Principal technical consultant, The Numerical Algorithms Group Ltd, Oxford, UK

and

Reinhold Bader

Scientific staff member at Leibniz Supercomputing Centre (LRZ) of the Bavarian Academy of Sciences, Germany

OXFORD
UNIVERSITY PRESS

Great Clarendon Street, Oxford, OX2 6DP,
United Kingdom

Oxford University Press is a department of the University of Oxford.
It furthers the University's objective of excellence in research, scholarship,
and education by publishing worldwide. Oxford is a registered trade mark of
Oxford University Press in the UK and in certain other countries

First edition published 1987 as *Fortran 8x Explained*
Second edition published 1989
Third edition published in 1990 *as Fortran 90 Explained*
Fourth edition published 1996 *as Fortran 90/95 Explained*
Fifth edition published 1999
Sixth edition published 2004 *as Fortran 95/2003 Explained*
Seventh edition published 2011 *as Modern Fortran Explained*
Eighth edition published *as Modern Fortran Explained: Incorporating Fortran 2018*
This edition published 2023

Published in the United States of America by Oxford University Press
198 Madison Avenue, New York, NY 10016, United States of America

British Library Cataloguing in Publication Data
Data available

Library of Congress Control Number: 2023946356

ISBN 9780198876571
ISBN 9780198876588 (pbk.)

DOI: 10.1093/oso/9780198876571.001.0001

Printed and bound by
CPI Group (UK) Ltd, Croydon, CR0 4YY

MIX
Paper | Supporting
responsible forestry
FSC
www.fsc.org   FSC® C013604

Preface

Fortran remains one of the principal languages used in the fields of scientific, numerical, and engineering programming, and a series of revisions to the standard defining successive versions of the language has progressively enhanced its power and kept it competitive with several generations of rivals.

Beginning in 1978, the technical committee responsible for the development of Fortran standards, X3J3 (now PL22.3 but still informally called J3), laboured to produce a new, much-needed modern version of the language, Fortran 90. Its purpose was to 'promote portability, reliability, maintainability, and efficient execution ... on a variety of computing systems'. That standard was published in 1991, and work began in 1993 on a minor revision, known as Fortran 95. Subsequently, and with the same purpose, a further major upgrade to the language was prepared by J3 and the international committee, WG5. That revision, which included object-oriented programming features, is now known as Fortran 2003. This was followed by a further revision, Fortran 2008, including coarrays, followed by a minor revision, Fortran 2018, including further coarray features; and, most recently, by a minor revision, Fortran 2023. Once again, we have prepared a definitive informal description of the language that this latest standard defines. This continues the series of editions of this book – the two editions of *Fortran 8x Explained* that described the two drafts of the standard (1987 and 1989), *Fortran 90 Explained* that described the Fortran 90 standard (1990), two editions of *Fortran 90/95 Explained* that included Fortran 95 as well (1996 and 1999) and *Fortran 95/2003* (2004), with its added chapters on Fortran 2003. In that endeavour, a third co-author was welcomed.

The first edition of *Modern Fortran Explained* (2011) described Fortran 2003 and included chapters on Fortran 2008; this was followed by a second edition (2018) in which all the additional material was consolidated into a description of Fortran 2008 as the basic language, with an additional four chapters describing the new features of Fortran 2018.

In this third edition the basic language becomes Fortran 2018. An initial chapter sets out the background to the work on new standards, Chapters 2 to 21 describe Fortran 2018 in a manner suitable both for grasping the implications of its features and for writing programs, and Chapters 22 and 23 describe the Fortran 2023 enhancements. Once again, we welcome a new co-author.

Some knowledge of programming concepts is assumed. In order to reduce the number of forward references and also to enable, as quickly as possible, useful programs to be written based on material already absorbed, the order of presentation does not always follow that of the standard. In particular, we have chosen to defer to appendices the description of features that are officially labelled as redundant (some of which have been deleted from the standard) and other features whose use we deprecate. They may be encountered in old programs, but are not needed in new ones.

Note that, apart from a small number of deletions, each of the languages Fortran 77, Fortran 90, Fortran 95, Fortran 2003, Fortran 2008, and Fortran 2018 is a subset of its successor.

In order to make the book a complete reference work it concludes with four appendices. They contain, successively, a description of various features whose use we deprecate and do not describe in the body of the book, a description of obsolescent and deleted features, extended examples illustrating the use of object orientation and of parallelism, and solutions to most of the exercises.

It is our hope that this book, by providing complete descriptions of Fortran 2018 and Fortran 2023, will continue the helpful role that earlier editions played for the corresponding versions of the standard, and that it will serve as a long-term reference work for the modern Fortran programming language.

Conventions used in this book

Fortran displayed text is set in typewriter font:

```
integer :: i, j
```

A line consisting of vertical dots (⋮):

```
subroutine sort
    ⋮
end subroutine sort
```

indicates omitted lines, and an ellipsis (. . .):

```
data_distance = ...
```

indicates omitted code.

Informal BNF terms are in italics:

```
if (scalar-logical-expr) action-stmt
```

Square brackets in italics indicate optional items:

```
end if [name]
```

here an ellipsis represents an arbitrary number of repeated items:

```
[ case selector [name]]
        block] ...
```

and *item-list* represents a comma-separated list with at least one item:

```
item [, item] ...
```

The italic letter *b* is sometimes used to signify a blank character.

Corrections to any significant errors detected in this book will be made available in the file *edits.pdf* accessible through the bullet for this book at

```
http://www.oup.com/fortran2023
```

Contents

1. Whence Fortran?

1.1 Introduction

This book is concerned with the Fortran 2018 and Fortran 2023 programming languages, setting out a reasonably concise description of the whole language. The form chosen for its presentation is that of a textbook intended for use in teaching or learning the language.

After this introductory chapter, the chapters are written so that simple programs can be coded after the first three (on language elements, expressions and assignments, and control) have been read. Successively more complex programs can be written as the information in each subsequent chapter is absorbed. Chapter 5 describes the important concept of the module and the many aspects of procedures. Chapters 6 and 7 complete the description of dynamic data and the powerful array features, Chapter 8 considers the details of specifying data objects and derived types, and Chapter 9 details the intrinsic procedures. Chapters 10, 11, and 12 cover all the input/output features in such a way that the reader can also approach this more difficult area feature by feature, but always with a useful subset already covered. Many, but by far not all, of the features described in Chapters 2 to 12 were available in Fortran 95.

Chapter 13 deals with parameterized data types, Chapter 14 with procedure pointers, Chapter 15 with object-oriented programming, and Chapter 16 with submodules. Coarrays, which are important for parallel processing, are described in Chapters 17 and 18. Chapter 19 describes floating-point exception handling, and Chapters 20 and 21 deal with interoperability with the C programming language. Finally, Chapter 23 describes the additional enhancements brought to the language by Fortran 2023. None of the features of Chapters 13 to 23 were available prior to Fortran 2003 or, in the cases of submodules and coarrays, Fortran 2008.

In Appendices A and B we describe features that are redundant in the language. Those of Appendix A are still fully part of the standard but their use is deprecated by us, while those of Appendix B are designated as obsolescent or deleted by the standard itself.

This introductory chapter has the task of setting the scene for those that follow. Section 1.2 presents the early history of Fortran, starting with its introduction over sixty years ago. Section 1.3 continues with the development of the Fortran 90 standard, summarizes its important new features, and outlines how standards are developed; Section 1.4 looks at the mechanism that has been adopted to permit the language to evolve. Sections 1.5 to 1.8 consider the development of Fortran 95 and its extensions, then of Fortran 2003, Fortran 2008, Fortran 2018, and Fortran 2023. The final section considers the requirements on programs and processors for conformance with the standard.

Modern Fortran Explained, 3rd Edition. M. Metcalf, J. Reid, M. Cohen, and R. Bader. Oxford University Press (2024). © M. Metcalf, J. Reid, M. Cohen, and R. Bader (2024). DOI 10.1093/oso/9780198876571.001.0001

1.2   Fortran's early history

Programming in the early days of computing was tedious in the extreme. Programmers required a detailed knowledge of the instructions, registers, and other aspects of the central processing unit (CPU) of the computer for which they were writing code. The **source code** itself was written in a numerical notation.[1] In the course of time mnemonic codes were introduced, a form of coding known as **machine** or **assembly code**. These codes were translated into the instruction words by programs known as assemblers. In the 1950s it became increasingly apparent that this form of programming was highly inconvenient, although it did enable the CPU to be used in a very efficient way.

These difficulties spurred a team led by John Backus of IBM to develop one of the earliest high-level languages, Fortran, which was first released commercially in 1957. Their aim was to produce a language which would be simple to understand but almost as efficient in execution as assembly language. In this they succeeded beyond their wildest dreams. The language was indeed simple to learn, as it was possible to write mathematical formulae almost as they are usually written in mathematical texts. (The name Fortran is a contraction of 'formula translation'.) This enabled working programs to be written more quickly than before, for only a small loss in efficiency, as a great deal of care was devoted to the generation of fast object code.

But Fortran was revolutionary as well as innovatory. Programmers were relieved of the tedious burden of using assembler language, and computers became accessible to any scientist or engineer willing to devote a little effort to acquiring a working knowledge of Fortran. No longer was it necessary to be an expert on computers to be able to write application programs.

Fortran spread rapidly as it fulfilled a real need. Inevitably, dialects of the language developed, which led to problems in exchanging programs between computers, and so in 1966 the then American Standards Association (later the American National Standards Institute, ANSI) brought out the first ever standard for a programming language, now known as Fortran 66.

But the proliferation of dialects remained a problem after the publication of the 1966 standard. There was widespread implementation of features which were essential for large-scale programs, but which were ignored by the standard. Different compilers implemented such features in different ways. These difficulties were partially resolved by the publication of a new standard in 1978, known as Fortran 77, which included several new features that were based on vendor extensions or preprocessors.

1.3   The drive for the Fortran 90 standard

After thirty years' existence, Fortran was far from being the only programming language available on most computers, but Fortran's superiority had always been in the area of numerical, scientific, engineering, and technical applications and so, in order that it be brought properly up to date, the ANSI-accredited technical committee X3J3 (subsequently

---

[1] An exhaustive history of the language is given in *The History of the Fortran Programming Language: Abstracting Away the Machine*, Mark Jones Lorenzo, SE Books, Philadelphia, PA, 2019 (ISBN 9781082395949).

known as J3 and now formally as PL22.3), working as a development body for the ISO committee ISO/IEC JTC1/SC22/WG5, once again prepared a new standard, now known as Fortran 90. We will use the abbreviations J3 and WG5 for these two committees.

J3 is composed of representatives of computer hardware and software vendors, users, and academia. It is now accredited to NCITS (National Council for Information Technology Standards). J3 acts as the development body for the corresponding international group, WG5, consisting of international experts responsible for recommending the choice of features. J3 maintains other close contacts with the international community by welcoming foreign members, including the present authors over many years.

What were the justifications for continuing to revise the definition of the Fortran language? As well as standardizing vendor extensions, there was a need to modernize it in response to requests from users, noting the developments in language design that had been exploited in other languages, such as APL, Algol 68, Pascal, Ada, C, and C++. Here, J3 could draw on the obvious benefits of concepts like data hiding. In the same vein was the need to begin to provide an alternative to dangerous storage association, to abolish the rigidity of the outmoded source form, and to improve further on the regularity of the language, as well as to increase the safety of programming in the language and to tighten the conformance requirements. Further, given its widespread use, the military standard MIL-STD-1753 needed to be incorporated into the standard. To preserve the vast investment in Fortran 77 code, the whole of Fortran 77 was retained as a subset. However, unlike the previous standard, which resulted almost entirely from an effort to standardize existing practices, the Fortran 90 standard was much more a development of the language, introducing features which were new to Fortran but were based on experience in other languages.

The main features of Fortran 90 were, first and foremost, the array language and data abstraction. The former is built on whole-array operations and assignments, array sections, intrinsic procedures for arrays, and dynamic storage. It was designed with optimization in mind. Data abstraction is built on modules and module procedures, derived data types, operator overloading, and generic interfaces, together with pointers. Also important were the new facilities for numerical computation, including a set of numeric inquiry functions, the parameterization of the intrinsic types, new control constructs such as `select case` and new forms of `do`, internal and recursive procedures, optional and keyword arguments, improved input/output facilities, and many new intrinsic procedures. Last but not least were the new free source form, an improved style of attribute-oriented specifications, the `implicit none` statement, and a mechanism for identifying redundant features for subsequent removal from the language. The requirement on compilers to be able to identify, for example, syntax extensions, and to report why a program has been rejected, were also significant. The resulting language was not only a far more powerful tool than its predecessor, but a safer and more reliable one too. Storage association, with its attendant dangers, was not abolished, but rendered unnecessary. Indeed, experience showed that compilers detected errors far more frequently than before, resulting in a faster development cycle. The array syntax and recursion also allowed quite compact code to be written, a further aid to safe programming.

1.4   Language evolution

The procedures under which J3 works require that a period of notice be given before any existing feature is removed from the language. This means, in practice, a minimum of one revision cycle, which for Fortran is at least five years. The need to remove features is evident: if the only action of the committee is to add new features, the language will become grotesquely large, with many overlapping and redundant items. The solution finally adopted by J3 was to publish as an appendix to a standard a set of two lists showing which items have been removed or are candidates for eventual removal.

One list contains the **deleted features**, those that have been removed. Since Fortran 90 contained the whole of Fortran 77, this list was empty for Fortran 90 and Fortran 2023. It was also empty for Fortran 2008 but was not for the others.

The second list contains the **obsolescent features**, those considered to be outmoded and redundant, and which are candidates for deletion in the next revision. This list was empty for Fortran 2003 and Fortran 2023, but was not for the others.

Apart from carriage control (Appendix B.2), the deleted features are still being supported by most compilers because of the demand for old, tried and tested programs to continue to work. Thus, the concept of obsolescence is really not working as intended, but at least it gives a clear signal that certain features are outmoded, and should be avoided in new programs and not taught to new programmers.

1.5   Fortran 95

Following the publication of the Fortran 90 standard in 1991, two further significant developments concerning the Fortran language occurred. The first was the continued operation of the two Fortran standards committees, J3 and WG5, and the second was the founding of the High Performance Fortran Forum (HPFF).

Early on in their deliberations, the standards committees decided on a strategy whereby a minor revision of Fortran 90 would be prepared by the mid-1990s and a major revision by about the year 2000.

The HPFF was set up in an effort to define a set of extensions to Fortran to make it possible to write portable code when using parallel computers for handling problems involving large sets of data that can be represented by regular grids. This version of Fortran was to be known as High Performance Fortran (HPF), and it was quickly decided, given the array features of Fortran 90, that it, and not Fortran 77, should be its base language. The final form of HPF[2] was a superset of Fortran 90, the main extensions being in the form of directives that take the form of Fortran 90 comment lines, and are thus recognized as directives only by an HPF processor. However, it also became necessary to add some additional syntax, as not all the desired features could be accommodated in the form of such directives.

The work of J3 and WG5 went on at the same time as that of the HPFF, and the bodies liaised closely. It was evident that, in order to avoid the development of divergent dialects of Fortran, it would be desirable to include the new syntax defined by the HPFF in Fortran 95 and, indeed, the HPF features were its most significant new features. Beyond this, a small

---

[2]*The High Performance Fortran Handbook*, C. Koebel et al., MIT Press, Cambridge, MA, 1994.

number of other pressing but minor language changes were made, mainly based on experience with the use of Fortran 90.

The details of Fortran 95 were finalized in 1995, and the new ISO standard, replacing Fortran 90, was adopted in 1997, following successful ballots, as ISO/IEC 1539-1:1997.

Further, in 1995, WG5 decided that three features,

- permitting allocatable arrays as structure components, dummy arguments, and function results;

- handling floating-point exceptions; and

- interoperability with C,

were so urgently needed in Fortran that it established development bodies to develop 'Technical Reports of Type 2'. The intent was that the material of these technical reports be implemented as extensions of Fortran 95 and integrated into the next revision of the standard, apart from any defects found in the field. It was essentially a beta-test facility for a language feature. In the event, the first two reports were completed, but not the third. The first was widely implemented in Fortran 95 compilers. The features of the two reports were incorporated in Fortran 2003. While there was no report for the third, features for interoperability with C were included in Fortran 2003. In this book, the three topics are included in Chapters 6, 19, and 20, respectively.

1.6 Fortran 2003

The next full language revision was published as ISO/IEC 1539-1:2004 in November 2004 and is known as Fortran 2003, since the details were finalized in 2003. Unlike Fortran 95, it was a major revision, its main new features being:

- Derived type enhancements: parameterized derived types, improved control of accessibility, improved structure constructors, and finalizers.

- Object-oriented programming support: type extension and inheritance, polymorphism, dynamic type allocation, and type-bound procedures.

- Data manipulation enhancements: allocatable components, deferred type parameters, the `volatile` attribute, explicit type specification in array constructors and `allocate` statements, pointer enhancements, extended initialization expressions (now called constant expressions), and enhanced intrinsic procedures.

- Input/output enhancements: asynchronous transfer, stream access, user-specified transfer operations for derived types, user-specified control of rounding during format conversions, named constants for preconnected units, the `flush` statement, regularization of keywords, and access to error messages.

- Procedure pointers.

- Support of IEC 60559 (IEEE 754) exceptions.

- Interoperability with the C programming language.

- Support for international usage: access to ISO 10646 four-byte characters and choice of decimal point or comma in numeric formatted input/output.

- Enhanced integration with the host operating system: access to command line arguments, environment variables, and processor error messages.

It was decided in 2001 that a feature to address problems with the maintenance of very large modules was too important to delay for the next revision and should be the subject of a further Technical Report. This was completed in 2005 as ISO/IEC TR 19767:2005 and defines submodules, see Chapter 16. Unfortunately, there were few early implementations.

The features for interoperability in Fortran 2003 provide a mechanism for sharing data between Fortran and C, but it was still necessary for users to implement a translation layer for a procedure with an argument that is optional, a pointer, allocatable, or of assumed shape. It was decided that adding features to address this was too important to wait for the next revision and so was made the subject of another Technical Report. Work began in 2006 but proved more difficult than expected and was not completed until 2012, so was not ready for incorporation in the next standard, Fortran 2008. It appeared as ISO/IEC TS 29113:2012 and is incorporated in Fortran 2018. ISO/IEC had meanwhile renamed these reports as Technical Specifications, but the intention was unchanged. We describe these features in Chapter 21.

1.7   Fortran 2008

Notwithstanding the fact that compilers for Fortran 2003 were very slow to appear, the standardization committees thought fit to plunge on with another standard. It was intended to be a small revision, but it became apparent that parallel programming was going to be universal, so coarrays were added and were its single most important new feature. Further, the `do concurrent` form of loop control and the `contiguous` attribute were introduced. The submodule feature, promised by the Technical Report mentioned in Section 1.6, was added. Other new features included various data enhancements, enhanced access to data objects, enhancements to input/output and to execution control, and more intrinsic procedures, in particular for bit processing. Fortran 2008 was published as ISO/IEC 1539-1:2010 in October 2010.

To avoid the extension of Fortran 2003 to Fortran 2008 becoming very large and to speed up agreement on all the details, WG5 decided to defer some of the coarray features to a Technical Specification. As for the Technical Reports, it was intended that the material would be implemented as extensions of Fortran 2008 and integrated into the next revision, apart from any defects found in the field. Work did not start until 2011 and was completed in 2015 as ISO/IEC TS 18508:2015.

1.8   Fortran 2018

Fortran 2018 is a less major extension of Fortran 2008. The features were chosen in 2015 and the language was originally called Fortran 2015. It was published in 2018 as ISO/IEC

1539-1:2018. It was decided to follow the more usual practice of naming the version by its date of publication, which was November 2018; this gives the impression that the gap from the previous version is larger than it really is.

The main changes concerned the features defined in the Technical Specifications for further coarray features, incorporated in Chapters 17 and 18, and further interoperability with C, see Chapter 21. It is the inclusion of these that led to this being a less minor extension than was originally intended.

Beyond these two major enhancements, there were quite a large number of small improvements, described throughout the book.

1.9   Fortran 2023

Fortran 2023 is a minor extension of Fortran 2018. It was purposely kept small to give vendors the opportunity to ensure that their compilers were up to date with the standard. The focus was on correcting errors and omissions from Fortran 2018 and adding small features that had been requested by users. The changes are described in Chapters 22 and 23.

1.10   Conformance

The standards are almost exclusively concerned with the rules for programs rather than for processors. A processor is required to accept a **standard-conforming program** and to interpret it according to the standard, subject to limits that the processor may impose on the size and complexity of the program. The processor is allowed to accept further syntax and to interpret relationships that are not specified in the standard, provided they do not conflict with the standard. In many places in this book we say '... is not permitted'. By this we mean that it is not permitted in a standard-conforming program. An implementation may nevertheless permit it as an extension. Of course, the programmer must avoid such syntax extensions if portability is desired.

The interpretation of some of the standard syntax is **processor dependent**, that is, may vary from processor to processor. For example, the set of characters allowed in character strings is processor dependent. Care must be taken whenever a processor-dependent feature is used in case it leads to the program not being portable to a desired processor.

A drawback of the Fortran 77 standard was that it made no statement about requiring processors to provide a means to detect any departure from the allowed syntax by a program, as long as that departure did not conflict with the syntax rules defined by the standard. The new standards are written in a different style. The syntax rules are expressed in a variant of BNF[3] with associated constraints, and the semantics are described by the text. This semi-formal style is not used in this book, so an example, from Fortran 95, is perhaps helpful:

---

[3]Backus–Naur form, a notation used to describe the syntax of a computer language.

R609	*substring*	**is**	*parent-string (substring-range)*
R610	*parent-string*	**is**	*scalar-variable-name*
		or	*array-element*
		or	*scalar-structure-component*
		or	*scalar-constant*
R611	*substring-range*	**is**	*[scalar-int-expr] : [scalar-int-expr]*

Constraint: *parent-string* shall be of type character.

The value of the first *scalar-int-expr* in *substring-range* is called the **starting point** and the value of the second one is called the **ending point**. The length of a substring is the number of characters in the substring and is $\mathrm{MAX}(\ell - f + 1, 0)$, where f and ℓ are the starting and ending points, respectively.

Here, the three production rules and the associated constraint for a character substring are defined, and the meaning of the length of such a substring explained.

The standards are written in such a way that a processor, at compilation time, may check that the program satisfies all the constraints. In particular, the processor must provide a capability to detect and report the use of any

- obsolescent feature;

- additional syntax;

- kind type parameter (Section 2.5) that it does not support;

- non-standard source form or character;

- name that is inconsistent with the scoping rules;

- non-standard intrinsic procedure;

- non-standard intrinsic module; or

- non-standard procedure of a standard intrinsic module; this includes a non-standard use of a standard procedure of a standard intrinsic module. For example, if a processor permits the use of c_loc (Section 20.3) on a polymorphic variable (forbidden by the standard), it must be capable of reporting such usage.

Furthermore, it must be able to report the reason for rejecting a program. These capabilities are of great value in producing correct and portable code.

2. Language elements

2.1 Introduction

Written prose in a natural language, such as an English text, is composed firstly of basic elements – the letters of the alphabet. These are combined into larger entities, words, which convey the basic concepts of objects, actions, and qualifications. The words of the language can be further combined into larger units, phrases and sentences, according to certain rules. One set of rules defines the grammar. This tells us whether a certain combination of words is correct in that it conforms to the **syntax** of the language; that is, those acknowledged forms which are regarded as correct renderings of the meanings we wish to express. Sentences can in turn be joined together into paragraphs, which conventionally contain the composite meaning of their constituent sentences, each paragraph expressing a larger unit of information. In a novel, sequences of paragraphs become chapters and the chapters together form a book, which usually is a self-contained work, largely independent of all other books.

2.2 Fortran character set

Analogies to these concepts are found in a programming language. In Fortran, the basic elements, or character set, are the twenty-six letters of the English alphabet, in both upper and lower case, the ten Arabic numerals, 0 to 9, the underscore, _, and the so-called special characters listed in Table 2.1. Within the Fortran syntax, the lower-case letters are equivalent to the corresponding upper-case letters; they are distinguished only when they form part of a character sequence. In this book, syntactically significant characters will always be written in lower case. The letters, numerals, and underscore are known as **alphanumeric** characters.

Except for the currency symbol, whose graphic may vary (for example, £ in the United Kingdom), the graphics are fixed, though their styles are not fixed. As shown in Table 2.1, some of the special characters have no specific meaning within Fortran 2018.[1]

In the course of this and the following chapters we shall see how further analogies with natural language may be drawn. The unit of Fortran information is the **lexical token**, which corresponds to a word or punctuation mark. Adjacent tokens are usually separated by spaces or the end of a line, but sensible exceptions are allowed just as for a punctuation mark in prose. Sequences of tokens form **statements**, corresponding to sentences. Statements, like

---

[1]In Fortran 2023, ? is used as a separator in a conditional expression or argument, and @ is used to make an array denote multiple subscripts or section subscripts.

Modern Fortran Explained, 3rd Edition. M. Metcalf, J. Reid, M. Cohen, and R. Bader. Oxford University Press (2024). © M. Metcalf, J. Reid, M. Cohen, and R. Bader (2024). DOI 10.1093/oso/9780198876571.001.0002

Table 2.1. The special characters of Fortran 2018.

Syntactic meaning		Syntactic meaning		No syntactic meaning	
=	Equals sign	:	Colon	\	Backslash
+	Plus sign		Blank	$	Currency symbol
−	Minus sign	!	Exclamation mark	?	Question mark
*	Asterisk	%	Percent	{	Left curly bracket
/	Slash	&	Ampersand	}	Right curly bracket
(	Left parenthesis	;	Semicolon	~	Tilde
)	Right parenthesis	<	Less than	`	Grave accent
[	Left square bracket	>	Greater than	^	Circumflex accent
]	Right square bracket	'	Apostrophe	|	Vertical line
,	Comma	"	Quotation mark	#	Number sign
.	Decimal point			@	Commercial at

sentences, may be joined to form larger units like paragraphs. In Fortran these are known as **program units**, and out of these may be built a **program**. A program forms a complete set of instructions to a computer to carry out a defined sequence of operations. The simplest program may consist of only a few statements, but programs of more than 100 000 statements are now quite common.

2.3   Tokens

Within the context of Fortran, alphanumeric characters (the letters, the underscore, and the numerals) may be combined into sequences that have one or more meanings. For instance, one of the meanings of the sequence 999 is a constant in the mathematical sense. Similarly, the sequence date might represent, as one possible interpretation, a variable quantity to which we assign the calendar date.

The special characters are used to separate such sequences and also have various meanings. We shall see how the asterisk is used to specify the operation of multiplication, as in x*y, and also has a number of other interpretations.

Basic significant sequences of alphanumeric characters or of special characters are referred to as **tokens**; they are labels, keywords, names, constants (other than complex literal constants), **operators** (listed in Table 3.4, Section 3.9), and **separators**, which are

$$/ \quad (\quad) \quad [\quad] \quad (/ \quad /) \quad , \quad = \quad => \quad : \quad :: \quad .. \quad ; \quad \%$$

For example, the expression x*y contains the three tokens x, *, and y.

Apart from within a character string or within a token, blanks may be used freely to improve the layout. Thus, whereas the variable date may not be written as d a t e, the sequence x * y is syntactically equivalent to x*y. In this context, multiple blanks are syntactically equivalent to a single blank.

A name, constant, or label must be separated from an adjacent keyword, name, constant, or label by one or more blanks or by the end of a line. For instance, in

```
    real x
    rewind 10
 30 do k=1,3
```

the blanks are required after `real`, `rewind`, 30, and `do`. Likewise, adjacent keywords must normally be separated, but some pairs of keywords, such as `else if`, are not required to be separated. Similarly, some keywords may be split; for example, `inout` may be written `in out`. We do not use these alternatives, but the exact rules are given in Table 2.2.

Table 2.2. Adjacent keywords where separating blanks are optional.

block data	double precision	else if	else where
end associate	end block	end block data	end critical
end do	end enum	end file	end forall
end function	end if	end interface	end module
end procedure	end program	end select	end submodule
end subroutine	end team	end type	end where
go to	in out	select case	select type

2.4 Source form

The statements of which a source program is composed are written on **lines**. Each line may contain up to 132 characters,[2] and usually contains a single statement. Since leading spaces are not significant, it is possible to start all such statements in the first character position, or in any other position consistent with the user's chosen layout. A statement may thus be written as

```
    x = (-y + root_of_discriminant)/(2.0*a)
```

In order to be able to mingle suitable comments with the code to which they refer, Fortran allows any line to carry a trailing comment field, following an exclamation mark (!). An example is

```
    x = y/a - b   ! Solve the linear equation
```

Any comment always extends to the end of the source line and may include processor-dependent characters (it is not restricted to the Fortran character set, Section 2.2). Any line whose first non-blank character is an exclamation mark, or which contains only blanks, or which is empty, is purely commentary and is ignored by the compiler. Such comment lines may appear anywhere in a program unit, including ahead of the first statement, and even after the final program unit. A **character context** (those contexts defined in Sections 2.6.5 and 11.2) is allowed to contain !, so the ! does not initiate a comment in this case; in all other cases it does.

---

[2]Lines containing characters of non-default kind (Section 2.6.5) are subject to a processor-dependent limit.

Since it is possible that a long statement might not be accommodated in the 132 positions allowed in a single line, up to 255 additional continuation lines are allowed.[3] The so-called **continuation mark** is the ampersand (&) character, and this is appended to each line that is followed by a continuation line. Thus, the first statement of this section (considerably spaced out) could be written as

```
x =                                                              &
   (-y + root_of_discriminant)                                   &
   /(2.0*a)
```

In this book the ampersands will normally be aligned to improve readability. On a non-comment line, if & is the last non-blank character or the last non-blank character ahead of the comment symbol !, the statement continues from the character immediately preceding the &. Continuation is normally to the first character of the next non-comment line, but if the first non-blank character of the next non-comment line is &, continuation is to the character following the &. For instance, the above statement may be written

```
x =                                                              &
   &(-y + root_of_discriminant)/(2.0*a)
```

In particular, if a token cannot be contained at the end of a line, the first non-blank character on the next non-comment line must be an & followed immediately by the remainder of the token.

Comments are allowed to contain any characters, including &, so they cannot be continued because a trailing & is taken as part of the comment. However, comment lines may be freely interspersed among continuation lines and do not count towards the limit of 255 lines.

In a character context, continuation must be from a line without a trailing comment and to a line with a leading ampersand. This is because both ! and & are permitted both in character contexts and in comments.

No line is permitted to have & as its only non-blank character, or as its only non-blank character ahead of !. Such a line is really a comment and becomes a comment if & is removed.

It can be convenient to write several statements on one line. The semicolon (;) character may always be used as a **statement separator**, for example:

```
a = 0; b = 0; c = 0
```

but must not appear as the first non-blank character of a line unless it is a continuation line. Adjacent semicolons, possibly separated by blanks, are interpreted as one.

Since commentary always extends to the end of the line, it is not possible to insert commentary between statements on a single line. In principle, it is possible to write several long statements one after the other in a solid block of lines, each 132 characters long and with the appropriate semicolons separating the individual statements. In practice, such code is unreadable, and the use of multiple-statement lines should be reserved for trivial cases such as the one shown in this example.

Any Fortran statement (that is not part of a compound statement) may be labelled, in order to be able to identify it. For some statements a label is mandatory. A statement **label** precedes the statement, and is regarded as a token. The label consists of from one to five digits, one of which must be nonzero. An example of a labelled statement is

---

[3]Such long statements might be generated automatically.

```
100 format(a10)
```

Leading zeros are not significant in distinguishing between labels. For example, 10 and 010 are equivalent.

2.5 Concept of type

In Fortran it is possible to define and manipulate various **types** of data. For instance, we may have available the value 10 in a program, and assign that value to an integer **scalar** variable denoted by i. Both 10 and i are of type integer; 10 is a fixed or **constant** value, whereas i is a **variable** which may be assigned other values. Integer expressions, such as i+10, are available too.

A **data type** consists of a set of data values, a means of denoting those values, and a set of operations that are allowed on them. For the integer data type the values are ..., -3, -2, -1, 0, 1, 2, 3, ... between some limits depending on the kind of integer and computer system being used. Such tokens as these are **literal constants**, and each data type has its own form for expressing them. Named scalar variables, such as i, may be established. During the execution of a program the value of i may change to any valid value, or may become **undefined**, that is, have no predictable value. The operations that may be performed on integers are those of usual arithmetic; we can write 1+10 or i-3 and obtain the expected results. Named constants may be established too; these have values that do not change during execution of the program.

Properties like those just mentioned are associated with all the data types of Fortran, and will be described in detail in this and the following chapters. The language itself contains five data types whose existence may always be assumed. These are known as the **intrinsic types**, whose literal constants form the subject of the next section. Of each intrinsic type there is a default kind and a processor-dependent number of other kinds. Each kind is associated with a non-negative integer value known as the **kind type parameter**. This is used as a means of identifying and distinguishing the various kinds available.

In addition, it is possible to define other data types based on collections of data of the intrinsic types, and these are known as **derived types**. The ability to define data types of interest to the programmer – matrices, geometrical shapes, lists, interval numbers – is a powerful feature of the language, one which permits a high level of **data abstraction**, that is, the ability to define and manipulate **data objects** without being concerned about their actual representation in a computer.

2.6 Literal constants of intrinsic type

2.6.1 The intrinsic types

The intrinsic data types are divided into two classes. The first class contains three **numeric** types which are used mainly for numerical calculations – integer, real, and complex. The second class contains the two **non-numeric** types which are used for such applications as text processing and program control – character and logical. The numerical types are used in conjunction with the usual operators of arithmetic, such as + and -, which will be described in Chapter 3. Each includes a zero, and the value of a signed zero is the same as that of an unsigned zero.[4] The non-numeric types are used with sets of operators specific to each type; for instance, character data may be concatenated. These too will be described in Chapter 3.

2.6.2   Integer literal constants

The first type of literal constant is the **integer literal constant**. The default kind is simply a signed or unsigned integer value, for example:

```
1
0
-999
32767
+10
```

The **range** of the default integers is not specified in the language but, on a computer with a word size of n bits, is often from -2^{n-1} to $+2^{n-1} - 1$. Thus, on a 32-bit computer the range is often from $-2\,147\,483\,648$ to $+2\,147\,483\,647$. However, the maximum integer size is required to have a range of at least 18 decimal digits, implying, on a binary machine (all modern computers), a 64-bit integer. Computers often also support 16-bit integers with range from $-32\,768$ to 32767, and even 8-bit integers with range from -128 to 127, but the standard does not require these.

To be sure that the range will be adequate on any computer requires the specification of the kind of integer by giving a value for the kind type parameter. This is best done through a named integer constant. For example, if the range $-999\,999$ to $999\,999$ is desired, k6 may be established as a constant with an appropriate value by the statement, fully explained later,

```
integer, parameter :: k6 = selected_int_kind(6)
```

and used in constants thus:

```
-123456_k6
+1_k6
 2_k6
```

Here, `selected_int_kind(6)` is an intrinsic inquiry function call, and it returns a kind parameter value that yields the range $-999\,999$ to $999\,999$ with the least margin (see Section 9.9.4).

On a given processor, it might be known that the kind value needed is 3. In this case, the first of our constants can be written

```
-123456_3
```

but this form is less portable. If we move the code to another processor, this particular value may be unsupported, or might correspond to a different range.

Many implementations use kind values that indicate the number of bytes of storage occupied by a value, but the standard allows greater flexibility. For example, a processor

---

[4]Although the representation of data is processor dependent, for the numeric data types the standard defines model representations and means to inquire about the properties of those models. The details are deferred to Section 9.9.

might have hardware only for four-byte integers, and yet support kind values 1, 2, and 4 with this hardware (to ease portability from processors that have hardware for one-, two-, and four-byte integers). However, the standard makes no statement about kind values or their order, except that the kind value is never negative.

The value of the kind type parameter for a given data type on a given processor can be obtained from the `kind` intrinsic function (Section 9.2):

```
kind(1)         ! for the default value
kind(2_k6)      ! for the example
```

and the decimal exponent range (number of decimal digits supported) of a given entity may be obtained from another function (Section 9.9.2), as in

```
range(2_k6)
```

which in this case would return a value of at least 6.

2.6.3   Real literal constants

The second type of literal constant is the **real literal constant**. The default kind is a floating-point form built of some or all of a signed or unsigned integer part, a decimal point, a fractional part, and a signed or unsigned exponent part. One or both of the integer part and fractional part must be present. The exponent part is either absent or consists of the letter e followed by a signed or unsigned integer. One or both of the decimal point and the exponent part must be present. An example is

```
-10.6e-11
```

meaning -10.6×10^{-11}, and other legal forms are

```
1.
-0.1
1e-1
3.141592653
```

The default real literal constants are representations of a subset of the real numbers of mathematics, and the standard specifies neither the allowed range of the exponent nor the number of significant digits represented by the processor. Many processors conform to the IEEE standard for floating-point arithmetic and have values of 10^{-37} to 10^{+37} for the range, and a precision of six decimal digits.

To guarantee obtaining a desired range and precision requires the specification of a kind parameter value. For example,

```
integer, parameter :: long = selected_real_kind(9, 99)
```

ensures that the constants

```
1.7_long
12.3456789e30_long
```

have a precision of at least nine significant decimals and an exponent range of at least 10^{-99} to 10^{+99}. The number of digits specified in the significand has no effect on the kind. In particular, it is permitted to write more digits than the processor can in fact use.

As for integers, many implementations use kind values that indicate the number of bytes of storage occupied by a value, but the standard allows greater flexibility. It specifies only that the kind value is never negative. If the desired kind value is known it may be used directly, as in the case

```
1.7_4
```

but the resulting code is then less portable.

The processor must provide at least one representation with more precision than the default, and this second representation may also be specified as double precision. A double precision real constant may be written with the letter d replacing the letter e, as in 1.73*d*0. We defer the description of this alternative but outmoded syntax to Appendix A.4.

The kind function is also valid for real values:

```
kind(1.0)          ! for the default value
kind(1.0_long)     ! for the example
```

In addition, there are two inquiry functions available which return the actual precision and range, respectively, of a given real entity (see Section 9.9.2). Thus, the value of

```
precision(1.7_long)
```

would be at least 9, and the value of

```
range(1.7_long)
```

would be at least 99.

2.6.4   Complex literal constants

Fortran, as a language intended for scientific and engineering calculations, has the advantage of having as a third literal constant type the **complex literal constant**. This is designated by a pair of literal constants, which are either integer or real, separated by a comma and enclosed in parentheses. Examples are

```
(1., 3.2)
(1, .99e-2)
(1.0, 3.7_8)
```

where the first constant of each pair is the real part of the complex number, and the second constant is the imaginary part. If one of the parts is integer, the kind of the complex constant is that of the other part. If both parts are integer, the kind of the constant is that of the default real type. If both parts are real and of the same kind, this is the kind of the constant. If both parts are real and of different kinds, the kind of the constant is that of one of the parts: the part with the greater decimal precision, or the part chosen by the processor if the decimal precisions are identical.

A default complex constant is one whose kind value is that of default real.

The kind, precision, and range functions are equally valid for complex entities.

Note that if an implementation uses the number of bytes needed to store a real as its kind value, the number of bytes needed to store a complex value of the corresponding kind is twice the kind value. For example, if the default real type has kind 4 and needs four bytes of storage, the default complex type has kind 4 but needs eight bytes of storage.

2.6.5 Character literal constants

The fourth type of literal constant is the **character literal constant**. The default kind consists of a string of characters enclosed in a pair of either apostrophes or quotation marks, for example

```
'Anything goes'
"Nuts & bolts"
```

The characters are not restricted to the Fortran set (Section 2.2). Any graphic character supported by the processor is permitted, but not control characters such as 'newline'. The apostrophes and quotation marks serve as **delimiters**, and are not part of the value of the constant. The value of the constant

```
'STRING'
```

is STRING. Note that in character constants the blank character is significant. For example

```
'a string'
```

is not the same as

```
'astring'
```

A problem arises with the representation of an apostrophe or a quotation mark in a character constant. Delimiter characters of one sort may be embedded in a string delimited by the other, as in the examples

```
'He said "Hello"'
"This contains an ' "
```

Alternatively, a doubled delimiter without any embedded intervening blanks is regarded as a single character of the constant. For example,

```
'Isn''t it a nice day'
```

has the value Isn't it a nice day.

The number of characters in a string is called its **length**, and may be zero. For instance, ' ' and "" are character constants of length zero.

We mention here the particular rule for the source form concerning character constants that are written on more than one line (needed because constants may include the characters ! and &): not only must each line that is continued be without a trailing comment, but each continuation line must begin with a continuation mark. Any blanks following a trailing ampersand or preceding a leading ampersand are not part of the constant, nor are the ampersands themselves part of the constant. Everything else, including blanks, is part of the constant. An example is

```
long_string =                                                      &
        'Were I with her, the night would post too soon;           &
      & But now are minutes added to the hours;                     &
      & To spite me now, each minute seems a moon;                  &
      & Yet not for me, shine sun to succour flowers!               &
      &    Pack night, peep day; good day, of night now borrow:     &
      &    Short, night, to-night, and length thyself tomorrow.'
```

On any computer the characters have a property known as their **collating sequence**. One may ask the question whether one character occurs before or after another in the sequence. This question is posed in a natural form such as 'Does C precede M?', and we shall see later how this may be expressed in Fortran terms. Fortran requires the computer's collating sequence to satisfy the following conditions:

- A is less than B is less than C ... is less than Y is less than Z;
- a is less than b is less than c ... is less than y is less than z;
- 0 is less than 1 is less than 2 ... is less than 8 is less than 9;
- blank is less than A and Z is less than 0, or blank is less than 0 and 9 is less than A;
- blank is less than a and z is less than 0, or blank is less than 0 and 9 is less than a.

Thus, we see that there is no rule about whether the numerals precede or succeed the letters, nor about the position of any of the special characters or the underscore, apart from the rule that blank precedes both partial sequences. Any given computer system has a complete collating sequence, and most computers nowadays use the collating sequence of the ASCII standard (also known as ISO/IEC 646:1991). However, Fortran is designed to accommodate other sequences, notably EBCDIC, so for portability no program should ever depend on any ordering beyond that stated above. Alternatively, Fortran provides access to the ASCII collating sequence on any computer through intrinsic functions (Section 9.6.1), but this access is not so convenient and is less efficient on some computers.

A processor is required to provide access to the default kind of character constant just described. In addition, it may support other kinds of character constants, in particular those of non-European languages, which may have more characters than can be provided in a single byte. For example, a processor might support Kanji with the kind parameter value 2; in this case, a Kanji character constant may be written

      2_'国内'

or

      kanji_"標準"

where kanji is an integer named constant with the value 2. We note that, in this case, the kind type parameter exceptionally precedes the constant.[5]

There is no requirement for a processor to provide more than one kind of character, and the standard does not require any particular relationship between the kind parameter values and the character sets and the number of bytes needed to represent them. In fact, all that is required is that each kind of character set includes a blank character. However, where other kinds are available, the intrinsic function selected_char_kind, fully described in Section 9.9.4, can be used to select a specific character set.

As for the other data types, the kind function gives the actual value of the kind type parameter, as in

      kind('ASCII')

Non-default characters are permitted in comments.

---

[5]This is to make it easier for a compiler to support multiple different character sets occurring within a single source file.

2.6.6 Logical literal constants

The fifth type of literal constant is the **logical literal constant**. The default kind has one of two values, `.true.` and `.false.`. These logical constants are normally used only to initialize logical variables to their required values, as we shall see in Section 3.6.

The default kind has a kind parameter value which is processor dependent. The actual value is available as `kind(.true.)`. As for the other intrinsic types, the kind parameter may be specified by an integer constant following an underscore, as in

```
.false._1
.true._long
```

Non-default logical kinds are useful for storing logical arrays compactly; we defer further discussion until Section 7.7.

2.6.7 Binary, octal, and hexadecimal constants

A binary, octal, or hexadecimal ('boz') constant is allowed as a literal constant in limited circumstances to provide the bit sequence of a stored integer or real value. It is known as a **'boz' constant**.

A 'boz' constant may take the form of the letter `b` followed by a binary integer in apostrophes, the letter `o` followed by an octal integer in apostrophes, or the letter `z` followed by a hexadecimal integer in apostrophes. Examples are `b'1011'`, `o'715'`, and `z'a2f'`. The letters `b`, `o`, and `z` may be in upper case, as may the hexadecimal digits `a` to `f`. A pair of quotation marks may be used instead of a pair of apostrophes. In all cases, it is an exact representation of a sequence of bits. The length must not exceed the largest storage size of a real or integer value.

A 'boz' constant is permitted only in a `data` statement (Section 8.5.2) and as an actual argument of a small number of intrinsic procedures (Sections 9.3.2, 9.10.3, 9.10.6, 9.10.7, 9.10.11, and A.4).

A binary, octal, or hexadecimal constant may also appear in an internal or external file as a digit string, without the leading letter and the delimiters (see Section 11.3.3).

2.7 Names

A Fortran program **references** many different entities by name. Such names must consist of between 1 and 63 alphanumeric characters – letters, underscores, and numerals – of which the first must be a letter. There are no other restrictions on the names; in particular, there are no reserved words in Fortran. We thus see that valid names are, for example,

```
a
a_thing
x1
mass
q123
real
time_of_flight
```

and invalid names are

```
1a            ! First character is not alphabetic
a thing       ! Contains a blank
$sign         ! Contains a non-alphanumeric character
```

Within the constraints of the syntax, it is important for program clarity to choose names that have a clear significance – these are known as **mnemonic names**. Examples are `day`, `month`, and `year` for variables to store the calendar date.

The use of names to refer to constants, already met in Section 2.6.2, will be fully described in Section 8.3.

2.8   Scalar variables of intrinsic type

We have seen in the section on literal constants (Section 2.6) that there exist five different intrinsic data types. Each of these types may have variables too. The simplest way by which a variable may be declared to be of a particular type is by specifying its name in a **type declaration statement** such as

```
integer   :: i
real      :: a
complex   :: current
logical   :: pravda
character :: letter
```

Here, all the variables have default kind, and `letter` has default length, which is 1. Explicit requirements may also be specified through **type parameters**, as in the examples

```
integer(kind=4)              :: i
real(kind=long)              :: a
character(len=20, kind=1)    :: english_word
character(len=20, kind=kanji) :: kanji_word
```

Character is the only type to have two parameters, and here the two character variables each have length 20. Where appropriate, just one of the parameters may be specified, leaving the other to take its default value, as in the cases

```
character(kind=kanji) :: kanji_letter
character(len=20)     :: english_word
```

The shorter forms

```
integer(4)         :: i
real(long)         :: a
character(20, 1)   :: english_word
character(20, kanji) :: kanji_word
character(20)      :: english_word
```

are available, but note that

```
character(kanji) :: kanji_letter        ! Beware
```

is not an abbreviation for

```
character(kind=kanji) :: kanji_letter
```

because a single unnamed parameter is taken as the length parameter.

2.9   Derived data types

When programming, it is often useful to be able to manipulate objects that are more sophisticated than those of the intrinsic types. Imagine, for instance, that we wished to specify objects representing persons. Each person in our application is distinguished by a name, an age, and an identification number. Fortran allows us to define a corresponding data type in the following fashion:

```
type person
   character(len=10) :: name
   real              :: age
   integer           :: id
end type person
```

This is the definition of the type and is known as a **derived-type definition**. A scalar object of such a type is called a **structure**. In order to create a structure of that type, we write an appropriate type declaration statement, such as

```
type(person) :: you
```

The scalar variable you is then a composite object of type person containing three separate **components**, one corresponding to the name, another to the age, and a third to the identification number.

As will be described in Sections 3.9 and 3.10, a variable such as you may appear in expressions and assignments involving other variables or constants of the same or different types. In addition, each of the components of the variable may be referenced individually using the **component selector** character percent (%). The identification number of you would, for instance, be accessed as

```
you%id
```

and this quantity is an integer variable which could appear in an expression such as

```
you%id + 9
```

Similarly, if there were a second object of the same type:

```
type(person) :: me
```

the differences in ages could be established by writing

```
you%age - me%age
```

It will be shown in Section 3.9 how a meaning can be given to an expression such as

```
you - me
```

Just as the intrinsic data types have associated literal constants, so too may literal constants of derived type be specified. Their simplest form is the name of the type followed by the constant values of the components, in order and enclosed in parentheses. Thus, the constant

```
person('Smith', 23.5, 2541)
```

may be written assuming the derived type defined at the beginning of this section, and could be assigned to a variable of the same type:

```
you = person('Smith', 23.5, 2541)
```

Any such **structure constructor**, fully described in Section 3.8, can appear only after the derived-type definition.

A derived type may have a component that is of a previously defined derived type. This is illustrated in Figure 2.1. A variable of type `triangle` may be declared thus

```
type(triangle) :: t
```

and t has components t%a, t%b, and t%c all of type point; t%a has components t%a%x and t%a%y of type real.

Figure 2.1 A derived type with a component of a previously defined derived type.

```
type point
   real :: x, y
end type point
type triangle
   type(point) :: a, b, c
end type triangle
```

The real and imaginary parts of a complex variable can be accessed as the pseudo-components re and im for the real and imaginary parts, respectively. For example, impedance%re and impedance%im are the real and imaginary parts of the complex variable impedance. They are also accessible, less conveniently, by the intrinsic functions real and aimag (Section 9.3.2).

2.10   Arrays

2.10.1   Array declarations

Another compound object supported by Fortran is the **array**. An array consists of a rectangular set of elements, all of the same type and type parameters. There are a number of ways in which arrays may be declared; for the moment we shall consider only the declaration of arrays of fixed sizes. To declare an array named a of ten real elements, we add the dimension **attribute** to the type declaration statement thus:

```
real, dimension(10) :: a
```

The successive elements of the **whole array** a are a(1), a(2), a(3), ..., a(10). The number of elements of an array is called its **size**. Each **array element** is a scalar.

Many problems require a more elaborate declaration than one in which the first element is designated 1, and it is possible in Fortran to declare a lower as well as an upper **bound**:

```
real, dimension(-10:5) :: vector
```

This is a vector of sixteen elements, vector(-10), vector(-9), ..., vector(5). We thus see that whereas we always need to specify the upper bound, the lower bound is optional, and by default has the value 1.

An array may extend in more than one dimension, and Fortran allows up to fifteen dimensions to be specified.[6] For instance,

```
real, dimension(5,4) :: b
```

declares an array with two dimensions, and

```
real, dimension(-10:5, -20:-1, 0:15, -15:0, 16, 16, 16) :: grid
```

declares seven dimensions, the first four with explicit lower bounds. It may be seen that the size of this second array is

$$16 \times 20 \times 16 \times 16 \times 16 \times 16 \times 16 = 335\,544\,320,$$

and that arrays of many dimensions can thus place large demands on the memory of a computer: an array

```
integer, dimension(10, 10, 10, 10, 10, 10, 10, 10, 10, 10) :: x
```

of four-byte integers requires 40 GB of memory. The number of dimensions of an array is known as its **rank**. Thus, `grid` has a rank of seven. Scalars are regarded as having rank zero. The number of elements along a dimension of an array is known as the **extent** in that dimension. Thus, `grid` has extents 16, 20, ...

The sequence of extents is known as the **shape**. For example, `grid` has the shape $(16, 20, 16, 16, 16, 16, 16)$.

A derived type may contain an array component. For example, the type

```
type triplet
   real              :: u
   real, dimension(3)    :: du
   real, dimension(3,3) :: d2u
end type triplet
```

might be used to hold the value of a variable in three dimensions and the values of its first and second derivatives. If `t` is of type `triplet`, `t%du` and `t%d2u` are arrays of type real.

2.10.2   Array elements and sections

Some statements treat the elements of an array one by one in a special order which we call the **array element order**. It is obtained by counting most rapidly in the early dimensions. Thus, the elements of `grid` in array element order are

```
grid(-10, -20,  0, -15,  1,  1,  1)
grid( -9, -20,  0, -15,  1,  1,  1)
   ⋮
grid(  5,  -1, 15,   0, 16, 16, 16)
```

This is illustrated for an array of two dimensions in Figure 2.2. Most implementations actually store arrays in **contiguous** storage in array element order, but we emphasize that the standard does not require this.

---

[6]If an array is also a coarray (Section 2.11), the limit applies to the sum of the rank and corank.

Figure 2.2 The ordering of elements in the array b(5,4).

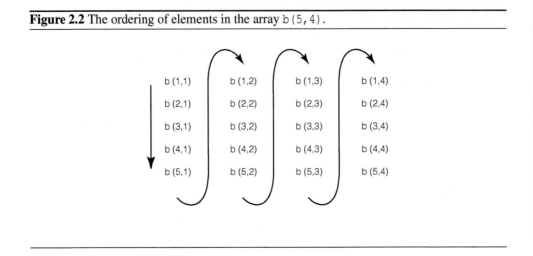

We reference an individual element of an array by specifying, as in the examples above, its **subscript** values. In the examples we used integer constants, but in general each subscript may be formed of a **scalar integer expression**, that is, any arithmetic expression whose value is scalar and of type integer. Each subscript must be within the corresponding ranges defined in the array declaration, and the number of subscripts must equal the rank. Examples are

```
a(1)
a(i*j)          ! i and j are of type integer
a(nint(x+3.))   ! x is of type real
t%d2u(i+1,j+2)  ! t is of derived type triplet
```

where nint is an intrinsic function to convert a real value to the nearest integer (see Section 9.3.2). In addition subarrays, called **sections**, may be referenced by specifying ranges and strides for one or more subscripts. The stride defaults to one if it is not specified. The following are examples of array sections:

```
a(i:j)          ! Rank-one array of size j-i+1
b(k, 1:n*2:2)   ! Rank-one array with elements b(k, 1), b(k, 3),
                ! b(k, 5), ... b(k, n*2-1)
c(1:i, 1:j, k)  ! Rank-two array with extents i and j
```

We describe array sections in more detail in Section 7.11. An array section is itself an array, but its individual elements must not be accessed through the section designator. Thus, b(k, 1:n)(1) cannot be written; it must be expressed as b(k, 1).

A further form of subscript is shown in

```
a(ipoint)       ! ipoint is a rank-one integer array
```

where ipoint is a rank-one array of indices, pointing to array elements. It may thus be seen that a(ipoint), which identifies as many elements of a as ipoint has elements, is an example of another **array-valued object**, and ipoint is referred to as a **vector subscript**. This will be met in greater detail in Section 7.11.

2.10.3 Array constants

It is often convenient to be able to define an array constant. In Fortran, a rank-one array may be constructed as a list of elements enclosed between the tokens (/ and /) or the characters [and]. A simple example is

```
(/ 1, 2, 3, 5, 10 /)
```

which is an array of rank one and size five. To obtain a series of values, the individual values may be defined by an expression that depends on an integer variable having values in a range, with an optional stride. Thus, the constructor

```
[1, 2, 3, 4, 5]
```

can be written as

```
[ (i, i = 1,5) ]
```

and

```
(/2, 4, 6, 8/)
```

as

```
(/ (i, i = 2,8,2) /)
```

and

```
(/ 1.1, 1.2, 1.3, 1.4, 1.5 /)
```

as

```
(/ (i*0.1, i=11,15) /)
```

An array constant of rank greater than one may be constructed by using the function reshape (see Section 9.15.3) to reshape a rank-one array constant.

A full description of array constructors is reserved for Section 7.15.

2.10.4 Declaring entities of differing shapes

In order to declare several entities of the same type but differing shapes, Fortran permits the convenience of using a single statement. Whether or not there is a dimension attribute present, arrays may be declared by placing the shape information after the name of the array:

```
integer :: a, b, c(10), d(10), e(8, 7)
```

If the dimension attribute is present, it provides a default shape for the entities that are not followed by their own shape information, and is ignored for those that are:

```
integer, dimension(10) :: c, d, e(8, 7)
```

2.11   Coarrays and images

For parallel programming, Fortran adopts the Single Program Multiple Data (SPMD) programming model. A single program is replicated a fixed number of times, each replication having its own set of data objects. Each replication of the program is called an **image**. Each image executes asynchronously, and the normal rules of Fortran apply within each image. Each image is identified by an **image index**, which is an integer between one and the number of images.

A scalar or array object may be declared as a **coarray**, which means that as well as being accessible in the usual way in its own image it is accessible from other images with cosubscripts enclosed in square brackets. Each valid set of cosubscripts maps to an image index in the same way as a valid set of array subscripts maps to a position in the array element order. A reference is always to a single image, that is, each cosubscript must be a scalar integer expression and the number of cosubscripts must match the declared corank. We call any object whose designator includes cosubscripts a **coindexed object**.

Each image executes on its own without regard to the execution of other images except when it encounters a special kind of statement called an **image control statement** that makes it delay further execution until one or more other images have executed corresponding image control statements. This is to ensure that one image altering the value of a coarray variable and another image accessing it are always ordered as the programmer intended. The simplest is the `sync all` statement that causes all images to delay until all have executed corresponding `sync all` statements.

It is anticipated that a coarray program might execute on a huge number of images. While the likelihood of a particular image failing during the execution of a program is small, the likelihood that one of them might fail may be significant. Therefore, the concept of continued execution in the presence of failed images has been introduced. Failure of an image is always treated as permanent because action on the other images might have started before the failed image resumes computation. The term **active** denotes an image that has neither failed nor stopped. It is not required that the system supports failed images.

Coarrays and images are described in detail in Chapters 17 and 18.

2.12   Character substrings

It is possible to build arrays of characters, just as it is possible to build arrays of any other type:

```
character, dimension(80) :: line
```

declares an array, called `line`, of 80 elements, each one character in length. Each character may be addressed by the usual reference, `line(i)` for example. In this case, however, a more appropriate declaration might be

```
character(len=80) :: line
```

which declares a scalar data object of 80 characters. These may be referenced individually or in groups using a **substring** notation

```
line(i:j)    ! i and j are of type integer
```

which references all the characters from i to j in line. The colon is used to separate the two substring subscripts, which may be any scalar integer expressions. The colon is obligatory in substring references, so that referencing a single character requires line(i:i). There are default values for the substring subscripts. If the lower one is omitted, the value 1 is assumed; if the upper one is omitted, a value corresponding to the character length is assumed. Thus,

line(:i) is equivalent to line(1:i)
line(i:) is equivalent to line(i:80)
line(:) is equivalent to line or line(1:80)

If i is greater than j in line(i:j), the value is a zero-sized string.

We may now combine the length declaration with the array declaration to build arrays of character objects of specified length, as in

```
character(len=80), dimension(60) :: page
```

which might be used to define storage for the characters of a whole page, with 60 elements of an array, each of length 80. To reference the line j on a page we may write page(j), and to reference character i on that line we could combine the array subscript and character substring notations into

```
page(j)(i:i)
```

A substring of a character constant or of a structure component may also be formed:

```
'ABCDEFGHIJKLMNOPQRSTUVWXYZ'(j:j)
you%name(1:2)
```

In this section we have used character variables with a declared maximum length. This is adequate for most character manipulation applications, but the limitation may be avoided with allocatable variables, as described in Section 15.4.2.

2.13 Dynamic memory management and aliasing

2.13.1 Allocatable objects

Sometimes an object is required to be of a size that is known only after some data have been read or some calculations performed. This can be implemented by the use of pointers (see Section 2.13.2), but there are significant advantages for memory management and execution speed in using **allocatable objects** when the added functionality of pointers is not needed.

Thus, for this more restricted purpose, an **allocatable** object may be declared by giving it the allocatable attribute:

```
real, dimension(:, :), allocatable :: a
character(len=:), allocatable :: s
```

The initial status of such an object is **unallocated**; while the rank of an allocatable array is determined by the declaration, the bounds and length type parameters of these unallocated allocatable objects, respectively, are undefined, and no memory is available for storing their data. In accordance with this, the declaration of these objects uses a colon, causing the bounds and length type parameters to be **deferred**.

The status can be changed to **allocated**

- either by explicit execution of an `allocate` statement, illustrated here for the array object a,

      allocate (a(n, 0:n+1)) ! n is of type integer, and is defined

  which, if successful, will supply the memory required for the object based on the specified bounds;
- or implicitly through execution of an assignment statement, shown here for the scalar string s,

      s = "Anything goes"

  for which the value of the length parameter is taken from the right-hand side.

In the case of arrays, the lower and upper bounds are specified just as for the `dimension` attribute (Section 2.10.1), except that any scalar integer expression is permitted.

The object will revert to the unallocated state either if it is an unsaved object and program execution causes it to go out of scope (Section 5.15), or when it is explicitly deallocated by execution of the statement

    deallocate (a)

Any memory resources previously assigned to the object will then be released.

Components of derived types are permitted to have the `allocatable` attribute. A simple example for the use of this is

```
type :: real_polynomial
    real, allocatable, dimension(:) :: coeff
end type
type(real_polynomial) :: poly_obj
⋮
degree = ...  ! only known at run time
allocate (poly_obj%coeff(0:degree))
```

This important feature is fully described in Chapter 6.

2.13.2 Pointers

In everyday language nouns are often used in a way that makes their meaning precise only because of the context. 'The chairman said that ...' will be understood precisely by the reader who knows that the context is the Fortran Committee developing Fortran 90 and that its chairman was then Jeanne Adams.

Similarly, in a computer program it can be very useful to be able to use a name that can be made to refer to different objects during execution of the program. One example is the multiplication of a vector by a sequence of square matrices. We might write code that calculates

$$y_i = \sum_{j=1}^{n} a_{ij} x_j, \quad i = 1, 2, \ldots, n$$

from the vector x_j, $j = 1, 2, \ldots, n$. In order to use this to calculate BCz, say, we might first make x refer to z and A refer to C, thereby using our code to calculate $y = Cz$, then make x refer to y and A refer to B so that our code calculates the result vector we finally want.

An object that can be made to refer to other objects in this way is called a **pointer**, and must be declared with the `pointer` attribute, for example

```
real, pointer              :: son
real, pointer, dimension(:)   :: x, y
real, pointer, dimension(:,:) :: a, c
```

In the case of an array, only the rank (number of dimensions) is declared, and the bounds (and hence shape) are taken from that of the object to which it points. Given such an array pointer declaration, the compiler arranges storage for a **descriptor** that will later hold the address of the actual object (known as the **target**) and holds, if it is an array, its bounds and strides.

A pointer may be made to point at an object (or a part of it) that is declared with the **target** attribute

```
real, target :: tx(n), ty(2*n), b(n, n)
```

by executing a **pointer assignment** statement:

```
x => tx ; y => ty(1:n:2)
a => b
```

Alternatively, an `allocate` statement such as

```
allocate (son, c(n, n))
```

can be used to create *anonymous* targets with appropriate bounds and length type parameters, and to implicitly set up the pointer association of the specified pointers with these targets. This use of pointers provides a means to access dynamic storage but, as seen in Section 2.13.1 and described later in Chapter 6, allocatable arrays provide a better way to to do this in cases where the 'pointing' property is not essential. Also, dynamically created pointer targets are not automatically deallocated when the pointer goes out of scope.

Finally, besides pointing to existing variables, a collection of pointers may be made explicitly to point at nothing:

```
nullify (son, x, y, a)
```

The `nullify` statement is described in Section 3.14.

By default, pointers are initially undefined (see also the final paragraph of Section 3.3). This is a very undesirable state since there is no way to test for it. However, it may be avoided by using the declaration:

```
real, pointer :: son => null()
```

(the function `null` is described in Section 9.17) and we recommend that this always be employed. Alternatively, pointers may be defined as soon as they come into scope by execution of a nullify statement or a pointer assignment.

Components of derived types are permitted to have the `pointer` attribute. This enables a major application of pointers: the construction of linked lists. As a simple example, we might decide to hold a sparse vector as a chain of variables of the type shown in Figure 2.3, which allows us to access the entries one by one; given

```
type(entry), pointer :: chain
```
where `chain` is a scalar of this type and holds a chain that is of length two, its entries are `chain%index` and `chain%next%index`, and `chain%next%next` will have been nullified. Additional entries may be created when necessary by an appropriate allocate statement. We defer the details to Section 3.13.

Figure 2.3 A type for holding a sparse vector as a chain of variables.

```
type entry
    real                   :: value
    integer                :: index
    type(entry), pointer :: next
end type entry
```

When a pointer is of derived type and a component such as `chain%index` is selected, it is actually a component of the pointer's target that is selected.

A subobject is not a pointer unless it has a final component selector for the name of a pointer component, for example, `chain%next` (whereas `chain%value` and `x(1)` are not).

Pointers will be discussed in detail in later chapters (especially Sections 3.13, 5.7.3, 7.12, 7.13, 8.5.3, 8.5.4, and 9.2).

2.14   Type extension

A new derived type may be constructed by extending an existing derived type using the `extends` attribute on the type definition statement. For example, the type
```
type person
    character(len=10) :: name
    real                :: age
    integer             :: id
end type person
```
can be extended to form a new type thus
```
type, extends(person) :: employee
    integer :: national_insurance_number
    real    :: salary
end type employee
```
The old type is known as the **parent type**. The new type inherits all the components of the parent type by a process known as **inheritance association** and usually has additional components. So an `employee` variable has the inherited components of `name`, `age`, and `id`, and the additional components of `national_insurance_number` and `salary`. Where the order matters, for example, in a structure constructor, the inherited components come first in their order, followed by the new components in their order.

A derived type is **extensible** unless it has the `sequence` (Appendix A.8) or `bind` (Section 20.4) attribute, or is `c_ptr` or `c_funptr` (Section 20.3). An extended type must not be given the `sequence` or `bind` attribute so it, too, is extensible. Type extension is discussed fully in Section 15.2.

2.15 Polymorphic variables

A pointer or allocatable variable may be declared using the `class` keyword in place of the `type` keyword to indicate that when it is allocated or associated with a target its type is its declared type or an extension of that type. We say that it is **polymorphic**. For example,

```
class(person), pointer :: p
```

declares a pointer p that may point to any object whose type is in the class of types consisting of `type(person)` and all of its extensions, including `employee` of Section 2.14.

We say that the polymorphic object is **type compatible** with any object of the same declared type or any of its extensions. Allocation and pointer assignment must have type-compatible allocations and targets, respectively.

The type named in the `class` attribute is called the **declared type** of the polymorphic variable, and the type of the object to which it refers is called the **dynamic type**. For example, execution of the typed allocation statement

```
allocate(employee :: p)
```

would create target memory sufficient to hold an object of type `employee`, and p becomes pointer associated with that target, causing its dynamic type to be `employee`.

However, when a polymorphic variable is referring to an object of an extended type, access via component notation is not available to components beyond those of the declared type. This is because the compiler only knows about the declared type of the object, and it cannot know about the dynamic type (which may vary at run time). Access to components that are in the dynamic type but not the declared type is provided by type-bound procedures (see Section 15.8) or by the `select type` construct (see Section 15.7).

Polymorphic entities are discussed fully in Section 15.3. Their use as dummy arguments is introduced in Section 5.7.4.

2.16 Objects and subobjects

We have seen that derived types may have components that are arrays, as in the type `triplet` of Section 2.10.1, and arrays may be of derived type as in the example

```
type(triplet), dimension(10) :: t
```

A single structure (for example, `t(2)`) is always regarded as a scalar, but it may have a component (for example, `t(2)%du`) that is an array. Derived types may have components of other derived types. This leads us to components of components, components of components of components, ..., continuing until we reach an intrinsic type, a pointer component, or an allocatable component. Each of these is called an **ultimate component** of the type. If we exclude pointer components but do not stop at allocatable components, we obtain the set of **potential subobject components** of the type. For example, given the types in Figure 2.4, the potential subobject components of type `two` are `ordinary`, `alloc`, `ordinary%comp`, and `alloc%comp`, but not `point` or `point%comp`.

Figure 2.4 Nested types.

```
type one
    integer :: comp
end type one
type two
    type(one)                 :: ordinary
    type(one), allocatable :: alloc(:)
    type(one), pointer     :: point
end type two
```

The ultimate components of type `two` are `ordinary%comp`, `alloc`, and `point`, but not `ordinary`, `alloc%comp`, or `point%comp`.

The terms 'ultimate component' and 'potential subobject component' are used for the corresponding subobjects of an object of derived type. A potential subobject component can be a subobject of an unallocated allocatable component. The concept is needed to describe language restrictions in the context of coarray programming that involve allocatable components that may be allocated or not.

An object referenced by an unqualified name (all characters alphanumeric) is called a **named object** and is not part of a bigger object. Its **subobjects** have **designators** that consist of the name of the **base object** followed by one or more qualifiers (for example, `t(1:7)` and `t(1)%du`). Each successive qualifier specifies a part of the base object specified by the name or designator that precedes it.

We note that the term **array** is used for any object that is not scalar, including an array section or an array-valued component of a structure. The term **variable** is used for any named object that is not specified to be a constant and for any part of such an object, including array elements, array sections, structure components, and substrings.

2.17   Summary

In this chapter we have introduced the elements of the Fortran language. The character set has been listed, and the manner in which sequences of characters form literal constants and names explained. In this context we have encountered the five intrinsic data types defined in Fortran, and seen how each data type has corresponding literal constants and named objects. We have seen how derived types may be constructed in various ways. We have introduced one method by which arrays may be declared, and seen how their elements may be referenced by subscript expressions. The concepts of the array section, character substring, and pointer have been presented. Allocatable and polymorphic objects have been introduced. Some important terms (including some relevant for parallel processing) have been defined. In the following chapter we shall see how these elements may be combined into expressions and statements, Fortran's equivalents of 'phrases' and 'sentences'.

Exercises

1. For each of the following assertions, state whether it is true, false, or not determined, according to the Fortran collating sequences:

    b is less than m
    8 is less than 2
    * is greater than T
    $ is less than /
    blank is greater than A
    blank is less than 6

2. Which of the Fortran lines in the code

    ```
    x = y
    a = b+c ! add
    word = 'string'
    a = 1.0; b = 2.0
    a = 15. ! initialize a; b = 22. ! and b
    song = "Life is just&
       & a bowl of cherries"
    chide = 'Waste not,
       want not!'
    c(3:4) = 'up"
    ```

    are correctly written according to the requirements of the Fortran source form? Which ones contain commentary? Which lines are initial lines and which are continuation lines?

3. Classify the following literal constants according to the five intrinsic data types of Fortran. Which are not legal literal constants?

-43	'word'
4.39	1.9-4
0.0001e+20	'stuff & nonsense'
4 9	(0.,1.)
(1.e3,2)	'I can''t'
'(4.3e9, 6.2)'	.true._1
e5	'shouldn' 't'
1_2	"O.K."
z10	z'10'

4. Which of the following names are legal Fortran names?

name	name32
quotient	123
a182c3	no-go
stop!	burn_
no_go	long__name

5. What are the first, tenth, eleventh, and last elements of the following arrays?

```
real, dimension(11)        :: a
real, dimension(0:11)      :: b
real, dimension(-11:0)     :: c
real, dimension(10,10)     :: d
real, dimension(5,9)       :: e
real, dimension(5,0:1,4)   :: f
```

Write an array constructor of eleven integer elements.

6. Given the array declaration

```
character(len=10), dimension(0:5,3) :: c
```

which of the following subobject designators are legal?

c(2,3)	c(4:3)(2,1)
c(6,2)	c(5,3)(9:9)
c(0,3)	c(2,1)(4:8)
c(4,3)(:)	c(3,2)(0:9)
c(5)(2:3)	c(5:6)
c(5,3)(9)	c(,)

7. Write derived-type definitions appropriate for:

i) a vehicle registration;

ii) a circle;

iii) a book (title, author, and number of pages).

Give an example of a derived type constant for each one.

8. Given the declaration for t in Section 2.16, which of the following objects and subobjects are arrays?

t	t(4)%du(1)
t(10)	t(5:6)
t(1)%du	t(5:5)

9. Write specifications for these entities:

i) an integer variable inside the range -10^{20} to 10^{20};

ii) a real variable with a minimum of twelve decimal digits of precision and a range of 10^{-100} to 10^{100};

iii) a Kanji character variable on a processor that supports Kanji with kind=2.

3. Expressions and assignments

3.1 Introduction

We have seen in the previous chapter how we are able to build the 'words' of Fortran – the constants, keywords, and names – from the basic elements of the character set. In this chapter we shall discover how these entities may be further combined into 'phrases' or **expressions**, and how these, in turn, may be combined into 'sentences' or **statements**.

In an expression, we describe a computation that is to be carried out by the computer. The result of the computation may then be assigned to a variable. A sequence of assignments is the way in which we specify, step by step, the series of individual computations to be carried out in order to arrive at the desired result. There are separate sets of rules for expressions and assignments, depending on whether the **operands** in question are numeric, logical, character, or derived in type, and whether they are scalars or arrays. There are also separate rules for pointer assignments. We shall discuss each set of rules in turn, including a description of the relational expressions that produce a result of type logical and are needed in control statements (see Chapter 4). To simplify the initial discussion, we commence by considering expressions and assignments that are intrinsically defined and involve neither arrays nor entities of derived data types.

An expression in Fortran is formed of operands and operators, combined in a way that follows the rules of Fortran syntax. A simple expression involving a **dyadic** (or **binary**) operator has the form

operand operator operand

an example being

    x+y

and a **unary** or **monadic** operator has the form

operator operand

an example being

    -y

The type and kind of the result are determined by the type and kind of the operands and do not depend on their values. The operands may be constants, variables, or functions (see Chapter 5), and an expression may itself be used as an operand. In this way we can build up more complicated expressions such as

operand operator operand operator operand

Modern Fortran Explained, 3rd Edition. M. Metcalf, J. Reid, M. Cohen, and R. Bader. Oxford University Press (2024). © M. Metcalf, J. Reid, M. Cohen, and R. Bader (2024). DOI 10.1093/oso/9780198876571.001.0003

where consecutive operands are separated by a single operator. Each operand must have a defined value.

The rules of Fortran state that the parts of expressions without parentheses are evaluated successively from left to right for operators of equal precedence, with the exception of `**` (exponentiation, see Section 3.2). If it is necessary to evaluate one part of an expression, or **subexpression**, before another, parentheses may be used to indicate which subexpression should be evaluated first. In

   operand operator (operand operator operand)

the subexpression in parentheses will be evaluated, and the result used as an operand to the first operator. The Fortran standard refers to an operand that contains no operators or is an expression in parentheses as a **primary**.

If an expression or subexpression has no parentheses, the processor is permitted to evaluate an equivalent expression; that is, an expression that always has the same value apart, possibly, from the effects of numerical round-off. For example, if a, b, and c are real variables, the expression

   `a/b/c`

might be evaluated as

   `a/(b*c)`

on a processor that can multiply much faster than it can divide. Usually, such changes are welcome to the programmer since the program runs faster, but when they are not (for instance, because they would lead to more round-off) parentheses should be inserted because the processor is required to respect them.

3.2   Scalar numeric expressions

A **numeric expression** is an expression whose operands are one of the three numeric types – integer, real, and complex – and whose operators are

   `**`   exponentiation
   `* /`   multiplication, division
   `+ -`   addition, subtraction

These operators are known as **numeric intrinsic** operators, and are listed here in their order of precedence. In the absence of parentheses, exponentiations will be carried out before multiplications and divisions, and these before additions and subtractions.

We note that the minus sign (–) and the plus sign (+) can be used as unary operators, as in

   `-tax`

Because it is not permitted in ordinary mathematical notation, a unary minus or plus must not follow immediately after another operator. When this is needed, as for x^{-y}, parentheses must be placed around the operator and its operand:

   `x**(-y)`

The type and kind type parameter of the result of a unary operation are those of the operand.

The exception to the left-to-right rule noted in Section 3.1 concerns exponentiations. Whereas the expression

```
-a+b+c
```

will be evaluated from left to right as

```
((-a)+b)+c
```

the expression

```
a**b**c
```

will be evaluated as

```
a**(b**c)
```

For integer data, the result of any division will be truncated towards zero, that is, to the integer value whose magnitude is equal to or just less than the magnitude of the exact result. Thus, the result of

6/3	is	2
8/3	is	2
-8/3	is	−2

This fact must always be borne in mind whenever integer divisions are written. Similarly, the result of

```
2**3   is   8
```

whereas the result of

```
2**(-3)   is   1/(2**3)
```

which is zero.

The rules of Fortran allow a numeric expression to contain numeric operands of differing types or kind type parameters. This is known as a **mixed-mode expression**. Except when raising a real or complex value to an integer power, the object of the weaker (or simpler) of the two data types will be converted, or **coerced**, into the type of the stronger one. The result will also be that of the stronger type. If, for example, we write

```
a*i
```

when a is of type real and i is of type integer, then i will be converted to a real data type before the multiplication is performed, and the result of the computation will also be of type real. The rules are summarized for each possible combination for the operations +, -, *, and / in Table 3.1, and for the operation ** in Table 3.2. The functions real and cmplx that they reference are defined in Section 9.3.2. In both tables, I stands for integer, R for real, and C for complex.

If both operands are of type integer, the kind type parameter of the result is that of the operand with the greater decimal exponent range, or is processor dependent if the kinds differ but the decimal exponent ranges are the same. If both operands are of type real or complex, the kind type parameter of the result is that of the operand with the greater decimal precision,

Table 3.1. Type of result of *a* .op. *b*, where .op. is +, −, *, or /.

Type of *a*	Type of *b*	Value of *a* used	Value of *b* used	Type of result
I	I	*a*	*b*	I
I	R	real(*a*, kind(*b*))	*b*	R
I	C	cmplx(*a*, 0, kind(*b*))	*b*	C
R	I	*a*	real(*b*, kind(*a*))	R
R	R	*a*	*b*	R
R	C	cmplx(*a*, 0, kind(*b*))	*b*	C
C	I	*a*	cmplx(*b*, 0, kind(*a*))	C
C	R	*a*	cmplx(*b*, 0, kind(*a*))	C
C	C	*a*	*b*	C

Table 3.2. Type of result of *a***b*.

Type of *a*	Type of *b*	Value of *a* used	Value of *b* used	Type of result
I	I	*a*	*b*	I
I	R	real(*a*, kind(*b*))	*b*	R
I	C	cmplx(*a*, 0, kind(*b*))	*b*	C
R	I	*a*	*b*	R
R	R	*a*	*b*	R
R	C	cmplx(*a*, 0, kind(*b*))	*b*	C
C	I	*a*	*b*	C
C	R	*a*	cmplx(*b*, 0, kind(*a*))	C
C	C	*a*	*b*	C

or is processor dependent if the kinds differ but the decimal precisions are the same. If one operand is of type integer and the other is real or complex, the type parameter of the result is that of the real or complex operand.

Note that a literal constant in a mixed-mode expression is held to its own precision, which may be less than that of the expression. For example, given a variable a of kind `long` (Section 2.6.3), the result of `a/1.7` will be less precise than that of `a/1.7_long`.

In the case of raising a complex value to a complex power, the principal value[1] is taken. Raising a negative real value to a real power is not permitted since the exact result probably has a nonzero imaginary part.

---

[1] The principal value of a^b is $\exp(b(\log|a| + i\arg a))$, with $-\pi < \arg a \leq \pi$.

3.3   Defined and undefined variables

In the course of the explanations in this and the following chapters, we shall often refer to a variable becoming **defined** or **undefined**. In the previous chapter, we showed how a scalar variable may be called into existence by a statement like

```
real :: speed
```

In this simple case, the variable `speed` has, at the beginning of the execution of the program, no defined value. It is undefined. No attempt must be made to reference its value since it has none. A common way in which it might become defined is for it to be assigned a value:

```
speed = 2.997
```

After the execution of such an **assignment statement** it has a value, and that value may be referenced, for instance in an expression:

```
speed*0.5
```

For a compound object, all of its subobjects that are not pointers must be individually defined before the object as a whole is regarded as defined. Thus, an array is said to be defined only when each of its elements is defined, an object of a derived data type is defined only when each of its non-pointer components is defined, and a character variable is defined only when each of its characters is defined.

A variable that is defined does not necessarily retain its state of definition throughout the execution of a program. As we shall see in Chapter 5, a variable that is local to a single subprogram usually becomes undefined when control is returned from that subprogram. In certain circumstances, it is even possible that a single array element becomes undefined and this causes the array considered as a whole to become undefined; a similar rule holds for entities of derived data type and for character variables.

A means to specify the initial value of a variable is explained in Section 8.5.

In the case of a pointer, the **pointer association** status may be **undefined** or **defined** by being **associated** with a target or being **disassociated**, which means that it is not associated with a target but has a definite status that may be tested by the function `associated` (Section 9.2). Even though a pointer is associated with a target, the target itself may be defined or undefined. Means to specify the initial status of disassociated are provided (see Section 8.5.3).

3.4   Scalar numeric assignment

The general form of a scalar numeric assignment is

   variable = expr

where *variable* is a scalar numeric variable and *expr* is a scalar numeric expression. If *expr* is not of the same type or kind as *variable*, it will be converted to that type and kind before the assignment is carried out, according to the set of rules given in Table 3.3 (the functions `int`, `real`, and `cmplx` are defined in Section 9.3.2).

We note that if the type of *variable* is integer but *expr* is not, then the assignment will result in a loss of precision unless *expr* happens to have an integral value. Similarly, assigning a real

Table 3.3. Numeric conversion for assignment statement *variable* = *expr*.

Type of *variable*	Value assigned
integer	int(*expr*, kind(*variable*))
real	real(*expr*, kind(*variable*))
complex	cmplx(*expr*, kind=kind(*variable*))

expression to a real variable of a kind with less precision will also cause a loss of precision to occur, and the assignment of a complex quantity to a non-complex variable involves the loss of the imaginary part. Thus, the values in i and a following the assignments

```
i = 7.3                  ! i of type default integer
a = (4.01935, 2.12372)   ! a of type default real
```

are 7 and 4.019 35, respectively. Also, if a literal constant is assigned to a variable of greater precision, the result will have the accuracy of the constant. For example, given a variable a of kind long (Section 2.6.3), the result of

```
a = 1.7
```

will be less precise than that of

```
a = 1.7_long
```

3.5　Scalar relational operators

It is possible in Fortran to test whether the value of one numeric expression bears a certain relation to that of another, and similarly for character expressions. The relational operators are

<	less than
<=	less than or equal
==	equal
/=	not equal
>	greater than
>=	greater than or equal

If either or both of the expressions are complex, only the operators == and /= are available.

The result of such a comparison is one of the default logical values .true. or .false., and we shall see in the next chapter how such tests areimportant in controlling the execution of a program. Examples of relational expressions (for i and j of type integer, a and b of type real, and char1 of type default character) are

```
i < 0            integer relational expression
a < b            real relational expression
a+b > i-j        mixed-mode relational expression
char1 == 'Z'     character relational expression
```

In the third expression above, we note that the two components are of different numeric types. In this case, and whenever either or both of the two components consist of numeric expressions, the rules state that the components are to be evaluated separately and converted to the type and kind of their sum before the comparison is made. Thus, a relational expression such as

```
a+b <= i-j
```

will be evaluated by converting the result of $(i-j)$ to type real.

For character comparisons, the kinds must be the same and the letters are compared from the left until a difference is found or the strings are found to be identical. The result of the comparison is determined by the positions in the collating sequence (see Section 2.6.5) of the first differing characters. If the lengths differ, the shorter one is regarded as being padded with blanks[2] on the right. Two zero-sized strings are considered to be identical. The effect is as for the order of words in a dictionary. For example, the relation `'ab' < 'ace'` is true.

No other form of mixed-mode relational operator is intrinsically available, though such an operator may be defined (Section 3.9). The numeric operators take precedence over the relational operators.

3.6 Scalar logical expressions and assignments

Logical constants, variables, and functions may appear as operands in logical expressions. The logical operators, in decreasing order of precedence, are:

unary operator:

    `.not.`   logical negation

binary operators:

`.and.`	logical intersection
`.or.`	logical union
`.eqv.` and `.neqv.`	logical equivalence and non-equivalence

If we assume a logical declaration of the form

```
logical :: i,j,k,l
```

then the following are valid logical expressions:

```
.not.j
j .and. k
i .or. l .and. .not.j
( .not.k .and. j .neqv. .not.l) .or. i
```

In the first expression we note the use of `.not.` as a unary operator. In the third expression, the rules of precedence imply that the subexpression `l.and..not.j` will be evaluated first, and the result combined with `i`. In the last expression, the two subexpressions

---

[2]Here and elsewhere, the blank padding character used for a non-default type is processor dependent.

.not.k.and.j and .not.l will be evaluated and compared for non-equivalence. The result of the comparison, .true. or .false., will be combined with i.

The kind type parameter of the result is that of the operand for .not., and for the others is that of one of the operands, but which is processor dependent.

We note that the .or. operator is an inclusive operator; the .neqv. operator provides an exclusive logical or (a.and..not.b .or. .not.a.and.b).

The result of any logical expression is the value .true. or .false., and this value may then be assigned to a logical variable such as element 3 of the logical array flag in the example

```
flag(3) = (.not. k .eqv. l) .or. j
```

The kind type parameter values of the variable and expression need not be identical.

A logical variable may be set to a predetermined value by an assignment statement:

```
flag(1) = .true.
flag(2) = .false.
```

In the foregoing examples, all the operands and results were of type logical – no other data type is allowed to participate in an intrinsic logical operation or assignment.

The results of several relational expressions may be combined into a logical expression, and assigned, as in

```
real    :: a, b, x, y
logical :: cond
   :
cond = a>b .or. x<0.0 .and. y>1.0
```

where we note the precedence of the relational operators over the logical operators. If the value of such a logical expression can be determined without evaluating a subexpression, a processor is permitted not to evaluate the subexpression. An example is

```
i<=10 .and. ary(i)==0    ! for a real array ary(10)
```

when i has the value 11. However, the programmer must not rely on such behaviour – an out-of-bounds subscript might be referenced if the processor chooses to evaluate the right-hand subexpression before the left-hand one. We return to this topic in Section 5.11.2.

3.7   Scalar character expressions and assignments

3.7.1   Concatenation and assignment

The only intrinsic operator for character expressions is the **concatenation** operator //, which has the effect of combining two character operands into a single character result. For example, the result of concatenating the two character constants AB and CD, written as

```
'AB'//'CD'
```

is the character string ABCD. The operands must have the same kind parameter values, but may be character variables, constants, or functions. For instance, if word1 and word2 are both of default kind and length 4, and contain the character strings LOOP and HOLE, respectively, the string POLE is the result of

```
word1(4:4)//word2(2:4)
```

The length of the result of a concatenation is the sum of the lengths of the operands. Thus, the length of the result of

```
word1//word2//'S'
```

is 9, which is the length of the string LOOPHOLES.

The result of a character expression may be assigned to a character variable of the same kind. Assuming the declarations

```
character(len=4) :: char1, char2
character(len=8) :: result
```

we may write

```
char1 = 'any '
char2 = 'book'
result = char1//char2
```

In this case, result will now contain the string any book. We note in these examples that the lengths of the left- and right-hand sides of the three assignments are in each case equal. If, however, the length of the result of the right-hand side is shorter than the length of the left-hand side, then the result is placed in the leftmost part of the left-hand side and the rest is filled with blank characters. Thus, in

```
character(len=5) :: fill
fill(1:4) = 'AB'
```

fill(1:4) will have the value AB*bb* (where *b* stands for a blank character). The value of fill(5:5) remains undefined, that is, it contains no specific value and should not be used in an expression. As a consequence, fill is also undefined. On the other hand, when the left-hand side is shorter than the result of the right-hand side, the right-hand end of the result is truncated. The result of

```
character(len=5) :: trunc8
trunc8 = 'TRUNCATE'
```

is to place in trunc8 the character string TRUNC. If a left-hand side is of zero length, no assignment takes place.

The left- and right-hand sides of an assignment may overlap. In such a case, it is always the old values that are used in the right-hand side expression. For example, the assignment

```
result(3:5) = result(1:3)
```

is valid, and if result began with the value ABCDEFGH it would be left with the value ABABCFGH.

Other means of manipulating characters and strings of characters, via intrinsic functions, are described in Sections 9.6 and 9.7.

3.7.2 ASCII character set

If the default character set for a processor is not ASCII, but ASCII is supported on that processor, intrinsic assignment is defined between them to convert characters appropriately. For example, on an EBCDIC machine, in

```
integer, parameter :: ascii = selected_char_kind('ASCII')
character          :: ce
character(kind=ascii) :: ca
ce = ascii_'X'
ca = 'X'
```

the first assignment statement will convert the ASCII upper-case X to an EBCDIC upper-case X, and the second assignment statement will do the reverse.

3.7.3   ISO 10646 character set

ISO/IEC 10646 UCS-4 is a four-byte character set designed to be able to represent every character in every language in the world, including all special characters in use in other coded character sets. It is a strict superset of seven-bit ASCII; that is, its first 128 characters are the same as those of ASCII.

Assignment of default characters or ASCII characters to ISO 10646 is allowed, and the characters are converted appropriately. Assignment of ISO 10646 characters to default or ASCII characters is also allowed; however, if any ISO 10646 character is not representable in the destination character set, the result is processor dependent (information will be lost).

For example, in

```
integer, parameter :: ascii = selected_char_kind('ASCII')
integer, parameter :: iso10646 = selected_char_kind('ISO_10646')
character(kind=ascii) :: x = ascii_'X'
character(kind=iso10646) :: y
y = x
```

the ISO 10646 character variable y will be set to the correct value for the upper-case letter X.

3.8   Structure constructors

A structure may be constructed from expressions for its components, just as a constant structure may be constructed from constants (Section 2.9). The general form of a **structure constructor** is

   derived-type-spec ([*component-spec-list*])

where *derived-type-spec* is the name of a derived type[3] and each *component-spec* is

   [*keyword* =] *expr*

or

   [*keyword* =] *target*

where *keyword* is the name of a component of the type, *expr* is a value for a non-pointer component, and *target* is a target for a pointer component.

In a simple case, the *component-spec*s are without *keyword*s and provide values for the components in the order they appear in the type declaration. For example, given the type

---

[3] A more general form will be discussed in Section 13.2.

```
type char10
   integer             :: length
   character(len=10) :: value
end type char10
```

and the variables

```
character(len=4) :: char1, char2
```

the following is a value of type `char10`:

```
char10(8, char1//char2)
```

The value `null()` indicates a disassociated pointer or an unallocated allocatable object. If *keyword*s are used they must appear after any *component-spec*s without keywords but may be in any order. No component may appear more than once. A non-pointer non-allocatable component may be absent if it has default initialization (Section 8.5.5). An absent allocatable component will be unallocated. A pointer component is not permitted to be absent.

Non-pointer components are treated as if the intrinsic assignment statement

derived-type-spec%keyword = expr

had been executed, and pointer components are treated as if the pointer assignment statement

derived-type-spec%keyword => target

had been executed. In both cases, the rules for intrinsic and pointer assignment apply.[4]

3.9   Scalar defined operators

No operators for derived types are automatically available. If a programmer defines a derived type and wishes operators to be available, he or she must define the operators, too. For a binary operator this is done by writing a function, with two intent `in` arguments, that specifies how the result depends on the operands, and an **interface block** that associates the function with the operator token (intent, functions, and interface blocks will be explained fully in Sections 5.10, 5.11, and 5.12.1). For example, given the type

```
type interval
   real :: lower, upper
end type interval
```

that represents intervals of numbers between a lower and an upper bound, we may define addition by a module (Section 5.5) containing the procedure

```
function add_interval(a,b)
   type(interval)              :: add_interval
   type(interval), intent(in) :: a, b
   add_interval%lower = a%lower + b%lower ! Production code would
   add_interval%upper = a%upper + b%upper ! allow for roundoff.
end function add_interval
```

and the interface block (see Section 5.18.1)

---

[4]In particular, *target* must not be a constant.

```
interface operator(+)
   module procedure add_interval
end interface
```

This function would be invoked in an expression such as

```
y + z
```

to perform this programmer-defined add operation for scalar variables `y` and `z` of type `interval`. A unary operator is defined by an interface block and a function with one intent in argument.

The operator token may be any of the tokens used for the intrinsic operators or may be a sequence of up to 63 letters enclosed in decimal points other than `.true.` or `.false.`. An example is

```
.sum.
```

In this case, the header line of the interface block would be written as

```
interface operator(.sum.)
```

and the expression as

```
y.sum.z
```

If an intrinsic token is used, the number of arguments must be the same as for the intrinsic operation, the precedence of the operation is as for the intrinsic operation, and a unary minus or plus must not follow immediately after another operator. Otherwise, it is of highest precedence for defined unary operators and lowest precedence for defined binary operators. The complete set of precedences is given in Table 3.4. Where another precedence is required within an expression, parentheses must be used.

Table 3.4. Relative precedence of operators (in decreasing order).

Type of operation when intrinsic	Operator
—	monadic (unary) defined operator
Numeric	`**`
Numeric	`*` or `/`
Numeric	monadic + or −
Numeric	dyadic + or −
Character	`//`
Relational	`==` `/=` `<` `<=` `>` `>=`
Logical	`.not.`
Logical	`.and.`
Logical	`.or.`
Logical	`.eqv.` or `.neqv.`
—	dyadic (binary) defined operator

Retaining the intrinsic precedences is helpful both to the readability of expressions and to the efficiency with which a compiler can interpret them. For example, if + is used for set union and * for set intersection, we can interpret the expression

```
i*j + k
```

for sets i, j, and k without difficulty.

Note that a defined unary operator not using an intrinsic token may follow immediately after another operator, as in

```
y .sum. .inverse. x
```

Operators may be defined for any types of operands, except where there is an intrinsic operation for the operator and types. For example, we might wish to be able to add an interval number to an ordinary real, which can be done with the extra procedure

```
function add_interval_real(a,b)
   type(interval)              :: add_interval_real
   type(interval), intent(in) :: a
   real, intent(in)           :: b
   add_interval_real%lower = a%lower + b ! Production code would
   add_interval_real%upper = a%upper + b ! allow for roundoff.
end function add_interval_real
```

changing the interface block to

```
interface operator(+)
   module procedure add_interval, add_interval_real
end interface
```

The result of such a **defined operation** may have any type. The type of the result, as well as its value, must be specified by the function.

Note that an operation that is defined intrinsically cannot be redefined; thus, in

```
real :: a, b, c
   ⋮
c = a + b
```

the meaning of the operation is always unambiguous.

3.10   Scalar defined assignments

Assignment of an expression of derived type to a variable of the same type is automatically available and takes place component by component. For example, if a is of the type interval defined at the start of Section 3.9, we may write

```
a = interval(0.0, 1.0)
```

(structure constructors were met in Section 3.8).

In other circumstances, however, we might wish to define a different action for an assignment involving an object of derived type, and indeed this is possible. An assignment may be redefined or another assignment may be defined by a subroutine with two arguments, the first having intent `out` or intent `inout` and corresponding to the variable, and the second having intent `in` and corresponding to the expression (subroutines will be dealt with fully in Chapter 5). In the case of an assignment involving an object of derived type and an object of a different type, such a definition must be provided. For example, assignment of reals to intervals and vice versa might be defined by a module containing the subroutines

```
subroutine real_from_interval(a,b)
    real, intent(out)          :: a
    type(interval), intent(in) :: b
    a = (b%lower + b%upper)/2
end subroutine real_from_interval
```

and

```
subroutine interval_from_real(a,b)
    type(interval), intent(out) :: a
    real, intent(in)            :: b
    a%lower = b
    a%upper = b
end subroutine interval_from_real
```

and the interface block

```
interface assignment(=)
    module procedure real_from_interval, interval_from_real
end interface
```

Given this, we may write

```
type(interval) :: a
a = 0.0
```

A **defined assignment** must not redefine the meaning of an **intrinsic assignment** for intrinsic types, that is, an assignment between two objects of numeric type, of logical type, or of character type with the same kind parameter, but may redefine the meaning of an intrinsic assignment for two objects of the same derived type. For instance, for an assignment between two variables of the type `char10` (Section 3.9) that copies only the relevant part of the character component, we might write

```
subroutine assign_string (left, right)
    type(char10), intent(out) :: left
    type(char10), intent(in)  :: right
    left%length = right%length
    left%value(1:left%length) = right%value(1:right%length)
end subroutine assign_string
```

Intrinsic assignment for a derived-type object always involves intrinsic assignment for all its non-pointer components, even if a component is of a derived type for which assignment has been redefined.

3.11   Array expressions

So far in this chapter we have assumed that all the entities in an expression are scalar. However, any of the unary intrinsic operations may also be applied to an array to produce another array of the same shape (identical rank and extents, see Section 2.10.1) and having each element value equal to that of the operation applied to the corresponding element of the operand. Similarly, binary intrinsic operations may be applied to a pair of arrays of the same shape to produce an array of that shape, with each element value equal to that of the operation applied to corresponding elements of the operands. One of the operands of a binary operation may be a scalar, in which case the result is as if the scalar had been broadcast to an array of the same shape as the array operand. Given the array declarations

```
real, dimension(10,20) :: a,b
real, dimension(5)     :: v
```

the following are examples of array expressions:

```
a/b           ! Array of shape [10,20], with elements a(i,j)/b(i,j)
v+1.          ! Array of shape [5], with elements v(i)+1.0
5/v+a(1:5,5)  ! Array of shape [5], with elements 5/v(i)+a(i,5)
a == b        ! Logical array of shape [10,20], with element (i,j)
              ! .true. if a(i,j) == b(i,j), and .false. otherwise
```

Two arrays of the same shape are said to be **conformable**, and a scalar is conformable with any array.

Note that the correspondence is by position in the extent and not by subscript value. For example,

```
a(2:9,5:10) + b(1:8,15:20)
```

has element values

```
a(i+1,j+4) + b(i,j+14)        where i = 1, 2, ..., 8, j = 1, 2, ..., 6
```

This is represented pictorially in Figure 3.1.

The order in which the scalar operations in any array expression are executed is not specified in the standard, thus enabling a compiler to arrange efficient execution on a vector or parallel computer.

Any scalar intrinsic operator may be applied in this way to arrays and array–scalar pairs. For derived operators, the programmer may define an elemental procedure with these properties (see Section 7.9). He or she may also define operators directly for certain ranks or pairs of ranks. For example, the type

```
type matrix
    real :: element
end type matrix
```

might be defined to have scalar operations that are identical to the operations for reals, but for arrays of ranks one and two the operator * defined to mean matrix multiplication. The type matrix would therefore be suitable for matrix arithmetic, whereas reals are not suitable because multiplication for real arrays is done element by element. This is discussed further in Section 7.5.

Figure 3.1 The sum of two array sections.

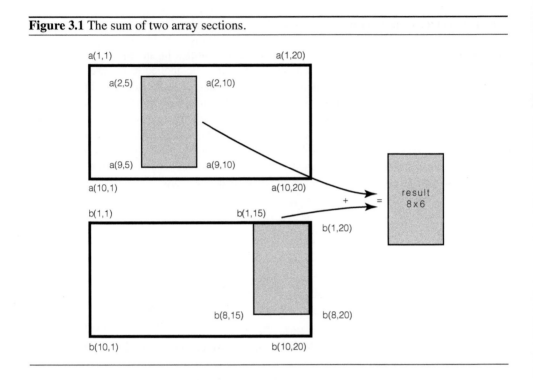

3.12   Array assignment

By intrinsic assignment, an array expression may be assigned to an array variable of the same shape,[5] which is interpreted as if each element of the expression were assigned to the corresponding element of the variable. For example, with the declarations of the beginning of the last section, the assignment

```
a = a + 1.0
```

replaces $a(i,j)$ by $a(i,j) + 1.0$ for $i = 1, 2, \ldots, 10$ and $j = 1, 2, \ldots, 20$. Note that, as for expressions, the element correspondence is by position within the extent rather than by subscript value. This is illustrated by the example

```
a(1,11:15) = v      ! a(1,j+10) is assigned from
                    ! v(j), j = 1, 2, ..., 5
```

A scalar expression may be assigned to an array, in which case the scalar value is broadcast to all the array elements.

If the expression includes a reference to the array variable or to a part of it, the expression is interpreted as being fully evaluated before the assignment commences. For example, the statement

```
v(2:5) = v(1:4)
```

---

[5]For allocatable arrays the shapes may differ, see Section 6.7.

results in each element v(i) for i = 2, 3, 4, 5 having the value that v(i-1) had prior to the commencement of the assignment. This rule exactly parallels the rule for substrings that was explained in Section 3.7.1. The order in which the array elements are assigned is not specified by the standard, to allow optimizations.

Sets of numeric and mathematical intrinsic functions, whose results may be used as operands in scalar or array expressions and in assignments, are described in Sections 9.3 and 9.4.

If a defined assignment (Section 3.10) is defined by an elemental subroutine (Section 7.9), it may be used to assign a scalar value to an array or an array value to an array of the same shape. A separate subroutine may be provided for any particular combination of ranks and will override the elemental assignment. If there is no elemental defined assignment, intrinsic assignment is still available for those combinations of ranks for which there is no corresponding defined assignment.

A form of array assignment under a mask is described in Section 7.6, and assignment expressed with the help of indices in Appendix B.1.13.

3.13   Pointers in expressions and assignments

A pointer may appear as a variable in the expressions and assignments that we have considered so far in this chapter, provided it has a valid association with a target. The target is accessed without any need for an explicit dereferencing symbol. In particular, if both sides of an assignment statement are pointers, data are copied from one target to the other target.

Sometimes the need arises for another sort of assignment. We may want the left-hand pointer to point to another target, rather than that its current target acquire fresh data. That is, we want the descriptor to be altered. This is called **pointer assignment** and takes place in a pointer assignment statement:

pointer => target

where *pointer* is the name of a pointer or the designator of a structure component that is a pointer, and *target* is usually a variable but may also be a reference to a pointer-valued function (see Section 5.11). For example, the statements

```
x => z
a => c
```

have variables as targets and are needed for the first matrix multiplication of Section 2.13.2, in order to make x refer to z and a to c. The statement

```
x => null()
```

(the function null is described in Section 9.17) nullifies x. Pointer assignment also takes place for a pointer component of a structure when the structure appears on the left-hand side of an ordinary assignment. For example, suppose we have used the type entry of Figure 2.3, Section 2.13.2, to construct a chain of entries and wish to add a fresh entry at the front. If first points to the first entry and current is a scalar pointer of type entry, the statements

```
allocate (current)
current = entry(new_value, new_index, first)
first => current
```

allocate a new entry and link it into the top of the chain. The assignment statement has the effect

```
current%next => first
```

and establishes the link. The pointer assignment statement gives `first` the new entry as its target without altering the old first entry. The ordinary assignment

```
first = current
```

would be incorrect because the target would be copied, destroying the old first entry, corresponding to the component assignments

```
first%value = current%value    ! Components of the
first%index = current%index    ! old first are lost.
first%next => current%next     !
```

In the case where the chain began with length two and consisted of

```
first :       (1.0, 10, associated)
first%next : (2.0, 15, null)
```

following the execution of the first set of statements it would have length three and consist of

```
first :            (4.0, 16, associated)
first%next :       (1.0, 10, associated)
first%next%next : (2.0, 15, null)
```

If the *target* in a pointer assignment statement is a variable that is not itself a pointer or a subobject of a pointer target, it must have the `target` attribute. For example, the statement

```
real, dimension(10), target :: y
```

declares `y` to have the `target` attribute. Any non-pointer subobject of an object with the `target` attribute also has the `target` attribute. The `target` attribute is required for the purpose of code optimization by the compiler. It is very helpful to the compiler to know that a variable that is not a pointer or a target may not be accessed as a pointer target.

The target in a pointer assignment statement may be a subobject of a pointer target. For example, given the declaration

```
character(len=80), dimension(:), pointer :: page
```

and an appropriate association, the following are all permitted targets:

```
page, page(10), page(2:4), page(2)(3:15)
```

Note that it is sufficient for the pointer to be at any level of component selection. For example, given the declaration

```
type(entry) :: node
```

which has a pointer component `next` (see Section 2.13.2) and an appropriate association, `node%next%value` is a permitted target.

If the *target* in a pointer assignment statement is itself a pointer target, then a straightforward copy of the descriptor takes place. If the pointer association status is undefined or disassociated, this state is copied.

If the *target* is a pointer or a subobject of a pointer target, the new association is with that pointer's target and is not affected by any subsequent changes to its pointer association status. This is illustrated by the following example. The sequence

```
b => c    ! c has the target attribute
a => b
nullify (b)
```

will leave a still pointing to c.

If the *target* is allocatable (Section 2.13.1), it must be allocated.

The type, type parameters, and rank of the *pointer* and *target* in a pointer assignment statement must each be the same. If the *pointer* is an array, it takes its shape and bounds from the *target*. The bounds are as would be returned by the functions lbound and ubound (Section 9.14.3) for the target, which means that an array section or array expression is usually taken to have the value 1 for a lower bound and the extent for the corresponding upper bound (but we shall see later how a lower bound may be specified, see Section 7.14).

Fortran is unusual in not requiring a special character for a reference to a pointer target, but requiring one for distinguishing pointer assignment from ordinary assignment. The reason for this choice was the expectation that most engineering and scientific programs will refer to target data far more often than they change targets.

3.14 The nullify statement

Pointers may be explicitly disassociated from their targets by executing a nullify statement. Its general form is

    nullify (*pointer-object-list*)

There must be no dependencies among the objects, in order to allow the processor to nullify the objects one by one in any order. The statement is also useful for giving the disassociated status to an undefined pointer. An advantage of nullifying pointers rather than leaving them undefined is that they may then be tested by the intrinsic function associated (Section 9.2). For example, the end of the chain of Section 3.13 will be flagged as a disassociated pointer if the statement

    nullify(first)

is executed initially to create a zero-length chain. Because there are often other ways to access a target (for example, through another pointer), the nullify statement does not deallocate the targets. If deallocation is also required, a deallocate statement (Section 6.6) should be executed instead.

3.15 Summary

In this chapter we have seen how scalar and array expressions of numeric, logical, character, and derived types may be formed, and how the corresponding assignments of the results may be made. The relational expressions and the use of pointers have also been presented. We now have the information required to write short sections of code forming a sequence of statements to be performed one after the other. In the following chapter we shall see how more complicated sequences may be built up.

Exercises

1. If all the variables are numeric scalars, which of the following are valid numeric expressions?

`a+b`	`-c`
`a+-c`	`d+(-f)`
`(a+c)**(p+q)`	`(a+c)(p+q)`
`-(x+y)**i`	`4.((a-d)-(a+4.*x)+1)`

2. In the following expressions, add the parentheses which correspond to Fortran's rules of precedence (assuming a, c–f are real scalars, i–n are logical scalars, and b is a logical array); for example, `a+d**2/c` becomes `a+((d**2)/c)`.

    ```
    c+4.*f
    4.*g-a+d/2.
    a**e**c**d
    a*e-c**d/a+e
    i .and. j .or. k
    .not. l .or. .not. i .and. m .neqv. n
    b(3).and.b(1).or.b(6).or..not.b(2)
    ```

3. What are the results of the following expressions?

`3+4/2`	`6/4/2`
`3.*4**2`	`3.**3/2`
`-1.**2`	`(-1.)**3`

4. A scalar character variable r has length eight. What are the contents of r after each of the following assignments?

    ```
    r = 'ABCDEFGH'
    r = 'ABCD'//'01234'
    r(:7) = 'ABCDEFGH'
    r(:6) = 'ABCD'
    ```

5. Which of the following logical expressions are valid if b is a logical array?

`.not.b(1).and.b(2)`	`.or.b(1)`
`b(1).or..not.b(4)`	`b(2)(.and.b(3).or.b(4))`

6. If all the variables are real scalars, which of the following relational expressions are valid?

`d <= c`	`p < t > 0`
`x-1 /= y`	`x+y < 3 .or. > 4.`
`d < c.and.3.0`	`q == r .and. s>t`

7. Write expressions to compute:

    i) the perimeter of a square of side *s*;

    ii) the area of a triangle of base *b* and height *h*;

    iii) the volume of a sphere of radius *r*.

8. An item costs *n* cents. Write a declaration statement for suitable variables and assignment statements which compute the change to be given from a $1 bill for any value of *n* from 1 to 99, using coins of denomination 1, 5, 10, and 25 cents.

9. Given the type declaration for `interval` in Section 3.9, the definitions of + given in Section 3.9, the definitions of assignment given in Section 3.10, and the declarations

```
type(interval) :: a,b,c,d
real           :: r
```

which of the following statements are valid?

```
a = b + c
c = b + 1.0
d = b + 1
r = b + c
a = r + 2
```

10. Given the type declarations

```
real, dimension(5,6) :: a, b
real, dimension(5)   :: c
```

which of the following statements are valid?

```
a = b               c = a(:,2) + b(5,:5)
a = c+1.0           c = a(2,:) + b(:,5)
a(:,3) = c          b(2:,3) = c + b(:5,3)
```

11. Write a definition suitable for a derived type holding a doubly-linked list (each entry has links to both the previous and the next entry). Write code for allocating an entry `new` and inserting it ahead of the entry `current`, assuming that this is not at the start of the list.

4. Control constructs

4.1 Introduction

We have learnt in the previous chapter how assignment statements may be written, and how these may be ordered one after the other to form a sequence of code which is executed step by step. In most computations, however, this simple sequence of **executable statements** (statements that can be executed, as opposed to those of an informative nature, such as declarations) is by itself inadequate for the formulation of the problem. For instance, we may wish to follow one of two possible paths through a section of code, depending on whether a calculated value is positive or negative. We may wish to sum 1000 elements of an array, and to do this by writing 1000 additions and assignments is clearly tedious; the ability to iterate over a single addition is required instead. We may wish to pass control from one part of a program to another, or even stop processing altogether.

For all these purposes, we have available in Fortran various facilities to enable the logical flow through the program statements to be controlled. The most important form is that of a **control construct**, which begins with an initial keyword statement, may have intermediate keyword statements, and ends with a matching terminal statement; it may be entered only at the initial statement. Each sequence of statements between keyword statements is called a **block**. A block may be empty, though such cases are rare.

Control constructs may be **nested**, that is, a block may contain another control construct. In such a case, the block must contain the whole of the inner construct. Execution of a block always begins with its first statement.

4.2 The `if` construct and statement

The `if` construct contains one or more sequences of statements (blocks), at most one of which is chosen for execution. The general form is shown in Figure 4.1. Here and throughout the book we use square brackets to indicate optional items, followed by dots if there may be any number (including zero) of such items. There can be any number (including zero) of `else if` statements, and zero or one `else` statements. Naming is optional, but an `else` or `else if` statement may be named only if the corresponding `if` and `end if` statements are named, and must be given the same name. The name may be any valid and distinct Fortran name (see Section 5.16 for a discussion on the scope of names).

Modern Fortran Explained, 3rd Edition. M. Metcalf, J. Reid, M. Cohen, and R. Bader. Oxford University Press (2024). © M. Metcalf, J. Reid, M. Cohen, and R. Bader (2024). DOI 10.1093/oso/9780198876571.001.0004

Figure 4.1 The if construct.

```
[ name : ] if  (scalar-logical-expr)  then
            block
          [ else if  (scalar-logical-expr)  then [ name ]
            block ] ...
          [ else [ name ]
            block ]
          end if [ name ]
```

An example of the if construct in its simplest form is

```
swap: if (x < y) then
          temp = x
          x = y
          y = temp
      end if swap
```

The block of three statements is executed if the condition is true; otherwise execution continues from the statement following the end if statement. Note that the block inside the if construct is indented. This is not obligatory, but makes the logic easier to understand, especially in nested if constructs as we shall see at the end of this section.

The next simplest form has an else block, but no else if blocks. Now there is an alternative block for the case where the condition is false. An example is

```
if (x < y) then
    x = -x
else
    y = -y
end if
```

in which the sign of x is changed if x is less than y, and the sign of y is changed if x is greater than or equal to y.

The most general type of if construct uses the else if statement to make a succession of tests, each of which has its associated block of statements. The tests are made one after the other until one is fulfilled, and the associated statements of the relevant if or else if block are executed. Control then passes to the end of the if construct. If no test is fulfilled, no block is executed, unless there is a final 'catch-all' else clause.

There is a useful shorthand form for the simplest case of all. An if construct of the form

```
if (scalar-logical-expr) then
    action-stmt
end if
```

where *action-stmt* is an executable statement, may be written as an if statement

```
if (scalar-logical-expr)  action-stmt
```

provided *action-stmt* is not itself an if statement (no nesting). Examples are

```
if (x-y > 0.0) x = 0.0
if (cond .or. p<q .and. r<=1.0) s(i,j) = t(j,i)
```

It is permitted to nest `if` constructs to an arbitrary depth, as shown to two levels in Figure 4.2, in which we see the necessity to indent the code in order to be able to understand the logic easily. For even deeper nesting, naming is to be recommended. The constructs must be properly nested; that is, each construct must be wholly contained in a block of the next outer construct.

Figure 4.2 A nested `if` construct.

```
if (i < 0) then
   if (j < 0) then
      x = 0.0
      y = 0.0
   else
      z = 0.0
   end if
else if (k < 0) then
   z = 1.0
else
   x = 1.0
   y = 1.0
end if
```

4.3   The `case` construct

Fortran provides another means of selecting one of several options, rather similar to that of the `if` construct. The principal differences between the two constructs are that, for the `case` construct, only **one** expression is evaluated for testing, and the evaluated expression may belong to no more than one of a series of predefined sets of values. The form of the `case` construct is shown by:

```
[ name: ]  select case (expr)
          [ case  selector [ name ]
               block ] ...
            end select [ name ]
```

As for the `if` construct, the leading and trailing statements must either both be unnamed or both bear the same name; a `case` statement within it may be named only if the leading statement is named and bears the same name. The expression *expr* must be scalar and of type character, logical, or integer, and the specified values in each *selector* must be of this type. In the character case, the lengths are permitted to differ, but not the kinds. In the logical and integer cases, the kinds may differ. The simplest form of *selector* is a scalar constant expression[1] in parentheses, such as in the statement

---

[1] A constant expression is a restricted form of expression that can be verified to be constant (the restrictions being chosen for ease of implementation). The details are tedious and are deferred to Section 8.4. In this section, all examples employ the simplest form of constant expression: the literal constant.

```
case (1)
```

For character or integer *expr*, a range may be specified by a lower and an upper scalar constant expression separated by a colon:

```
case (low:high)
```

Either *low* or *high*, but not both, may be absent; this is equivalent to specifying that the case is selected whenever *expr* evaluates to a value that is less than or equal to *high*, or greater than or equal to *low*, respectively. An example is shown in Figure 4.3.

Figure 4.3 A case construct.

```
select case (number)     ! number is of type integer
case (:-1)               ! all values below 0
   n_sign = -1
case (0)                 ! only 0
   n_sign = 0
case (1:)                ! all values above 0
   n_sign = 1
end select
```

The general form of *selector* is a list of non-overlapping values and ranges, all of the same type as *expr*, enclosed in parentheses, such as

```
case (1, 2, 7, 10:17, 23)
```

The form

```
case default
```

is equivalent to a list of all the possible values of *expr* that are not included in the other selectors of the construct. Though we recommend that the values be in order, as in this example, this is not required. Overlapping values are not permitted within one *selector*, nor between different ones in the same construct.

There can only be a single case default *selector* in a given case construct, as shown in Figure 4.4. The case default clause does not necessarily have to be the last clause of the case construct.

Figure 4.4 A case construct with a case default selector.

```
select case (ch)          ! ch of type character
case ('c', 'd', 'r':)
   ch_type = .true.
case ('i':'n')
   int_type = .true.
case default
   real_type = .true.
end select
```

Since the values of the selectors are not permitted to overlap, at most one selector may be satisfied; if none is satisfied, control passes to the next executable statement following the end select statement.

Like the if construct, case constructs may be nested inside one another.

4.4   The do construct

Many problems in mathematics require the ability to iterate. If we wish to sum the elements of an array a of length 10, we could write

```
sum = a(1)
sum = sum+a(2)
  ⋮
sum = sum+a(10)
```

which is clearly laborious. Fortran provides a facility known as the do construct which allows us to reduce these ten lines of code to

```
sum = 0.0
do  i = 1,10 ! i is of type integer
    sum = sum+a(i)
end do
```

In this fragment of code we first set sum to zero, and then require that the statement between the do statement and the end do statement shall be executed ten times. For each iteration there is an associated value of an index, kept in i, which assumes the value 1 for the first iteration through the loop, 2 for the second, and so on up to 10. The variable i is a normal integer variable, but is subject to the rule that it must not be explicitly modified within the do construct.

The do statement has more general forms. If we wished to sum the fourth to ninth elements we would write

```
do  i = 4, 9
```

thereby specifying the required first and last values of i. If, alternatively, we wished to sum all the odd elements, we would write

```
do  i = 1, 9, 2
```

where the third of the three **loop parameters**, namely the 2, specifies that i is to be incremented in steps of 2, rather than by the default value of 1, which is assumed if no third parameter is given. In fact, we can go further still, as the parameters need not be constants at all, but integer expressions, as in

```
do  i = j+4, m, -k(j)**2
```

in which the first value of i is j+4, and subsequent values are decremented by k(j)**2 until the value of m is reached. Thus, do indices may run 'backwards' as well as 'forwards'. If any of the three parameters is a variable or is an expression that involves a variable, the value of the variable may be modified within the loop without affecting the number of iterations, as the **initial** values of the parameters are used for the control of the loop.

The general form of this type of bounded do construct control clause is

[*name* :] do [,] *variable* = *expr*$_1$, *expr*$_2$ [, *expr*$_3$]
      block
   end do [*name*]

where *variable* is a named scalar integer variable, *expr*$_1$, *expr*$_2$, and *expr*$_3$ (*expr*$_3$ must be nonzero when present) are any valid scalar integer expressions, and *name* is the optional construct name. The do and end do statements must be unnamed or bear the same *name*.

The number of iterations of a do construct is given by the formula

$$\max\left(\frac{expr_2 - expr_1 + expr_3}{expr_3}, 0\right),$$

where max is a function which we shall meet in Section 9.3.3 and which here returns either the value of the first expression or zero, whichever is the larger. There is a consequence following from this definition, namely that if a loop begins with the statement

```
do   i = 1, n
```

then its body will not be executed at all if the value of n on entering the loop is zero or less. This is an example of the **zero-trip loop**, and results from the application of the max function.

A very simple form of the do statement is the unbounded

[*name* :] do

which specifies an endless loop. In practice, a means to exit from an endless loop is required, and this is provided in the form of the exit statement:

exit [*name*]

where *name* is optional and is used to specify from which do construct the exit should be taken in the case of nested constructs.[2] Execution of an exit statement causes control to be transferred to the next executable statement after the end do statement to which it refers. If no name is specified, it terminates execution of the innermost do construct in which it is enclosed. As an example of this form of do, suppose we have used the type entry of Section 2.13.2 to construct a chain of entries in a sparse vector, and we wish to find the entry with index 10, known to be present. If first points to the first entry, the code in Figure 4.5 is suitable.

Figure 4.5 Searching a linked list.

```
type(entry), pointer :: first, current
   ⋮
current => first
do
    if (current%index == 10) exit
    current => current%next
end do
```

---

[2] In fact, a named exit can be used to exit from nearly any construct, not just a loop; see Section 4.5.

The `exit` statement is also useful in a bounded loop when all iterations are not always needed.

A related statement is the `cycle` statement

```
cycle [ name ]
```

which transfers control to the `end do` statement of the corresponding construct. Thus, if further iterations are still to be carried out, the next one is initiated.

The value of a `do` construct index (if present) is incremented at the end of every loop iteration for use in the subsequent iteration. As the value of this index is available outside the loop after its execution, we have three possible situations, each illustrated by the following loop:

```
do  i = 1, n
    ⋮
    if (i==j) exit
    ⋮
end do
l = i
```

The situations are as follows:

i) If, at execution time, n has the value zero or less, i is set to 1 but the loop is not executed, and control passes to the statement following the `end do` statement.

ii) If n has a value which is greater than or equal to j, an exit will be taken at the `if` statement, and l will acquire the last value of i, which is of course j.

iii) If the value of n is greater than zero but less than j, the loop will be executed n times, with the successive values of i being 1, 2, ..., n. When reaching the end of the loop for the nth time, i will be incremented a final time, acquiring the value n+1, which will then be assigned to l.

We see how important it is to make careful use of loop indices outside the `do` block, especially when there is the possibility of the number of iterations taking on the boundary value of the maximum for the loop.

The `do` block, just mentioned, is the sequence of statements between the `do` statement and the `end do` statement. It is prohibited to jump into a `do` block or to its `end do` statement from anywhere outside the block.

It is similarly invalid for the block of a `do` construct (or any other construct, such as an `if` or `case` construct), to be only partially contained in a block of another construct. The construct must be completely contained in the block. The following two sequences are valid:

```
do i = 1, n
    if (scalar-logical-expr) then
        ⋮
    end if
end do
```

and

```
if (scalar-logical-expr) then
    do i = 1, n
        ⋮
    end do
else
    ⋮
end if
```

Any number of do constructs may be nested. We may thus write a matrix multiplication as shown in Figure 4.6.

Figure 4.6 Matrix multiplication as a triply nested do construct.

```
do  i = 1, n
    do  j = 1, m
        a(i,j) = 0.0
        do  l = 1, k
            a(i,j) = a(i,j) + b(i,l)*c(l,j)
        end do
    end do
end do
```

A further form of do construct, the do concurrent construct, is described in Section 7.16, and additional, but redundant, forms of do syntax in Appendix A.11.

Finally, it should be noted that many short do loops can be expressed alternatively in the form of array expressions and assignments. However, this is not always possible, and a particular danger to watch for is where one iteration of the loop depends upon a previous one. Thus, the loop

```
do i = 2, n
    a(i) = a(i-1) + b(i)
end do
```

cannot be replaced by the statement

```
a(2:n) = a(1:n-1) + b(2:n)          ! Beware
```

4.5 Further uses of the `exit` statement

The simplest way of imposing a change on the execution flow is by use of an exit statement inside a block construct. Figure 4.7 shows how a named block construct can be used to deal with error conditions in a structured manner. For a detailed description of the block construct, see Section 8.12.

The exit statement can, in fact, be used to complete the execution of any construct except the do concurrent construct (Section 7.16). An example is shown in Figure 4.8.

Note that an exit statement without a construct name exits the innermost do construct. Since the different behaviours can easily confuse the reader, we recommend that if an exit

---

Figure 4.7 Use of an `exit` statement inside a `block` construct.

```
integer, parameter :: err_singular = -1
integer :: ierr
real :: x

try_recipe : block
   ⋮                 ! calculate x
   if (abs(x) < threshold) then
      ierr = err_singular
      exit try_recipe ! terminate block construct
   end if
   ierr = 0
   ⋮                 ! code that needs sufficiently large x
end block try_recipe
⋮                     ! check ierr and act appropriately
```

Figure 4.8 Exit from an `if` construct.

```
adding_to_set: if (add_x_to_set) then
   find_position: do i=1, size(set)
      if (x==set(i)) exit adding_to_set
      if (x>set(i)) exit find_position
   end do find_position
   set = [set(:i-1), x, set(i:)]   !set is reallocated (see Section 6.7)
end if adding_to_set
```

---

from a non-do construct is used in proximity to an `exit` from a do construct (as in Figure 4.8), both `exit` statements have construct labels.

Note that exit from a `do concurrent` construct is prohibited, as is exit from an outer construct from within a `do concurrent`, `critical` (Section 17.14), or `change team` (Section 18.3) construct.

4.6   Summary

In this chapter we have introduced the four main features by which control in Fortran code may be programmed – the `if` statement and construct, the `case` construct, and the `do` construct. The `block` construct has also been mentioned. The effective use of these features is key to sound code.

We have touched upon the concept of a program unit as being like the chapter of a book. Just as a book may have only one chapter, so a complete program may consist of just one program unit, which is known as a main program. In its simplest form it consists of a series of statements of the kinds we have been dealing with so far, and terminates with an end statement, which acts as a signal to the computer to stop processing the current program. In order to test whether a program unit of this type works correctly, we need to be able to output, to a terminal or printer, the values of the computed quantities. This topic will be fully explained in Chapter 10, and for the moment we need to know only that this can be achieved by a statement of the form

```
print * , ' var1 = ', var1 , ' var2 = ', var2
```

which will output a line such as

```
var1 = 1.0   var2 = 2.0
```

Similarly, input data can be read by statements like

```
read *, val1, val2
```

This is sufficient to allow us to write simple programs like that in Figure 4.9, which outputs the converted values of a temperature scale between specified limits, and Figure 4.10, which constructs a linked list. Sample inputs are shown at the end of each example.

Figure 4.9 Print a conversion table.

```
!    Print a conversion table of the Fahrenheit and Celsius
!    temperature scales between specified limits.
!
   real      :: celsius, fahrenheit
   integer   :: low_temp, high_temp, temperature
   character :: scale
!
read_loop: do
!
!   Read scale and limits
      read *, scale, low_temp, high_temp
!
!   Check for valid data
      if (scale /= 'C' .and. scale /= 'F') exit read_loop
!
!   Loop over the limits
      do  temperature = low_temp, high_temp
!
!   Choose conversion formula
         select case (scale)
         case ('C')
            celsius = temperature
            fahrenheit = 9.0/5.0*celsius + 32.0
!   Print table entry
            print *, celsius, ' degrees C correspond to',    &
               fahrenheit, ' degrees F'
         case ('F')
            fahrenheit = temperature
            celsius = 5.0/9.0*(fahrenheit-32.0)
!   Print table entry
            print *, fahrenheit, ' degrees F correspond to',&
               celsius, ' degrees C'
         end select
      end do
   end do read_loop
!
!   Termination
   print *, ' End of valid data'
   end
C  90    100
F  20    32
*   0    0
```

Figure 4.10 Constructing and printing a linked list.

```
type entry ! Type for sparse matrix
    real                :: value
    integer             :: index
    type(entry), pointer :: next
end type entry

type(entry), pointer :: first, current
integer              :: key
real                 :: value
!
! Create a null list
    nullify (first)
!
! Fill the list
    do
        read *, key, value
        if (key <= 0) exit
        allocate (current)
        current = entry(value, key, first)
        first => current
    end do
!
! Print the list
    current => first
    do
        if (.not.associated(current)) exit
        print *, current%index, current%value
        current => current%next
    end do
end

1 4
2 9
0 0
```

Exercises

1. Write a program which

    i) defines an array to have 100 elements;

    ii) assigns to the elements the values 1, 2, 3, ..., 100;

    iii) reads two integer values in the range 1 to 100;

    iv) reverses the order of the elements of the array in the range specified by the two values.

2. The first two terms of the Fibonacci series are both 1, and all subsequent terms are defined as the sum of the preceding two terms. Write a program which reads an integer value `limit` and which computes and prints the values of the first `limit` terms of the series.

3. The coefficients of successive orders of the binomial expansion are shown in the normal Pascal triangle form as

Write a program which reads an integer value `limit` and prints the coefficients of the first `limit` lines of the Pascal triangle.

4. Define a character variable of length 80. Write a program which reads a value for this variable. Assuming that each character in the variable is alphabetic, write code which sorts them into alphabetic order and prints out the frequency of occurrence of each letter.

5. Write a program to read an integer value `limit` and print the first `limit` prime numbers, by any method.

6. Write a program which reads a value x and calculates and prints the corresponding value x/(1.+x). The case x = −1. should produce an error message and be followed by an attempt to read a new value of x.

7. Given a chain of entries of the type `entry` of Section 2.13.2, modify the code in Figure 4.5 (Section 4.4) so that it removes the entry with index 10, and makes the previous entry point to the following entry.

5. Program units and procedures

5.1 Introduction

As we saw in the previous chapter, it is possible to write a complete Fortran program as a single unit, but it is preferable to break the program down into manageable units. Each such **program unit** corresponds to a program task that can be readily understood and, ideally, can be written, compiled, and tested in isolation. We will discuss three kinds of program unit: the main program, external subprogram, and module. Submodules (Chapter 16) enable the subdivision of very large modules, which assists their maintenance.

A complete program must, as a minimum, include one **main program**. This may contain statements of the kinds that we have met so far in examples, but normally its most important statements are invocations or calls to subsidiary programs, each of which is known as a **subprogram**. A subprogram defines a **function** or a **subroutine**. These differ in that a function returns a single object and usually does not alter the values of its arguments (so that it represents a function in the mathematical sense), whereas a subroutine usually performs a more complicated task, returning several results through its arguments and by other means. Functions and subroutines are known collectively as **procedures**.

There are various kinds of subprograms. A subprogram may be a program unit in its own right, in which case it is called an **external subprogram** and defines an **external procedure**. External procedures may also be defined by means other than Fortran. A subprogram may be a member of a collection in a program unit called a **module**, in which case we call it a **module subprogram** and it defines a **module procedure**. A subprogram may be placed inside a module subprogram, an external subprogram, or a main program, in which case we call it an **internal subprogram** and it defines an **internal procedure**. Internal subprograms may not be nested, that is, they may not contain further subprograms, and we expect them normally to be short sequences of code, say up to about twenty lines. We illustrate the nesting of subprograms in program units in Figure 5.1. If a program unit or subprogram contains a subprogram, it is called the **host** of that subprogram.

Besides containing a collection of subprograms, a module may contain data definitions, derived-type definitions, interface blocks (Section 5.12.1), and namelist groups (Section 8.19). This possibly-large collection may provide facilities associated with some particular task, such as providing matrix arithmetic, a library facility, or a database.

In this chapter we will describe program units and the statements that are associated with them. Within a complete program they may appear in any order, but many compilers require a module to precede other program units that use it.

Modern Fortran Explained, 3rd Edition. M. Metcalf, J. Reid, M. Cohen, and R. Bader. Oxford University Press (2024). © M. Metcalf, J. Reid, M. Cohen, and R. Bader (2024). DOI 10.1093/oso/9780198876571.001.0005

Figure 5.1 Nesting of subprograms in program units.

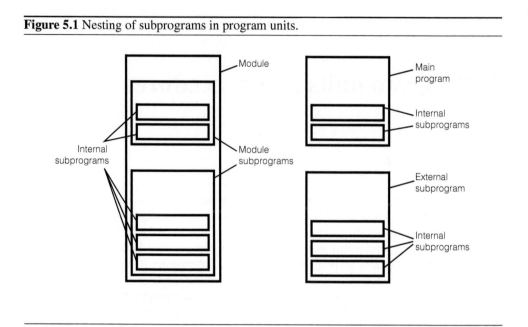

5.2   Main program

Every complete program must have one, and only one, main program. Optionally, it may contain calls to subprograms. A main program has the following form:

> *[* program *program-name]*
> *   [specification-stmts]*
> *   [executable-stmts]*
> *[* contains
> *   [internal-subprogram] ...]*
> end *[* program *[program-name]]*

The program statement is optional, but we recommend its use. The *program-name* may be any valid Fortran name such as model. The only non-optional statement is the end statement, which has two purposes. It acts as a signal to the compiler that it has reached the end of the program unit and, when executed, it causes the program to stop. This is **normal termination**. If the end program statement includes *program-name*, this must be the name on the program statement. We recommend using the full form so that it is clear both to the reader and to the compiler exactly what is being ended.

A main program without calls to subprograms is usually used only for short tests, as in

```
program test
    print *, 'Hello world!'
end program test
```

The specification statements define the environment for the executable statements. So far, we have met the type declaration statement (integer, real, complex, logical, character, and type(*type-name*)) that specifies the type and other properties of the entities that it lists, and the type definition block (bounded by type *type-name* and end type statements). We will meet other specification statements in this and the next three chapters.

The executable statements specify the actions that are to be performed. So far, we have met the assignment statement, the pointer assignment statement, the if statement and construct, the do and case constructs, the block construct, and the read and print statements. We will meet other executable statements in this and later chapters. Execution of a program always commences with the first executable statement of the main program.

The contains statement separates any internal subprograms from the body of the main program. We will describe internal subprograms in Section 5.6. They are excluded from the sequence of executable statements of the main program; if the last executable statement before a contains is executed without branching (Appendix A.12.3), the next statement executed will be the end statement. Note that although the syntax permits a contains statement without any following internal subprograms, it serves no purpose and so should be avoided. The end statement may be the target of a branch from one of the executable statements. If the end statement is executed, further execution stops with normal termination.

5.3 Program termination

Another way to stop program execution with normal termination is to execute a stop statement. This statement may appear in the main program or any subprogram. A well-designed program normally returns control to the main program for program termination, so the stop statement should appear there. However, in applications where several stop statements appear in various places in a complete program, it is possible to distinguish which of the stop statements has caused the termination by adding to each one a *stop code* consisting of a default integer or default character scalar expression (Sections 3.2 and 3.7.1). This might be used by a given processor to indicate the origin of the stop in a message. Examples are

```
stop
stop 12345
stop -2**20
stop 'Incomplete data. Program terminated.'
stop 'load_data_type_1'//': value out of range'
```

The Fortran standard requires that on termination by a stop statement, if any IEEE floating-point exception is signaling (Chapter 19), a warning message be written to the error file unit error_unit of the module iso_fortran_env (Section 9.24.2).

The standard also recommends that any stop code be written to the same unit and, if it is an integer, that it be used as the 'process exit status' if the operating system has such a concept. It further recommends that an exit status of zero be supplied if the stop code is of type character or the program is terminated by an end program statement. However, these are only recommendations, and in any case operating systems often have only a limited range for the process exit status, so values outside the range 0–127 should be avoided for this purpose.

There is also an `error stop` statement that causes **error termination** of program execution. The main differences between the `stop` and `error stop` statements are that

- normal termination properly closes all files, waiting for any input/output operation in progress to complete, but error termination has no such requirement (this could cause data loss if files are still being written);
- in a coarray program with multiple images, the entire computation is terminated with error termination, rather than a single image (see Sections 17.18 and 17.19 for details).

Error termination has the same IEEE exception and stop code reporting requirements, and the recommended process exit status with a numeric stop code is again the stop code value, but otherwise the exit status should be nonzero, in accordance with typical operating system conventions where zero indicates success and nonzero indicates failure. (This is also the recommended exit code for other error-termination situations, such as an unhandled input/output error or allocation failure.)

Even though the normal and error termination exit code values are merely recommendations of the Fortran standard, it rarely makes sense to second-guess the processor's choices here. For these reasons, we recommend the use of an informative message rather than an integer for both the `stop` and `error stop` statements.

Finally, we note that output of the stop code and exception summary from a `stop` or `error stop` statement can be controlled with a `quiet=` specifier, for example

```
stop failure_message, quiet=no_messages
```

If the `quiet=` specifier is true, no stop code or exception summary is output. Any scalar logical expression may be used for the `quiet=` specifier.

5.4   External subprograms

External subprograms are called from a main program or elsewhere, usually to perform a well-defined task within the framework of a complete program. Apart from the leading statement, they have a form that is very like that of a main program:

subroutine-stmt
    [specification-stmts]
    [executable-stmts]
[contains
    [internal-subprogram] ... *]*
end *[* subroutine *[subroutine-name]]*

or

function-stmt
    [specification-stmts]
    [executable-stmts]
[contains
    [internal-subprogram] ... *]*
end *[* function *[function-name]]*

The `contains` statement plays exactly the same role as within a main program (see Section 5.2). As before, although the syntax permits a `contains` statement without any following internal subprograms, we do not recommend doing that. The effect of executing an `end` statement in a subprogram is to return control to the caller, rather than to stop execution. As for the `end program` statement, we recommend using the full form for the `end` statement so that it is clear both to the reader and to the compiler exactly what it terminates.

The simplest form of external subprogram defines a subroutine without any arguments and has a *subroutine-stmt* of the form

```
subroutine subroutine-name
```

Such a subprogram is useful when a program consists of a sequence of distinct phases, in which case the main program consists of a sequence of `call` statements that invoke the subroutines as in the example

```
program game        ! Main program to control a card game
    call shuffle    ! First shuffle the cards.
    call deal       ! Now deal them.
    call play       ! Play the game.
    call display    ! Display the result.
end program game    ! Cease execution.
```

But how do we handle the flow of information between the subroutines? How does `play` know which cards `deal` has dealt? There are, in fact, two methods by which information may be passed. The first is via data held in a module (Section 5.5) and accessed by the subprograms, and the second is via arguments (Section 5.7) in the procedure calls.

5.5 Modules

The third type of program unit, the module, provides a means of packaging global data, derived types and their associated operations, subprograms, interface blocks (Section 5.12.1), and namelist groups (Section 8.19). Everything associated with some task (such as interval arithmetic, see later in this section) may be collected into a module and accessed whenever it is needed. Those parts that are associated with the internal working and are of no interest to the user may be made 'invisible' to the user, which allows the internal design to be altered without the need to alter the program that uses it and prevents accidental alteration of internal data. Fortran libraries often consist of sets of modules.

A module has the form

```
module module-name
    [ specification-stmts ]
[ contains
    [ module-subprogram ] ... ]
end [ module [ module-name ] ]
```

As for other program units, although the syntax permits a `contains` statement without any following module subprograms, we do not recommend doing that. As for the `end program`, `end subroutine`, and `end function` statements, we recommend using the full form for the `end` statement.

In its simplest form, the body consists only of data specifications. For example

```
module state
   integer, dimension(52) :: cards
end module state
```

might hold the state of play of the game of Section 5.4. It is accessed by the statement

```
use state
```

appearing at the beginnings of the main program game and subprograms shuffle, deal, play, and display. The array cards is set by shuffle to contain the integer values 1 to 52 in a random order, where each integer value corresponds to a predefined playing card. For instance, 1 might stand for the ace of clubs, 2 for the two of clubs, etc. up to 52 for the king of spades. The array cards is changed by the subroutines deal and play, and finally accessed by subroutine display.

A further example of global data in a module would be the definitions of the values of the kind type parameters (Section 2.6) that might be required throughout a program. They can be placed in a module and used wherever they are required. On a processor that supports all the kinds listed, an example might be:

```
module numeric_kinds
   ! named constants for 4, 2, and 1 byte integers:
   integer, parameter ::                               &
         i4b = selected_int_kind(9),                   &
         i2b = selected_int_kind(4),                   &
         i1b = selected_int_kind(2)
   ! and for single, double and quadruple precision reals:
   integer, parameter ::                               &
         sp = kind(1.0),                               &
         dp = selected_real_kind(2*precision(1.0_sp)), &
         qp = selected_real_kind(2*precision(1.0_dp))
end module numeric_kinds
```

A very useful role for modules is to contain definitions of types and their associated operators. For example, a module might contain the type interval of Section 3.9, as shown in Figure 5.2. Given this module, any program unit needing this type and its operators need only include the statement

```
use interval_arithmetic
```

at the head of its specification statements.

A module subprogram has exactly the same form as an external subprogram. It has access to other entities of the module, including the ability to call other subprograms of the module, rather as if it contained a use statement for its module.

A module may contain use statements that access other modules. It must not access itself directly or indirectly through a chain of use statements, for example a accessing b and b accessing a. No overall ordering of modules is required by the standard, but no module may precede a module that it uses.

It is possible within a module to specify that some of the entities are private to it and cannot be accessed from other program units. Also, there are forms of the use statement that allow

---

Figure 5.2 A module for interval arithmetic.

```
module interval_arithmetic
   type interval
      real :: lower, upper
   end type interval
   interface operator(+)
      module procedure add_intervals
   end interface
   ⋮
contains
   function add_intervals(a,b)
      type(interval)               :: add_intervals
      type(interval), intent(in) :: a, b
      add_intervals%lower = a%lower + b%lower
      add_intervals%upper = a%upper + b%upper
   end function add_intervals
   ⋮
end module interval_arithmetic
```

---

access to only part of a module and forms that allow renaming of the entities accessed. These features will be explained in Sections 8.6.1 and 8.13. For the present, we assume that the whole module is accessed without any renaming of the entities in it.

5.6   Internal subprograms

We have seen that internal subprograms may be defined inside main programs and external subprograms, and within module subprograms. They have the form

> *subroutine-stmt*
>   *[specification-stmts]*
>   *[executable-stmts]*
> end *[* subroutine *[subroutine-name]]*

or

> *function-stmt*
>   *[specification-stmts]*
>   *[executable-stmts]*
> end *[* function *[function-name]]*

that is, the same form as a module subprogram, except that they may not contain further internal subprograms. An internal subprogram automatically has access to all the host's entities, including the ability to call its other internal subprograms. This is known as **host association**. Internal subprograms must be preceded by a contains statement in the host. As for other end statements, we recommend using the full form of the end statement for internal subprograms.

In the rest of this chapter we describe several properties of subprograms that apply to external, module, and internal subprograms. We therefore do not need to describe internal subprograms separately. An example is given in Figure 5.13 (Section 5.16).

5.7   Arguments of procedures

5.7.1   Argument association

Procedure arguments provide an alternative means for two program units to access the same data. Returning to our card game example, instead of placing the array `cards` in a module, we might declare it in the main program and pass it as an actual argument to each subprogram, as shown in Figure 5.3.

Figure 5.3 Subroutine calls with actual arguments.

```
program game          ! Main program to control a card game
   integer, dimension(52) :: cards
   call shuffle(cards)        ! First shuffle the cards.
   call deal(cards)           ! Now deal them.
   call play(cards)           ! Play the game.
   call display(cards)        ! Display the result.
end program game              ! Cease execution.
```

Each subroutine receives `cards` as a dummy argument. For instance, `shuffle` has the form shown in Figure 5.4.

Figure 5.4 A subroutine with a dummy argument.

```
subroutine shuffle(cards)
   ! Subroutine that places the values 1 to 52 in cards
   ! in random order.
   integer, dimension(52) :: cards
   ! Statements that fill cards
   ⋮
end subroutine shuffle    ! Return to caller.
```

We can, of course, imagine a card game in which `deal` is going to deal only three cards to each of four players. In this case, it would be a waste of time for `shuffle` to prepare a deck of 52 cards when only the first 12 cards are needed. This can be achieved by requesting `shuffle` to limit itself to a number of cards that is transmitted in the calling sequence thus:

```
call shuffle(3*4, cards(1:12))
```

Inside `shuffle`, we would define the array to be of the given length and the algorithm to fill `cards` would be placed in a do construct with this number of iterations, as seen in Figure 5.5.

We have seen how it is possible to pass an array and a constant expression between two program units. An actual argument may be any variable or expression (or a procedure

Figure 5.5 A subroutine with two dummy arguments.

```
subroutine shuffle(ncards, cards)
   integer                   :: ncards, icard
   integer, dimension(ncards) :: cards
   do icard = 1, ncards
      :
      cards(icard) = ...
   end do
end subroutine shuffle
```

name, see Section 5.13). Each dummy argument of the called procedure must agree with the corresponding actual argument in type, type parameters, and shape (except as described in Section 20.5). However, the names do not have to be the same. For instance, if two decks had been needed, we might have written the code thus:

```
program game
   integer, dimension(52) :: acards, bcards
   call shuffle(acards)       ! First shuffle the a deck.
   call shuffle(bcards)       ! Next shuffle the b deck.
   :
end program game
```

The important point is that subprograms can be written independently of one another, the association of the dummy arguments with the actual arguments occurring each time the call is executed. This is known as **argument association**. We can imagine shuffle being used in other programs which use other names. In this manner, libraries of subprograms may be built up.

Being able to have different names for actual and dummy arguments provides useful flexibility, but it should only be used when it is actually needed. When the same name can be used, the code is more readable.

As the type of an actual argument and its corresponding dummy argument must agree, care must be taken when using component selection within an actual argument. Thus, supposing the derived-type definitions point and triangle of Figure 2.1 (Section 2.9) are available in a module def, we might write

```
      use def
      type(triangle) :: t
      :
      call sub(t%a)
      :
   contains
      subroutine sub(p)
         type(point) :: p
         :
```

A dummy argument of a procedure may be used as an actual argument in a procedure call. The called procedure may use its dummy argument as an actual argument in a further procedure call. A chain is built up with argument association at every link. The chain ends at an object that is not a dummy argument, which is known as the **ultimate argument** of the original dummy argument.

5.7.2   Assumed-shape arrays

Outside Appendix B, we require that the shapes of actual and dummy arguments agree, and so far we have achieved this by passing the extents of the array arguments as additional arguments. However, it is possible to require that the shape of the dummy array be taken automatically to be that of the corresponding actual array argument. Such an array is said to be an **assumed-shape array**. When the shape is declared by the dimension clause, each dimension has the form

   [lower-bound] :

where *lower-bound* is an integer expression that may depend on module data or the other arguments (see Section 8.17 for the exact rules). If *lower-bound* is omitted, the default value is one. Note that it is the shape that is passed, and not the upper and lower bounds. For example, if the actual array is a, declared thus:

```
real, dimension(0:10, 0:20) :: a
```

and the dummy array is da, declared thus:

```
real, dimension(:, :) :: da
```

then a(i,j) corresponds to da(i+1,j+1); to get the natural correspondence, the lower bound must be declared:

```
real, dimension(0:, 0:) :: da
```

In order that the compiler knows that additional information is to be supplied, the interface must be explicit (Section 5.12) at the point of call. A dummy array with the pointer or allocatable attribute is not regarded as an assumed-shape array because its shape is not necessarily assumed.

If the size or shape is needed within the code, the intrinsic functions size and shape are available, for example

```
subroutine sub (da)
real, dimension(:, :) :: da
integer :: size_da, shape_da(2)
size_da = size(da)
shape_da = shape(da)
   ⋮
```

5.7.3 Pointer arguments

A dummy argument is permitted to have the attribute `pointer`. Unless it has intent `in`, see Section 7.17.3, the actual argument must also have the attribute `pointer`. When the subprogram is invoked, the rank of the actual argument must match that of the dummy argument, and its pointer association status is passed to the dummy argument. On return, the actual argument normally takes its pointer association status from that of the dummy argument, but it becomes undefined if the dummy argument is associated with a target that becomes undefined when the return is executed (for example, if the target is a local variable that does not have the `save` attribute, Section 5.9).

In the case of a module or internal procedure, the compiler knows when the dummy argument is a pointer. In the case of an external or dummy procedure, the compiler assumes that the dummy argument is not a pointer unless it is told otherwise in an interface block (Section 5.12.1).

A pointer actual argument is also permitted to correspond to a non-pointer dummy argument. In this case, the pointer must have a target and the target is associated with the dummy argument, as in

```
real, pointer :: a(:,:)
   :
allocate ( a(80,80) )
call find (a)
   :
subroutine find (c)
   real :: c(:,:) ! Assumed-shape array
```

The term **effective argument** is used to refer to the entity that is associated with a dummy argument, that is, the actual argument if it is not a pointer, and its target otherwise.

5.7.4 Polymorphic arguments

A dummy argument that is neither allocatable nor a pointer is permitted to be polymorphic. The corresponding actual argument must be type-compatible with it and the dummy argument assumes its dynamic type from the actual argument. This provides a convenient means of writing a function that applies to any extension of a type, for example

```
real function distance(a, b)
  class(point) :: a, b
  distance = sqrt((a%x-b%x)**2 + (a%y-b%y)**2)
end function distance
```

This function will work unchanged, for example, not only on a scalar of type `point` but also on a scalar of type

```
type, extends(point) :: data_point
   real, allocatable :: data_value(:)
end type data_point
```

As for allocatable and pointer polymorphic variables (Section 2.15), access to components that are in the dynamic type but not the declared type is provided by type-bound procedures (see Section 15.8) or by the `select type` construct (see Section 15.7).

A polymorphic dummy argument that is allocatable or a pointer is permitted to be associated only with an actual argument of the same declared type. This is to ensure that allocations or pointer assignments within the procedure involve objects that are type-compatible with the actual argument.

Polymorphic arguments are discussed fully in Chapter 15.

5.7.5   Restrictions on actual arguments

There are two important restrictions on actual arguments, which are designed to allow the compiler to optimize on the assumption that the dummy arguments are distinct from each other and from other entities that are accessible within the procedure. For example, a compiler may arrange for an array to be copied to a local variable on entry, and copied back on return. While an actual argument is associated with a dummy argument the following statements hold:

i) Action that affects the allocation status or pointer association status of the argument or any part of it (any pointer assignment, allocation, deallocation, or nullification) must be taken through the dummy argument. If this is done, then throughout the execution of the procedure, the argument may be referenced only through the dummy argument.

ii) Action that affects the value of the argument or any part of it must be taken through the dummy argument unless

    a) the dummy argument has the `pointer` attribute;

    b) the part is all or part of a pointer subobject; or

    c) the dummy argument has the `target` attribute, the dummy argument does not have intent in (Section 5.10), the dummy argument is scalar or an assumed-shape array (Section 5.7.2), and the actual argument is a target other than an array section with a vector subscript.

If the value of the argument or any part of it is affected through a dummy argument for which neither a), b), nor c) holds, then throughout the execution of the procedure, the argument may be referenced only through that dummy argument.

An example of i) is a pointer that is nullified (Section 3.14) while still associated with the dummy argument. As an example of ii), consider

```
call modify(a(1:5), a(3:9))
```

Here, a(3:5) may not be changed through either dummy argument since this would violate the rule for the other argument. However, a(1:2) may be changed through the first argument and a(6:9) may be changed through the second. Another example is an actual argument that is an object being accessed from a module; here, the same object must not be accessed from the module by the procedure and redefined. As a third example, suppose an internal procedure call associates a host variable h with a dummy argument d. If d is defined during the call, then at no time during the call may h be referenced directly.

5.7.6   Arguments with the target attribute

In most circumstances an implementation is permitted to make a copy of an actual argument on entry to a procedure and copy it back on return. This may be desirable on efficiency grounds, particularly when the actual argument is not held in contiguous storage (for example `a(1:n:3)`). In any case, if a dummy argument has neither the `target` nor `pointer` attribute, any pointers associated with the actual argument do not become associated with the corresponding dummy argument but remain associated with the actual argument.

However, copy-in copy-out is not allowed when

i) a dummy argument has the `target` attribute and is either scalar or is an assumed-shaped array; and

ii) the actual argument is a target other than an array section with a vector subscript.

In this case, the dummy and actual arguments must have the same shape, any pointer associated with the actual argument becomes associated with the dummy argument on invocation, and any pointer associated with the dummy argument on return remains associated with the actual argument.

When a dummy argument has the `target` attribute, but the actual argument is not a target or is an array section with a vector subscript, any pointer associated with the dummy argument obviously becomes undefined on return.

In other cases where the dummy argument has the `target` attribute, whether copy-in copy-out occurs is processor dependent. No reliance should be placed on the pointer associations with such an argument after the invocation.

5.8   The `return` statement

We saw in Section 5.2 that if the `end` statement of a main program is executed, further execution stops. Similarly, if the `end` statement in a subprogram is executed, control returns to the point of invocation. Just as the `stop` statement is an executable statement that provides an alternative means of stopping execution, so the `return` statement provides an alternative means of returning control from a subprogram. It has the form

```
return
```

and must not appear in a main program.

5.9   Local variables

There is often a need for a procedure to employ additional variables to assist the calculation. A simple example is `icard` in Figure 5.5. The value is usually needed only while the procedure is in execution and compilers often allocate storage for it on entry and deallocate it on return. If it is required for the value to be retained between calls, the programmer can ensure that it has the `save` attribute by declaring it with this attribute explicitly,

```
real, save :: a
```

or by giving it an initial value,

```
integer :: count = 0
```

We illustrate with code to compute mean values of positive numbers in Figure 5.6. A negative value is used as a sequence terminator. The local variables count and sum are given initial values, which makes them have the save attribute.

Figure 5.6 Computing mean values of positive numbers.

```
subroutine average (x, mean)
  real :: x, mean
  integer :: count = 0 ! Local variable with save attribute
  real :: sum = 0.0    ! Local variable with save attribute
  if (x<0.0) then
    count = 0
    sum = 0.0
  else
    sum = sum + x
    count = count + 1
    mean = sum/count
  end if
end subroutine
```

5.10   Argument intent

In Figure 5.5, the dummy argument cards was used to pass information out from shuffle and the dummy argument ncards was used to pass information in; a third possibility is for a dummy argument to be used for both input and output variables. We can specify such intent on the type declaration statement for the argument, for example:

```
subroutine shuffle(ncards, cards)
  integer, intent(in)                        :: ncards
  integer, intent(out), dimension(ncards) :: cards
```

For input/output arguments, intent inout may be specified.

If a dummy argument is specified with intent in, it (or any part of it) must not be redefined by the procedure, say by appearing on the left-hand side of an assignment or by being passed on as an actual argument to a procedure that redefines it. For the specification intent inout, the corresponding actual argument must be a variable because the expectation is that it will be redefined by the procedure. For the specification intent out, the corresponding actual argument must again be a variable; in this case, the intention is that it be used only to pass information out, so it becomes undefined on entry to the procedure, apart from any components with default initialization (Section 8.5.5).

If a function specifies a defined operator (Section 3.9), its dummy arguments must have intent in. If a subroutine specifies defined assignment (Section 3.10), its first argument must have intent out or inout, and its second argument must have intent in.

If a dummy argument has no intent, the corresponding actual argument may be a variable or an expression, but the actual argument must be a variable if the dummy argument is redefined. In this context we note that an actual argument variable, say x, is transformed into an expression if it is enclosed in parentheses, (x); its value is then passed and cannot be redefined. The compiler will need to make a copy if the variable x could be changed during execution of the procedure, e.g. if x is not a local variable or is accessible via a pointer. We recommend that all dummy arguments be given a declared intent, allowing compilers to make more checks at compile time and as good documentation.

If a dummy argument has the `pointer` attribute, any intent refers to its pointer association (and **not** to the value of the target); that is, it refers to the descriptor. An intent `out` pointer has undefined association status on entry to the procedure; an intent `in` pointer cannot be nullified or associated during execution of the procedure; and the actual argument for an intent `inout` pointer must be a pointer variable (that is, it cannot be a reference to a pointer-valued function).

Note that, although an intent `in` pointer cannot have its pointer association status changed inside the procedure, if it is associated with a target the value of its target may be changed. For example,

```
subroutine maybe_clear(p)
   real, pointer, intent(in) :: p(:)
   if (associated(p)) p = 0.0
end subroutine maybe_clear
```

Likewise, if a dummy argument is of a derived type with a pointer component, its `intent` attribute refers to the pointer association status of that component (and **not** to the target of the component). For example, if the intent is `in`, no pointer assignment, allocation, or deallocation is permitted.

5.11 Functions

5.11.1 Function results

Functions are similar to subroutines in many respects, but they are invoked within an expression and return a value that is used within the expression. For example, the subprogram in Figure 5.7 returns the distance between two points in space; the statement

```
if (distance(a, c) > distance(b, c)) then
```

invokes the function twice in the logical expression that it contains.

Note the type declaration for the function result. The result behaves just like a dummy argument with intent `out`. It is initially undefined, but once defined it may appear in an expression and it may be redefined. The type may also be defined on the `function` statement:

```
real function distance(p, q)
```

It is permissible to write functions that change the values of their arguments, modify values in modules, rely on local data saved from a previous invocation, or perform input/output operations. However, these are known as **side-effects** and conflict with good programming

Figure 5.7 A function that returns the distance between two points in space. The intrinsic function sqrt is defined in Section 9.4.

```
function distance(p, q)
   real                        :: distance
   real, intent(in), dimension(3) :: p, q
   distance = sqrt( (p(1)-q(1))**2 + (p(2)-q(2))**2 +    &
                    (p(3)-q(3))**2 )
end function distance
```

practice. Where they are needed, a subroutine should be used. It is reassuring to know that when a function is called, nothing else goes on 'behind the scenes', and it may be very helpful to an optimizing compiler, particularly for internal and module subprograms. A formal mechanism for avoiding side-effects is provided, but we defer its description to Section 7.8.

A function result may be an array, in which case it must be declared as such.

A function result may also be a pointer. The result is initially undefined. Within the function, it must become associated or defined as disassociated. We expect the function reference usually to be such that a pointer assignment takes place for the result, that is, the reference occurs as the right-hand side of a pointer assignment (Section 3.13), for example,

```
real          :: x(100)
real, pointer :: y(:)
   :
y => compact(x)
   :
```

or as a pointer component of a structure constructor. The reference may also occur as a primary of an expression or as the right-hand side of an ordinary assignment, in which case the result must become associated with a target that is defined and the value of the target is used. We do not recommend this practice, however, since it is likely to lead to memory leakage, discussed at the end of Section 6.6.

The value returned by a non-pointer function must always be defined.

As well as being a scalar or array value of intrinsic type, a function result may also be a scalar or array value of a derived type, as we have seen already in Section 3.9. When the function is invoked, the function value must be used as a whole, that is, it is not permitted to be qualified by substring, array-subscript, array-section, or structure-component selection.

A function is permitted to have an empty argument list. In this case, the brackets are obligatory both within the function statement and at every invocation.

5.11.2   Prohibited side-effects

In order to assist an optimizing compiler, the standard prohibits reliance on certain side-effects. It specifies that it is not necessary for a processor to evaluate all the operands of an expression, or to evaluate entirely each operand, if the value of the expression can be determined otherwise. For example, in evaluating

```
x>y .or. l(z)   ! x, y, and z are real; l is a logical function
```

the function reference need not be made if x is greater than y. Since some processors will make the call and others will not, any variable (for example z) that is redefined by the function is regarded as undefined following such an expression evaluation. Similarly, it is not necessary for a processor to evaluate any subscript or substring expressions for an array of zero size or character object of zero character length.

Another prohibition is that a function reference must not redefine the value of a variable that appears in the same statement or affect the value of another function reference in the same statement. For example, in

```
d = max(distance(p,q), distance(q,r))
```

distance is required not to redefine its arguments. This rule allows any expressions that are arguments of a single procedure call to be evaluated in any order. With respect to this rule, an if statement,

```
if (lexpr) stmt
```

is treated as the equivalent if construct

```
if (lexpr) then
    stmt
end if
```

and the same is true for the where statement (Section 7.6).

5.12 Explicit interfaces

5.12.1 Interface blocks

A call to an internal subprogram must be from a statement within the same program unit. It may be assumed that the compiler will process the program unit as a whole and will therefore know all about any internal subprogram. In particular, it will know about its **interface**, that is, whether it defines a function or a subroutine, the names and properties of the arguments, and the properties of the result if it defines a function. This, for example, permits the compiler to check whether the actual and dummy arguments match in the way that they should. We say that the interface is **explicit**.

A call to a module subprogram must either be from another statement in the module or from a statement following a use statement for the module. In both cases the compiler will know all about the subprogram, and again we say that the interface is explicit. Similarly, intrinsic procedures (Chapter 9) always have explicit interfaces.

As will be explained later (in Section 14.1), for external and dummy procedures an explicit interface can be specified by the procedure declaration statement. However, another mechanism is provided for the interface to be specified. It may be done through an interface block of the form

```
interface
    [interface-body]...
end interface
```

Normally, each *interface-body* is an exact copy of the subprogram's header, the specifications of its arguments and function result, and its end statement. However,

- the names of the arguments may be changed;

- other specifications may be included (for example, for a local variable), but not internal procedures, data statements, or format[1] statements;

- the information may be given by a different combination of statements;

- in the case of an array argument or function result, the expressions that specify a bound may differ as long as their values can never differ; and

- a recursive procedure (Sections 5.17.1 and 5.17.2) or a pure procedure (Section 7.8) need not be specified as such if it is not called as such.

Because the procedure itself will not have access to the declarations in the host of the interface, the interface does not have access to these declarations. If they are wanted an import statement (Section 5.12.2) is needed.

An *interface-body* may be provided for a call to an external procedure defined by means other than Fortran (usually C or assembly language).

An external or dummy procedure without an interface block is said to have an **implicit** interface. Such an interface is error-prone and we do not recommend it. Hereafter, we will assume that all interfaces are explicit except that we discuss them further in Appendix A.13, where we detail the situations in which explicit interfaces are required.

The interface block is placed in a sequence of specification statements. Perhaps the most convenient way to do this is to place the interface block among the specification statements of a module and then use the module. Libraries can be written as sets of external subprograms together with modules holding interface blocks for them. This keeps the modules of modest size.

Interface blocks may also be used to allow procedures to be called as defined operators (Section 3.9), as defined assignments (Section 3.10), or under a single generic name. We therefore defer description of the full generality of the interface block until Section 5.18, where overloading is discussed.

5.12.2   The import statement

As we have seen, an interface body for an external or dummy procedure does not access its environment by host association, and therefore cannot use any named constants and derived types defined therein. In particular, it might be desirable in a module procedure to be able to describe a dummy procedure that uses types and kind type parameters defined in the module, but without host association this is impossible, as a module cannot 'use' itself. This problem

---

[1]Section 10.2.

is solved by the `import` statement. This statement can be used only in an interface body, and gives access to named entities of the containing scoping unit.

For example, in Figure 5.8 the interface body would be invalid without the `import` statement shown in bold, because it would have no means of accessing either type `t` or the constant `wp`.

Figure 5.8 Using an `import` statement to provide an explicit interface using types and constants defined in the module.

```
module m
    integer, parameter :: wp = kind(0.0d0)
    type t
        ⋮
    end type t
contains
    subroutine apply(fun,...)
        interface
            function fun(f)
                import  :: t, wp
                type(t)  :: fun
                real(wp) :: f
            end function fun
        end interface
    end subroutine apply
end module m
```

The statement must be placed after any `use` statements but ahead of any other statements of the body. It has the basic form

    `import [ [::] ` *import-name-list* `]`

as well as

    `import, only:`   *import-name-list*
    `import, none`
    `import, all`

where each *import-name* is that of an entity that is accessible in the containing scoping unit.

If one `import` statement in a scoping unit is an `import, only` statement, they must all be, and only the entities listed become accessible by host association.

If an `import, none` statement appears in a scoping unit, no entities are accessible by host association and it must be the only `import` statement in the scoping unit; `import, none` is not permitted in the specification part of a submodule except within an interface body.

If an `import, all` statement appears in a scoping unit, all entities of the host are accessible by host association and it must be the only `import` statement in the scoping unit.

Each *import-name* must be the name of a host entity and must not appear in a context that makes the host entity inaccessible, such as being declared as a local entity. For `import, all` no entity of the host may be inaccessible for this reason. This restriction does not apply to the simple form of import.

If an imported entity is defined in the containing scoping unit, it must be explicitly declared prior to the interface body.

An import statement without a list imports all entities from the containing scoping unit that are not declared to be local entities of the interface body; this works the same way as normal host association.

5.13   Procedures as arguments

So far, we have taken the actual arguments of a procedure invocation to be variables and expressions, but another possibility is for them to be procedures. Let us consider the case of a library subprogram for function minimization. It needs to receive the user's function, just as the subroutine shuffle in Figure 5.5 needs to receive the required number of cards. The minimization code might look like the code in Figure 5.9. Notice the way the procedure argument is declared by an interface block playing a similar role to that of the type declaration statement for a data object.

Figure 5.9 A library subprogram for function minimization.

```
real function minimum(a, b, func) ! Returns the minimum
        ! value of the function func(x) in the interval (a,b)
   real, intent(in) :: a, b
   interface
      real function func(x)
         real, intent(in) :: x
      end function func
   end interface
   real :: f,x
   :
   f = func(x)    ! invocation of the user function.
   :
end function minimum
```

Just as the type and shape of actual and dummy data objects must agree, so must the properties of the actual and dummy procedures. The agreement is exactly as for a procedure and an interface body for that procedure (see Section 5.12.1). It would make no sense to specify an intent attribute (Section 5.10) for a dummy procedure, and this is not permitted.

On the user side, the code may look like that in Figure 5.10. Notice that the structure is rather like a sandwich: user-written code invokes the minimization code which in turn invokes user-written code. An external procedure here would instead require the presence of an interface block and reference to an abstract interface (Section 14.1).

The procedure that is passed can be an external, internal, or module procedure, or an associated procedure pointer (Section 14.2).

Figure 5.10 Invoking the library code of Figure 5.9.

```
module code
contains
   real function fun(x)
      real, intent(in) :: x
         ⋮
   end function fun
end module code
program main
   use code
   use library ! Contains function minimum
   real :: f
      ⋮
   f = minimum(1.0, 2.0, fun)
      ⋮
end program main
```

When an internal procedure is passed as an actual argument, the environment of the host procedure is passed with it. That is, when it is invoked via the corresponding dummy argument, it has access to the variables of the host procedure as if it had been invoked there. For example, in Figure 5.11, invocations of the function fun from integrate will use the values for the variables freq and alpha from the host procedure.

Figure 5.11 Quadrature using internal procedures.

```
subroutine s(freq, alpha, lower, upper, ...)
   use library ! Contains function integrate
   real(wp), intent(in) :: freq, alpha, lower, upper
      ⋮
   z = integrate(fun, lower, upper)
      ⋮
contains
   real(wp) function fun(x)
      real(wp), intent(in) :: x
      fun = x*sin(freq*x)/sqrt(1-alpha*x**2)
   end function
end subroutine
```

If the host procedure is recursive (Sections 5.17.1 and 5.17.2), the instance that called the procedure with its internal procedure as an actual argument is known as the **host instance** of the internal procedure. It is the data in this instance that can be accessed by the internal procedure.

Apart from the convenience, this code can in principle safely be part of a multi-threaded program because the data for the function evaluation are not being passed by global variables.

5.14   Keyword and optional arguments

In practical applications, argument lists can get long and actual calls may need only a few arguments. For example, a subroutine for constrained minimization might have the form

```
subroutine mincon(n, f, x, upper, lower,                          &
                   equalities, inequalities, convex, xstart)
```

For many calls there may be no upper bounds, or no lower bounds, or no equalities, or no inequalities, or it may not be known whether the function is convex, or a sensible starting point may not be known. All the corresponding dummy arguments may be declared optional (see also Section 8.9). For instance, the bounds might be declared by the statement

```
real, optional, dimension(n) :: upper,lower
```

If the first four arguments are the only ones wanted, we may use the statement

```
call mincon(n, f, x, upper)
```

but usually the wanted arguments are scattered. In this case, we may follow a (possibly empty) ordinary positional argument list for leading arguments by a keyword argument list, as in the statement

```
call mincon(n, f, x, equalities=q, xstart=x0)
```

The keywords are the dummy argument names and there must be no further positional arguments after the first keyword argument.

This example also illustrates the merits of both positional and keyword arguments as far as readability is concerned. A small number of leading positional arguments (for example, n, f, and x) are easily linked in the reader's mind to the corresponding dummy arguments. Beyond this, the keywords are very helpful to the reader in making these links. We recommend their use for long argument lists even when there are no gaps caused by optional arguments that are not present.

A non-optional argument must appear exactly once, either in the positional list or in the keyword list. An optional argument may appear at most once, either in the positional list or in the keyword list. An argument must not appear in both lists.

The called subprogram needs some way to detect whether an argument is present so that it can take appropriate action when it is not. This is provided by the intrinsic function present (see Section 9.2). For example,

```
present(xstart)
```

returns the value .true. if the current call has provided a starting point and .false. otherwise. When it is absent, the subprogram might, for example, use a random number generator to provide a starting point.

A slight complication occurs if an optional dummy argument is used within the subprogram as an actual argument in a procedure invocation. For example, our minimization subroutine mincon might start by calling a subroutine that handles the corresponding equality problem by the call

```
call mineq(n, f, x, equalities, convex, xstart)
```

In such a case, an absent optional argument is also regarded as absent in the second-level subprogram. For instance, when `convex` is absent in the call of `mincon`, it is regarded as absent in `mineq` too. Such absent arguments may be propagated through any number of calls, provided the dummy argument is optional in each case. An absent argument further supplied as an actual argument must be specified as a whole, and not as a subobject. Furthermore, an absent pointer is not permitted to be associated with a non-pointer dummy argument (the target is doubly absent).

5.15   Use and scope of labels

Fortran statement labels have an impact on the execution of a program in the following situations:

- the label appears on a `format` statement and is referenced by a data transfer statement (Sections 10.7 and 10.8);

- the label appears on a statement that is permitted as the target for branched execution (Appendices A.12, B.1.5, and B.1.11);

- the labelled do construct (Appendix B.1.14).

To avoid ambiguities in referencing labels, each subprogram has its own independent set of labels. This includes the case of a host subprogram with several internal subprograms. The same label may be referenced in the host and the internal subprograms without ambiguity.

This is our first encounter with **scope**. The scope of a label, known as **inclusive scope**, is a main program or a subprogram, excluding any internal subprograms that it contains. The label may be referenced unambiguously anywhere among the executable statements of its scope. Notice that the host `end` statement may be labelled and be a branch target from a host statement; that is, the internal subprograms leave a hole in the scope of the host.

5.16   Scope of names

In the case of a named entity, there is a similar set of statements within which the name may always be used to refer to the entity. Here, derived-type definitions and interface blocks as well as subprograms can knock holes in scopes. This leads us to regard each program unit as consisting of a set of non-overlapping scoping units. A **scoping unit** is one of the following:

- a derived-type definition;
- a procedure interface body, excluding any derived-type definitions and interface bodies contained within it;
- a program unit or subprogram, excluding derived-type definitions, interface bodies, and subprograms contained within it; or
- a block construct (Section 8.12).

Figure 5.12 An example of nested scopes.

```
module scope1              ! scope 1
    ⋮                      ! scope 1
contains                   ! scope 1
    subroutine scope2      ! scope 2
        type scope3        ! scope 3
            ⋮              ! scope 3
        end type scope3    ! scope 3
        interface          ! scope 2
            ⋮              ! scope 4
        end interface      ! scope 2
        ⋮                  ! scope 2
    contains               ! scope 2
        function scope5(...) ! scope 5
            ⋮              ! scope 5
        end function scope5 ! scope 5
    end subroutine scope2  ! scope 2
end module scope1          ! scope 1
```

An example containing five scoping units is shown in Figure 5.12.

Once an entity has been declared in a scoping unit, its name may be used to refer to it in that scoping unit. An entity declared in another scoping unit is always a different entity even if it has the same name and exactly the same properties.[2] Each is known as a **local entity**. This is very helpful to the programmer, who does not have to be concerned about the possibility of accidental name clashes. Note that this is true for derived types, too. Even if two derived types have the same name and the same components, entities declared with them are treated as being of different types.[2]

A use statement of the form

use *module-name*

is regarded as a redeclaration of all the module entities inside the local scoping unit, with exactly the same names and properties. The module entities are said to be accessible by **use association**. Names of entities in the module may not be used to declare local entities (but see Section 8.13 for a description of further facilities provided by the use statement when greater flexibility is required).

In the case of a derived-type definition, a module subprogram, or an internal subprogram, the scoping unit that immediately contains it is known as the **host scoping unit**. The name of an entity in the host scoping unit (including an entity accessed by use association) is treated as being automatically redeclared with the same properties, provided no entity with

---

[2] Apart from the effect of storage association, which is not discussed until Appendix A and whose use we strongly discourage.

this name is declared locally, is a local dummy argument or function result, or is accessed by use association. The host entity is said to be accessible by **host association**. For example, in the subroutine `inner` of Figure 5.13, x is accessible by host association, but y is a separate local variable and the y of the host is inaccessible. We note that `inner` calls another internal procedure that is a function, f; it must not contain a type specification for that function, as the interface is already explicit. Such a specification would, in fact, declare a different, external function of that name. The same remark applies to a module procedure calling a function in the same module.

Figure 5.13 Examples of host association.

```
subroutine outer
   real :: x, y
   ⋮
contains
   subroutine inner
      real :: y
      y = f(x) + 1. ! x and f accessed by host association
      ⋮
   end subroutine inner
   function f(z)
      real             :: f
      real, intent(in) :: z
      ⋮
   end function f
end subroutine outer
```

Note that the host has no access to the local entities of a subroutine that it contains.

Host association does not extend to interface blocks unless an `import` statement (Section 5.12.2) is used. This allows an interface body to be constructed mechanically from the specification statements of an external procedure.

Within a scoping unit, each named data object, procedure, derived type, named construct, and namelist group (Section 8.19) must have a distinct name, with the one exception of generic names of procedures (to be described in Section 5.18). Note that this means that any appearance of the name of an intrinsic procedure in another role makes the intrinsic procedure inaccessible by its name (the renaming facility described in Section 8.13 allows an intrinsic procedure to be accessed from a module and renamed). Within a derived-type definition, each component of the type, each intrinsic procedure referenced, and each derived type or named constant accessed by host association must have a distinct name. Apart from these rules, names may be reused. For instance, a name may be used for the components of two types, or the arguments of two procedures referenced with keyword calls.

The names of program units and external procedures are **global**, that is, available anywhere in a complete program. Each must be distinct from the others and from any of the local entities of the program unit.

At the other extreme, the do variable of an implied-do in a data statement (Section 8.5.2) or an array constructor (Section 7.15) has a scope that is just the implied-do. It is different from any other entity with the same name.

5.17   Recursion

5.17.1   Direct recursion

Unless it is declared non-recursive (Section 5.17.3), a subprogram may invoke itself, either directly or indirectly, through a sequence of invocations; this may optionally be indicated by having the keyword recursive prefixed to its leading statement. Where the subprogram is a function that calls itself directly in this fashion, the function name cannot be used for the function result and another name is needed. This is done by adding a further clause to the function statement as in Figure 5.14, which illustrates the use of a recursive function to sum the entries in a chain (see Section 2.13.2).

Figure 5.14 Summing the entries in a linked list.

```
recursive function sum(top) result(s)
   type(entry), pointer :: top
   real                 :: s
   if (associated(top)) then
      s = top%value + sum(top%next)
   else
      s = 0.0
   end if
end function sum
```

The type of the function (and its result) may be specified on the function statement, either before or after the token recursive:

```
integer recursive function factorial(n) result(res)
```

or

```
recursive integer function factorial(n) result(res)
```

or in a type declaration statement for the result name (as in Figure 5.14). In fact, the result name, rather than the function name, must be used in any specification statement. In the executable statements, the function name refers to the function itself and the result name must be used for the **result variable**. If there is no result clause, the function name is used for the result, and is not available for a recursive function call.

The result clause may also be used in a non-recursive function.

Just as in Figure 5.14, any recursive procedure that calls itself directly must contain a conditional test that terminates the sequence of calls at some point, otherwise it will call itself indefinitely.

Each time a recursive procedure is invoked, a fresh set of local data objects is created, which ceases to exist on return. They consist of all data objects declared in the procedure's specification statements or declared implicitly (see Section 8.2.1), but excepting those with the `save` attribute and any dummy arguments.

5.17.2 Indirect recursion

A procedure may also be invoked by indirect recursion, that is, it may call itself through calls to other procedures. To illustrate that this may be useful, suppose we wish to perform a two-dimensional integration but have only a procedure for one-dimensional integration. For example, suppose that it is desired to integrate a function $f(x,y)$ over a rectangle. We might write a Fortran function in a module to receive the value of x as an argument and the value of y from the module itself by host association, as shown in Figure 5.15. We can then integrate over x for a particular value of y, as shown in Figure 5.16, where `integrate` might be as shown in Figure 5.17. We may now integrate over the whole rectangle thus:

```
volume = integrate(fy, ybounds)
```

Note that `integrate` calls `fy`, which in turn calls `integrate`.

Figure 5.15 A two-dimensional function to be integrated.

```
module func
    real                :: yval
    real, dimension(2) :: xbounds, ybounds
contains
    function f(xval)
        real            :: f
        real, intent(in) :: xval
        f = ...      ! Expression involving xval and yval
    end function f
end module func
```

Figure 5.16 Integrate over *x*.

```
function fy(y)
    use func
    real            :: fy
    real, intent(in) :: y
    yval = y
    fy = integrate(f, xbounds)
end function fy
```

Figure 5.17 Library code for one-dimensional integration.

```
recursive function integrate(f, bounds)
   ! Integrate f(x) from bounds(1) to bounds(2)
   real :: integrate
   interface
      function f(x)
         real              :: f
         real, intent(in) :: x
      end function f
   end interface
   real, dimension(2), intent(in) :: bounds
   :
end function integrate
```

5.17.3   Non-recursive procedures

All procedures, apart from functions whose result is of type character with an asterisk character length,[3] are recursive by default but the keyword non_recursive allows a procedure to be specified as non-recursive. It can be used wherever recursive can be used. Using a non-recursive procedure recursively is likely to produce invalid results on a system that does not detect such use, so it is safer to require the keyword non_recursive for the case where the user is sure that this is all that is wanted.

5.18   Overloading

5.18.1   Generic interfaces

We saw in Section 5.12 how to use a simple interface block to provide an explicit interface to an external or dummy procedure. Another use is for overloading, that is, being able to call several procedures by the same **generic identifier**. Here, the interface block contains several interface bodies and the interface statement specifies the generic identifier. For example, the code in Figure 5.18 permits both the functions sgamma and dgamma to be invoked using the generic name gamma.

A specific name for a procedure may be the same as its generic name. For example, the procedure sgamma could be renamed gamma without invalidating the interface block.

Furthermore, a generic name may be the same as another accessible generic name. In such a case, all the procedures that have this generic name may be invoked through it. This capability is important, since a module may need to extend the intrinsic functions such as sin to a new type such as interval (Section 3.9).

If it is desired to overload a module procedure, the interface is already explicit so it is inappropriate to specify an interface body. Instead, the statement

---

[3]An obsolescent feature, see Section B.1.10 – such functions cannot be recursive.

Figure 5.18 A generic interface block.

```
interface gamma
   function sgamma(x)
      real (selected_real_kind( 6))                 :: sgamma
      real (selected_real_kind( 6)), intent(in) :: x
   end function sgamma
   function dgamma(x)
      real (selected_real_kind(12))                 :: dgamma
      real (selected_real_kind(12)), intent(in) :: x
   end function dgamma
end interface
```

```
[ module ] procedure [::] procedure-name-list
```
is included in the interface block in order to name the module procedures for overloading; if the functions sgamma and dgamma were defined in a module, the interface block becomes

```
interface gamma
   module procedure sgamma, dgamma
end interface
```
It is probably most convenient to place such a block in the module itself.

The generic name on an interface statement may be repeated on the corresponding end interface statement, for example,

```
end interface gamma
```
As for other end statements, we recommend use of this fuller form.

Another form of overloading occurs when an interface block specifies a defined operation (Section 3.9), a defined assignment (Section 3.10), or defined derived-type input/output (Section 11.6). The scope of the defined operation, defined assignment, or defined input/output is the scoping unit that contains the interface block, but it may be accessed elsewhere by use or host association. If an intrinsic operator is extended, the number of arguments must be consistent with the intrinsic form (for example, it is not possible to define a unary * operator).

The general form of the interface block is

```
interface [generic-spec]
   [ interface-body ]...
   [ [ module ] procedure [ :: ] procedure-name-list ]...
   ! Interface bodies and module procedure statements
   ! may appear in any order.
end interface [generic-spec]
```
where *generic-spec* is one of
> *generic-name*
> operator (*defined-operator*)
> assignment (=)
> *defined-io-spec*

and *defined-io-spec* is one of

```
read(formatted)
write(formatted)
read(unformatted)
write(unformatted)
```

A `module procedure` statement is permitted only when a *generic-spec* is present, and all the procedures must be accessible module procedures (as shown in the complete module in Figure 5.23). No procedure name may be given a particular *generic-spec* more than once in the interface blocks accessible within a scoping unit. An interface body must be provided for an external or dummy procedure.

5.18.2   Distinguishing generic invocations

If `operator` is specified on the interface statement, all the procedures in the block must be functions with one or two non-optional arguments having intent `in`. If `assignment` is specified, all the procedures must be subroutines with two non-optional arguments, the first having intent `out` or `inout` and the second intent `in`. In order that invocations are always unambiguous, if two procedures have the same generic operator and the same number of arguments or both define assignment, one must have a dummy argument that corresponds by position in the argument list to a dummy argument of the other that has a different type, different kind type parameter, or different rank.

All procedures that have a given generic name must be subroutines or all must be functions, including the intrinsic ones when an intrinsic procedure is extended. Any two non-intrinsic procedures with the same generic name must have arguments that are **distinguishable** in order that any invocation will be unambiguous. Two dummy arguments are distinguishable if

- one is a procedure and the other is a data object;

- they are both data objects or known to be functions but have different types,[4] kind type parameters, or rank;

- one is allocatable and the other is a pointer; or

- one is a function with nonzero rank and the other is not known to be a function.

The rules for any two non-intrinsic procedures with the same generic name are that either

i) one of them has a non-optional data-object dummy argument such that this procedure's number of non-optional data-object dummy arguments that are not distinguishable from it differs from the other procedure's number of such dummy arguments; or

ii) at least one of them has both

- a non-optional dummy argument that corresponds by position in the argument list to a dummy argument that is distinguishable from it, or for which no dummy argument corresponds by position; and

---

[4]Or, if either or both are polymorphic (Section 15.3), they are not type-compatible.

Figure 5.19 An example of a broken overloading rule.

```
interface f  ! Invalid interface block
    function fxi(x,i)
        real             :: fxi
        real, intent(in) :: x
        integer          :: i
    end function fxi
    function fix(i,x)
        real             :: fix
        real, intent(in) :: x
        integer          :: i
    end function fix
end interface
```

- a non-optional dummy argument with the same name as a dummy argument that is distinguishable from it, or for which there is no dummy argument of that name;

where these two arguments must either be the same or the argument that corresponds by position must occur earlier in the dummy argument list; or

iii) one procedure has more non-optional dummy procedures than the other has dummy procedures (including optional ones).

For case ii), both rules are needed in order to cater for both keyword and positional dummy argument lists. For instance, the interface in Figure 5.19 is invalid because the two functions are always distinguishable in a positional call, but not on a keyword call such as f(i=int, x=posn). If a generic invocation is ambiguous between a non-intrinsic and an intrinsic procedure, the non-intrinsic procedure is invoked.

Figure 5.20 Generic disambiguation based on procedureness.

```
interface g1
    subroutine s1(a)
        real a
    end subroutine
    subroutine s2(a)
        interface
            real function a()
            end function
        end interface
    end subroutine
end interface
```

The interface block in Figure 5.20 illustrates disambiguation based on procedureness: since the compiler always knows whether an actual argument is a procedure, no reference

---

Figure 5.21 Generic disambiguation based on `pointer` vs. `allocatable`.

```
interface log_deallocate
  subroutine log_deallocate_real_pointer_2(a)
    real, pointer, intent(inout) :: a(:, :)
  end subroutine
  subroutine log_deallocate_real_allocatable_2(a)
    real, allocatable, intent(inout) :: a(:, :)
  end subroutine
end interface
```

---

to g1 could ever be ambiguous. The interface block in Figure 5.21 illustrates disambiguation based on whether the argument is a pointer or allocatable: in this case its purpose is to allow switching between using allocatable and pointer, without having to change the name of the deallocation procedure. The reason that `allocatable` and `pointer` are only considered to be mutually distinguishable when the pointer does not have intent `in` is that there is an interaction with the automatic targetting feature (see Section 7.17.3) that would have made it possible to write an ambiguous reference.

Note that the presence or absence of the pointer attribute is insufficient to ensure an unambiguous invocation since a pointer actual argument may be associated with a non-pointer dummy argument, see Section 5.7.3.

As already shown, the keyword `module` is optional; for example, the interface block in Section 5.18.1 may be written

```
interface gamma
   procedure :: sgamma, dgamma
end interface
```

If the keyword `module` is omitted, the named procedures need not be module procedures but may also be external procedures, internal procedures, dummy procedures, or procedure pointers. A procedure is allowed to appear in more than one interface block. For example, the code in Figure 5.22 allows the use of both the `*` and `.and.` operators for 'bitwise and' on values of type `bitstring`.

A generic name is permitted to be the same as a type name and takes precedence over the type name; a structure constructor for the type is interpreted as such only if it cannot be interpreted as a reference to the generic procedure.

There are many scientific applications in which it is useful to keep a check on the sorts of quantities involved in a calculation. For instance, in dimensional analysis, whereas it might be sensible to divide length by time to obtain velocity, it is not sensible to add time to velocity. There is no intrinsic way to do this, but we conclude this section with an outline example, see Figures 5.23 and 5.24, of how it might be achieved using derived types. Note that definitions for operations between like entities are also required, as shown by `time_plus_time`. Similarly, any intrinsic function that might be required, here `sqrt`, must be overloaded appropriately. Of course, this can be avoided if the components of the variables are referenced directly, as in

```
t%seconds = t%seconds + 1.0
```

Figure 5.22 Explicit interface for external procedure.

```
type bitstring
   ⋮
end type
⋮
interface operator(*)
   elemental type(bitstring) function bitwise_and(a, b)
      import :: bitstring
      type(bitstring), intent(in) :: a, b
   end function bitwise_and
end interface
interface operator(.and.)
   procedure :: bitwise_and
end interface
```

5.19 Assumed character length

A character dummy argument may be declared with an asterisk for the value of the length type parameter, in which case it automatically takes the value from the actual argument. For example, a subroutine to sort the elements of a character array might be written thus

```
subroutine sort(n,chars)
   integer, intent(in)                      :: n
   character(len=*), dimension(n), intent(in) :: chars
   ⋮
end subroutine sort
```

If the length of the associated actual argument is needed within the procedure, the intrinsic function `len` (Section 9.7.1) may be invoked, as in Figure 5.25.

An asterisk must not be used for a kind type parameter value. This is because a change of character length is analogous to a change of an array size and can easily be accommodated in the object code, whereas a change of kind probably requires a different machine instruction for every operation involving the dummy argument. A different version of the procedure would need to be generated for each possible kind value of each argument. The overloading feature (previous section) gives the programmer an equivalent functionality with explicit control over which versions are generated.

Figure 5.23 A module for distinguishing real entities.

```
module sorts
   type time
      real :: seconds
   end type time
   type velocity
      real :: metres_per_second
   end type velocity
   type length
      real :: metres
   end type length
   type length_squared
      real :: metres_squared
   end type length_squared
   interface operator(/)
      module procedure length_by_time
   end interface
   interface operator(+)
      module procedure time_plus_time
   end interface
   interface sqrt
      module procedure sqrt_metres_squared
   end interface
contains
   function length_by_time(s, t)
      type(length), intent(in) :: s
      type(time), intent(in)   :: t
      type(velocity)           :: length_by_time
      length_by_time%metres_per_second = s%metres / t%seconds
   end function length_by_time
   function time_plus_time(t1, t2)
      type(time), intent(in)   :: t1, t2
      type(time)               :: time_plus_time
      time_plus_time%seconds = t1%seconds + t2%seconds
   end function time_plus_time
   function sqrt_metres_squared(l2)
      type(length_squared), intent(in) :: l2
      type(length)                     :: sqrt_metres_squared
      sqrt_metres_squared%metres = sqrt(l2%metres_squared)
   end function sqrt_metres_squared
end module sorts
```

Figure 5.24 Use of the module of Figure 5.23.

```
program test
   use sorts
   type(length)         :: s  = length(10.0), l
   type(length_squared) :: s2 = length_squared(10.0)
   type(velocity)       :: v
   type(time)           :: t  = time(3.0)
   v = s / t
   ! Note: v = s + t   or   v = s * t  would be invalid.
   t = t + time(1.0)
   l = sqrt(s2)
   print *, v, t, l
end program test
```

Figure 5.25 A function with an argument of assumed character length.

```
   ! Count the number of occurrences of letter in string.
   integer function count (letter, string)
      character (1), intent(in) :: letter
      character (*), intent(in) :: string
      count = 0
      do i = 1, len(string)
         if (string(i:i) == letter) count = count + 1
      end do
   end function count
```

5.20   The subroutine and function statements

We finish this chapter by giving the syntax of the subroutine and function statements, which have so far been explained through examples. It is

   [prefix]... subroutine *subroutine-name [([dummy-argument-list])] [bindC]*

and

   [prefix]... function *function-name ([dummy-argument-list]) [suffix]...*

where *prefix* is type, recursive, non_recursive, module, pure, impure, or elemental; and *bindC* is

   bind(c *[,*name=*character-string]*)

and *suffix* is *bindC* or result *(result-name)*. A *prefix* or *suffix* must not be repeated. For details of *type*, see Section 8.16; *type* must not be present on a subroutine statement.

   Apart from pure, elemental, impure, and *bindC*, which will be explained in Sections 7.8, 7.9, and 20.9, each feature has been explained separately and the meanings are the same in the combinations allowed by the syntax.

5.21   Requirements on statement ordering

The standard places certain requirements on the order in which different classes of statements may appear in a program, and these have been observed as each new class has been introduced. In Table 5.1, we give our own recommendations on ordering; it includes the `implicit none` statement that we will meet later in Section 8.2, and the `format` statement (Chapter 10).[5]

Table 5.1: Recommended statement ordering.

`program`, `function`, `subroutine`, `module`, or `submodule` statement
`use` statements
`import` statements
`implicit none`
Specification part
Executable part
`format` statements
`contains` statement
Internal subprograms or module subprograms
`end` statement

The standard is less restrictive: it permits `data` statements (a specification statement described in Section 8.5.2) in the executable part, and `format` statements anywhere in the specification or executable parts. Such mixing can give programs a messy appearance.

5.22   Summary

A program consists of a sequence of program units. It must contain exactly one main program but may contain any number of modules and external subprograms. We have described each kind of program unit. Modules contain data definitions, derived-type definitions, namelist groups, interface blocks and the import statement, and module subprograms, all of which may be accessed in other program units with the `use` statement. The program units may be in any order, but many compilers require modules to precede their use.

Subprograms define procedures, which may be functions or subroutines. They may also be defined intrinsically (Chapter 9), and external procedures may be defined by means other than Fortran. We have explained how information is passed between program units and to procedures through argument lists and through the use of modules. Procedures may be called recursively provided they are correspondingly specified.

---

[5]A number of obsolescent statements that are described in Appendix B are omitted.

A procedure may be called with keyword arguments, and the procedure may have optional arguments. Interface blocks permit procedures to be invoked as operations or assignments, or by a generic name. The character lengths of dummy arguments may be assumed.

We have also explained about the scope of labels and Fortran names, and introduced the concepts of inclusive scope and scoping unit.

Exercises

1. A subroutine receives a rank-one array of values, x, as an argument. If the mean and variance of the values in x are estimated by

$$\text{mean} = \frac{1}{n} \sum_{i=1}^{n} x(i)$$

and

$$\text{variance} = \frac{1}{n-1} \sum_{i=1}^{n} (x(i) - \text{mean})^2,$$

write a subroutine which returns these calculated values as arguments.

2. A subroutine `matrix_mult` multiplies together two matrices A and B, whose dimensions are $i \times j$ and $j \times k$, respectively, returning the result in a matrix C, dimensioned $i \times k$. Write `matrix_mult`, given that each element of C is defined by

$$C(m,n) = \sum_{\ell=1}^{j} (A(m,\ell) \times B(\ell,n))$$

The matrices should appear as arguments to `matrix_mult`.

3. The subroutine `random_number` (Section 9.18.4) returns a random number in the range 0.0 to 1.0, that is

    ```
    call random_number(r)     ! 0≤r<1
    ```

    Using this function, write the subroutine `shuffle` of Figure 5.4.

4. A character string consists of a sequence of letters. Write a function to return that letter of the string which occurs earliest in the alphabet; for example, the result of applying the function to `DGUMVETLOIC` is C.

5. Write an internal procedure to calculate the volume, $\pi r^2 \ell$, of a cylinder of radius r and length ℓ, using as the value of π the result of `acos(-1.0)`, and reference it in a host procedure.

6. For a simple card game of your own choice, and using the random number procedure (Section 9.18.4), write the subroutines `deal` and `play` of Section 5.4, using data in a module to communicate between them.

7. Objects of the intrinsic type `character` are of a fixed length. Write a module containing a definition of a variable-length character string type, of maximum length 80, and also the procedures necessary to:

    i) assign a character variable to a string;

    ii) assign a string to a character variable;

    iii) return the length of a string;

    iv) concatenate two strings.

6. Allocation of data

6.1 Introduction

There is an underlying assumption in Fortran that the processor supplies a mechanism for managing heap storage. (A heap is a memory management mechanism whereby fresh storage may be established and old storage may be discarded in any order. Mechanisms to deal with the progressive fragmentation of the memory are usually required.) The statements described in this chapter are the user interface to that mechanism.

6.2 The `allocatable` attribute

As we have seen in Section 2.13.1, sometimes the required size of an object is known only after some data have been read or some calculations performed. For this purpose, an object may be given the `allocatable` attribute by a statement such as

```
real, dimension(:, :), allocatable :: a
```

Its rank is specified when it is declared, but the bounds (if it is an array) are undefined until an `allocate` statement has been executed for it. Its initial status is unallocated and it becomes allocated following successful execution of an `allocate` statement. Only then does the shape of the array become defined: the array is said to have a **deferred shape**.

Although the Fortran standard does not mention descriptors, it is very helpful to think of an allocatable object as being held as a **descriptor** that records whether it is allocated and, if so, its address and its bounds in each dimension. This is like a descriptor for a pointer, but no strides need be held since these are always unity. As for pointers, the expectation is that the array itself is held separately.

An important example is shown in Figure 6.1. The array `work` is placed in a module and is allocated at the beginning of the main program to a size that depends on input data. The array is then available throughout program execution in any subprogram that has a `use` statement for `work_array`.

When an allocatable object `a` is no longer needed, it may be deallocated by execution of the statement

```
deallocate (a)
```

following which the object is unallocated. The `deallocate` statement is described in more detail in Section 6.6.

Modern Fortran Explained, 3rd Edition. M. Metcalf, J. Reid, M. Cohen, and R. Bader. Oxford University Press (2024). © M. Metcalf, J. Reid, M. Cohen, and R. Bader (2024). DOI 10.1093/oso/9780198876571.001.0006

Figure 6.1 An allocatable array in a module.

```
module work_array
   integer                           :: n
   real, dimension(:,:,:), allocatable :: work
end module work_array
program main
   use work_array
   read *, n
   allocate (work(n, 2*n, 3*n))
   ⋮
```

If it is required to make any change to the bounds of an allocatable array, the array must be deallocated and then allocated afresh. This can be carried out either by explicit deallocation and reallocation, or automatically (see Section 6.7). It is an error to allocate an allocatable array that is already allocated, or to deallocate an allocatable array that is unallocated, but one that can easily be avoided by the use of the `allocated` intrinsic function (Section 9.2) to inquire about the allocation status.

An undefined allocation status cannot occur. On return from a subprogram, an allocated allocatable object without the `save` attribute (Section 5.9) is automatically deallocated if it is local to the subprogram.

6.3 Deferred type parameters

A `len` type parameter value is permitted to be a colon in a type declaration statement such as

```
character(len=:), pointer :: varchar
```

for a pointer or an allocatable entity. It indicates a **deferred type parameter**; such a type parameter has no defined value until it is given one by allocation or pointer assignment. For example, in

```
character(:), pointer :: varchar
character(100), target :: name
character(200), target :: address
⋮
varchar => name
⋮
varchar => address
```

the character length of `varchar` after each pointer assignment is the same as that of its target; that is, 100 after the first pointer assignment and 200 after the second.

For intrinsic types, only character length may be deferred. Derived types that are parameterized may have type parameters which can be deferred, see Section 13.2.2.

Deferred type parameters can be given values by the `allocate` statement, see Section 6.5. For allocatable variables, they can also be given values by assignment, see Section 6.7.

Further, any array that has a deferred type parameter must also be of deferred shape. Thus, what is permitted or not can be illustrated by:

```
character(:), pointer :: b(:,:)    ! Allowed
character(3), pointer :: c(:,:)    ! Allowed
character(:), pointer :: d(3)      ! Not permitted.
```

6.4   Allocatable scalars

The `allocatable` attribute (and hence the `allocated` function) may also be applied to scalar variables and components. This is particularly useful when combined with deferred type parameters, for example, in

```
character(:), allocatable :: chdata
integer                   :: unit, reclen
   ⋮
read *, reclen
allocate (character(reclen) :: chdata)
read *, chdata
```

where `reclen` allows the length of `character` to be specified at run time.

Another relevant use of allocatable scalars is for polymorphic objects (Section 2.15).

6.5   The `allocate` statement

The general form of the `allocate` statement is

allocate ([*type-spec* ::] *allocation-list* [, *alloc-spec*] ...)

where *allocation-list* is a list of allocations of the form

allocate-object [(*array-bounds-list*)]

each *array-bound* has the form

[*lower-bound* :] *upper-bound*

and *alloc-spec* is one of

```
errmsg=erm
mold=expr
source=expr
stat=stat
```

where no specifier may appear more than once, *stat* is a scalar integer variable, and *erm* is a scalar default character variable. Neither *stat* nor *erm* may be part of an object being allocated.

The optional *type-spec*, and `mold=` and `source=` specifiers, are discussed in Section 15.4.

If the `stat=` specifier is present, *stat* is given either the value zero after a successful allocation or a positive value after an unsuccessful allocation (for example, if insufficient storage is available). It is recommended that any `stat=` variable should have a decimal exponent range of at least four to ensure that the error code is representable in the variable. After an unsuccessful execution, each array that was not successfully allocated retains its previous allocation or pointer association status. If `stat=` is absent and the allocation is unsuccessful, error termination occurs.

If the `errmsg=` specifier is present and an error during allocation occurs, an explanatory message is assigned to the variable. For example,

```
character(200) :: error_message ! Probably long enough.
        ! In Fortran 2023, error_message may be an allocatable scalar.
   ⋮
allocate (x(n), stat=allocate_status, errmsg=error_message)
if (allocate_status > 0) then
   print *, 'Allocation of X failed:', trim(error_message)
   ⋮
end if
```

where the function `trim` is described in Section 9.7.2. This is helpful, because the error codes available in the *stat* variable are processor dependent.

Each *allocate-object* is allocatable or a pointer. It is permitted to have zero character length.

Each *lower-bound* and each *upper-bound* is a scalar integer expression. The default value for the lower bound is 1. The number of *array-bound*s in a list must equal the rank of the *allocate-object*. They determine the array bounds, which do not alter if the value of a variable in one of the expressions changes subsequently. An array may be allocated to be of size zero.

The bounds of all the arrays being allocated are regarded as undefined during the execution of the `allocate` statement, so none of the expressions that specify the bounds may depend on any of the bounds or on the value of the `stat=` variable. For example,

```
allocate (a(size(b)), b(size(a)))    ! invalid
```

or even

```
allocate (a(n), b(size(a)))          ! invalid
```

is not permitted, but

```
allocate (a(n))
allocate (b(size(a)))
```

is valid. This restriction allows the processor to perform the allocations in a single `allocate` statement in any order.

In contrast to the case with an allocatable object, a pointer may be allocated a new target even if it is currently associated with a target. In this case the previous association is broken. If the previous target was created by allocation, it becomes inaccessible unless another pointer is associated with it. Linked lists are normally created by using a single pointer in an `allocate` statement for each node of the list. There is an example in Figure 4.10.

6.6   The `deallocate` statement

When an allocatable object or pointer target is no longer needed, its storage may be recovered by using the `deallocate` statement. Its general form is

  `deallocate (` *allocate-object-list* [`,stat=`*stat*] [`, errmsg=`*erm*] `)`

where each *allocate-object* is an allocatable object that is allocated or a pointer that is associated with the whole of a target that was allocated through a pointer in an `allocate` statement.[1] A pointer becomes disassociated (Section 3.3) following successful execution of the statement. Here, *stat* is a scalar integer variable that must not be deallocated by the statement nor depend on an object that is deallocated by the statement. If `stat=` is present, *stat* is given either the value zero after a successful execution or a positive value after an unsuccessful execution (for example, if a pointer is disassociated). If an error during deallocation occurs, an explanatory message is assigned to the optional `errmsg=` variable. After an unsuccessful execution, each object that was not successfully deallocated retains its previous allocation or pointer association status. If `stat=` is absent and the deallocation is unsuccessful, error termination occurs.

If there is more than one object in the list, there must be no dependencies among them, to allow the processor to deallocate the objects one by one in any order. An object must not be deallocated while it or a subobject of it is associated with a dummy argument; this can happen, for example, if the object is in a module.

A danger in using the `deallocate` statement is that an object may be deallocated while pointers are still associated with it. Such pointers are left 'dangling' in an undefined state, and must not be reused until they are nullified or again associated with an actual target.

In order to avoid an accumulation of unused and unusable storage, all allocated storage should be explicitly deallocated when it is no longer required (this is automatic for unsaved local allocatable variables, see the end of Section 6.2). This explicit management is required in order to avoid a potentially significant overhead on the part of the processor in handling arbitrarily complex allocation and reference patterns.

Note also that the standard does not specify whether the processor recovers storage that held a target allocated through a pointer but is no longer accessible through this or any other pointer. This failure to recover storage is known as **memory leakage**. It might be important where, for example, a pointer function is referenced within an expression – the programmer cannot rely on the compiler to arrange for deallocation. To ensure that there is no memory leakage, it is necessary to use such functions only on the right-hand side of pointer assignments or as pointer component values in structure constructors, and to deallocate the pointer target when it is no longer needed.

Allocatable objects have the advantage that they do not leak memory, because an allocatable object cannot be allocated if already allocated, and is automatically deallocated on return from a subprogram if it is local to the subprogram and does not have the `save` attribute.

---

[1]Note that this excludes a pointer that is associated with an allocatable array.

6.7 Automatic reallocation in intrinsic assignment

Intrinsic assignment

    variable = *expr*

for an allocatable *variable* may involve automatic reallocation. If *expr* is a scalar and *variable* is an array, *variable* must be allocated (to define the bounds). If *variable* is allocated but differs from *expr* in shape, length type parameter values, or dynamic type and type parameters, it is deallocated. Following deallocation or if already unallocated, it is allocated to have the type and type parameter values of *expr* and the bounds of *expr* if it is an array or its previous bounds if it is a scalar.

Automatic reallocation simplifies the use of array functions which return a variable-sized result (such as the intrinsic functions `pack` and `unpack`). For example, in

```
subroutine process(x)
    real(wp), intent(inout) :: x(:)
    real(wp), allocatable   :: nonzero_values(:)
    nonzero_values = pack(x, x/=0)
```

the array `nonzero_values` is automatically allocated to be of the correct length to contain the results of the intrinsic function `pack`, instead of the user having to allocate it manually (which would necessitate counting the number of nonzeros separately).

Automatic reallocation also permits a simple extension of an existing allocatable array whose lower bounds are all 1. To add some extra values to such an integer array a of rank 1, it is sufficient to write, for example,

```
a = [ a, 5, 6 ]
```

An example where a deferred type parameter differs is

```
character(:), allocatable :: quotation
   ⋮
quotation = 'Now is the winter of our discontent.'
   ⋮
quotation = "This ain't the summer of love."
```

In each of the assignments to `quotation`, it is reallocated to be the right length (unless it is already of that length) to hold the desired quotation. If instead the normal truncation or padding is required in an assignment to an allocatable-length character, substring notation can be used to suppress the automatic reallocation. For example,

```
quotation(:) = ''
```

leaves `quotation` at its current length, setting all of it to blanks.

Automatic reallocation only occurs for normal intrinsic assignment, and not for defined assignment or for `where` constructs (Section 7.6).

6.8 Transferring an allocation

The intrinsic subroutine `move_alloc` moves an allocation from one allocatable object to another. Unless coarrays are involved, which we discuss in Section 17.6, the subroutine is pure (Section 7.8). It has the form

```
call move_alloc( from, to[ , stat ][ , errmsg ])
```

> **from** is allocatable and of any type. It has intent `inout`.

> **to** has intent `out`, is allocatable, of the same rank as `from`, and type compatible with `from`.[2] If `from` is polymorphic, `to` must be polymorphic and have the same declared type parameters unless deferred (Section 13.2.2).

> **stat** is an optional integer scalar with a decimal exponent range of at least 4. It has intent `out`. If it is not present and an error occurs, error termination is initiated. If the execution is successful, `stat` is assigned the value zero. If an error occurs, it is assigned a nonzero value.

> **errmsg** is an optional intent `inout` scalar of type default character. When present, it provides an explanatory message in the event of an error condition, and is unchanged otherwise.

After the call, the allocation status of `to` is that of `from` beforehand and `from` becomes unallocated. If `to` has the `target` attribute, any pointer that was associated with `from` will be associated with `to`; otherwise, such a pointer will become undefined.

This subroutine provides what is essentially the allocatable equivalent of pointer assignment: allocation transfer. However, unlike pointer assignment, this maintains the allocatable semantics of having at most one allocated object for each allocatable variable. For example,

```
real, allocatable :: a1(:), a2(:)
allocate (a1(0:10))
a1(3) = 37
call move_alloc(from=a1, to=a2)
! a1 is now unallocated,
! a2 is allocated with bounds (0:10) and a2(3)==37.
```

The subroutine `move_alloc` can be used to minimize the amount of copying required when one wishes to expand or contract an allocatable array; the canonical sequence for this is

```
real, allocatable :: a(:,:), temp(:,:)
    :
! Increase size of a to (n, m)
allocate (temp(n, m))
temp(1:size(a,1), 1:size(a,2)) = a
call move_alloc(temp, a)
! a now has shape (/ n, m /), and temp is unallocated
```

This sequence only requires one copying operation instead of the two that would have been required without `move_alloc`. Because the copy is controlled by the user, pre-existing values will end up where the user wants them (which might be at the same subscripts, or all at the beginning, or all at the end, etc.).

---

[2]Same type if neither is polymorphic.

6.9 Allocatable dummy arguments

A dummy argument is permitted to have the allocatable attribute. In this case, the corresponding actual argument must be allocatable and of the same type, kind parameters, and rank. The dummy argument always receives the allocation status (descriptor) of the actual argument on entry and the actual argument receives that of the dummy argument on return. In both cases, this may be unallocated. If allocated, the bounds are also received (as for an array pointer but not an assumed-shape array).

Our expectation is that some compilers will perform copy-in copy-out of the descriptor. Rule i) of Section 5.7.5 is applicable and is designed to permit compilers to do this. In particular, this means that no reference to the actual argument (for example, through it being a module variable) is permitted from the invoked procedure if the dummy array is allocated or deallocated there.

For the object itself, the situation is just like the case when the actual and dummy arguments are both explicit-shape arrays (see Section 5.7.6). Copy-in copy-out is permitted unless both objects have the `target` attribute.

An allocatable dummy argument is permitted to have intent and this applies both to the allocation status (the descriptor) and to the object itself. If the intent is `in`, the object is not permitted to be allocated or deallocated and the value is not permitted to be altered. If the intent is `out` and the object is allocated on entry, it becomes deallocated. An example of the application of an allocatable dummy argument to reading arrays of variable bounds is shown in Figure 6.2.

Figure 6.2 Reading arrays whose size is not known beforehand.

```
subroutine load(array, unit)
   real, allocatable, intent(out), dimension(:, :, :) :: array
   integer, intent(in)          :: unit
   integer                      :: n1, n2, n3
   read (unit,*) n1, n2, n3     ! read from specified unit
   allocate (array(n1, n2, n3))
   read (unit,*) array
end subroutine load
```

6.10 Allocatable functions

A function result is permitted to have the allocatable attribute, which is very useful when the size of the result depends on a calculation in the function itself, as illustrated in Figure 6.3. The allocation status on each entry to the function is unallocated. The result may be allocated and deallocated any number of times during execution of the procedure, but it must be allocated and have a defined value on return.

The result is automatically deallocated after execution of the statement in which the reference occurs, even if it has the `target` attribute. If the reference to an allocatable function is as an actual argument, it is just its value that is referenced (which is not allocatable).

Figure 6.3 An allocatable function to remove duplicate values.

```
program no_leak
   real, allocatable, dimension(:) :: x, y
      ⋮
   y = compact(x)**2
      ⋮
contains
   function compact(x) ! To remove duplicates from the array x
      real, allocatable, dimension(:) :: compact
      real, dimension(:), intent(in)  :: x
      integer                         :: n
         ⋮                ! Find the number of distinct values, n
      allocate (compact(n))
         ⋮                ! Copy the distinct values into compact
   end function compact
end program no_leak
```

6.11   Allocatable components

Components of a derived type are permitted to have the allocatable attribute. For example, a lower-triangular matrix may be held by using an allocatable array for each row. Consider the type

```
type row
   real, dimension(:), allocatable :: r
end type row
```

and the arrays

```
type(row), dimension(n) :: s, t      ! n of type integer
```

Storage for the rows can be allocated thus

```
do i = 1, n                 ! i of type integer
   allocate (t(i)%r(1:i)) ! Allocate row i of length i
end do
```

The array assignment

```
s = t
```

would then be equivalent to the assignments

```
s(i)%r = t(i)%r
```

for all the components.

For an object of a derived type that has a component of derived type, we need the concept of an **ultimate allocatable component**, which is an ultimate component (Section 2.16) that is allocatable. Just as for an ordinary allocatable object, the initial state of an ultimate allocatable component is unallocated. Hence, there is no need for default initialization of allocatable components. In fact, initialization in a derived-type definition of an allocatable component is not permitted, see Section 8.5.5.

In a structure constructor (Section 3.8), an expression corresponding to an allocatable component must be an object or a reference to the intrinsic function `null` with no arguments. If it is an allocatable object, the component takes the same allocation status and, if allocated, the same bounds and value. If it is an object, but not allocatable, the component is allocated with the same bounds and is assigned the same value. If it is a reference to the intrinsic function `null` with no arguments, the component receives the allocation status of unallocated.

Allocatable components are illustrated in Figure 6.4, where code to manipulate polynomials with variable numbers of terms is shown.

Figure 6.4 Using allocatable components for adding polynomials.

```
module real_polynomial_module
    type real_polynomial
        real, allocatable, dimension(:) :: coeff
    end type real_polynomial
    interface operator(+)
        module procedure rp_add_rp
    end interface operator(+)
contains
    function rp_add_rp(p1, p2)
        type(real_polynomial)                :: rp_add_rp
        type(real_polynomial), intent(in) :: p1, p2
        integer                           :: m, m1, m2
        m1 = ubound(p1%coeff,1)
        m2 = ubound(p2%coeff,1)
        allocate (rp_add_rp%coeff(max(m1,m2)))
        m = min(m1,m2)
        rp_add_rp%coeff(:m) = p1%coeff(:m) +p2%coeff(:m)
        if (m1 > m) rp_add_rp%coeff(m+1:) = p1%coeff(m+1:)
        if (m2 > m) rp_add_rp%coeff(m+1:) = p2%coeff(m+1:)
    end function rp_add_rp
end module real_polynomial_module
program example
    use real_polynomial_module
    type(real_polynomial) :: p, q, r
    p = real_polynomial((/4.0, 2.0, 1.0/))   ! Set p to 4+2x+x**2
    q = real_polynomial((/-1.0, 1.0/))
    r = p + q
    print *, 'Coefficients are: ', r%coeff
end program example
```

An object of a type having an ultimate allocatable component is permitted to have the `parameter` attribute (be a constant). In this case the component is always unallocated. It is not permitted to appear in an `allocate` statement.

When a variable of derived type is deallocated, any ultimate allocatable component that is allocated is also deallocated, as if by a `deallocate` statement. The variable may be a pointer or allocatable, and the rule applies recursively, so that all allocated allocatable components at all levels (apart from any lying beyond pointer components) are deallocated. Such deallocations of components also occur when a variable is associated with an intent `out` dummy argument.

Intrinsic assignment

   variable = expr

for a type with an ultimate allocatable component (as in `r = p + q` in Figure 6.4) consists of the following steps for each such component.

  i) If the component of *variable* is allocated, it is deallocated.

  ii) If the component of *expr* is allocated, the component of *variable* is allocated with the same bounds and the value is then transferred using intrinsic assignment.

If the allocatable component of *expr* is unallocated, nothing happens in step ii), so the component of *variable* is left unallocated. Note that if the component of *variable* is already allocated with the same shape, the compiler may choose to avoid the overheads of deallocation and reallocation unless finalization (Section 15.11.1) is involved. Note also that if the compiler can tell that there will be no subsequent reference to *expr*, because it is a function reference or a temporary variable holding the result of expression evaluation, no allocation or assignment is needed – all that has to happen is the deallocation of any allocated ultimate allocatable components of *variable* followed by the copying of the descriptor.

If a component is itself of a derived type with an allocatable component, the intrinsic assignment in step ii) will involve these rules, too. In fact, they are applied recursively at all levels, and copying occurs in every case. This is known as **deep copying**, as opposed to **shallow copying**, which occurs for pointer components, where the descriptor is copied and nothing is done for components of a target of a pointer component.

If an actual argument and the corresponding dummy argument have an ultimate allocatable component, rule i) of Section 5.7.5 is applicable and requires all allocations and deallocations of the component to be performed through the dummy argument, in case copy-in copy-out is in effect.

If a statement contains a reference to a function whose result is of a type with an ultimate allocatable component, any allocated ultimate allocatable components of the function result are deallocated after execution of the statement. This parallels the rule for allocatable function results (Section 6.10).

6.11.1   Allocatable components of recursive type

An allocatable component is permitted to be of any derived type, including the type being defined or a type defined later in the program unit. This can be used to define

Figure 6.5 Allocatable list example. The read statement here is explained in Section 10.7.

```
type my_real_list
  real value
  type(my_real_list), allocatable :: next
end type
type(my_real_list), allocatable, target :: list
type(my_real_list), pointer :: last
real :: x
 ⋮
last => null()
do
  read (unit, *, iostat=ios) x
  if (ios/=0 .or. x==0.) exit
  if (.not.associated(last)) then
    list = my_real_list(x)
    last => list
  else
    last%next = my_real_list(x)
    last => last%next
  end if
end do
! list now contains all the input values, in order of reading.
 ⋮
deallocate (list) ! deallocates every element in the list.
```

dynamic structures without involving pointers, thus gaining the usual benefits of allocatable variables: no aliasing (except where the target attribute is used), contiguity, and automatic deallocation. Automatic deallocation means that deallocating the parent variable (or returning from the procedure in which it is defined) will completely deallocate the entire dynamic structure. Figure 6.5 shows how this can be used to build a list. For building up the list in that example, it was convenient to use a pointer whose target is the end of the list. If, on the other hand, we want to insert a new value somewhere else (such as at the beginning of the list), careful use of the move_alloc intrinsic (Section 6.8) is recommended to avoid making temporary copies of the entire list. We illustrate this with the subroutine push for adding an element to the top of a stack in Figure 6.6. Similar comments apply to element deletion, illustrated by subroutine pop in Figure 6.6.

One might imagine that the compiler would produce similar code (that is, code avoiding deep copies) for the much simpler statements

```
list = my_real_list(newvalue, list)
```

and

```
list = list%next
```

Figure 6.6 Allocatable stack procedures.

```
subroutine push(list, newvalue)
  type(my_real_list), allocatable :: list, temp
  real, intent(in)                :: newvalue
  call move_alloc(list, temp)
  list = my_real_list(newvalue)
  call move_alloc(temp, list%next)
end subroutine
subroutine pop(list)
  type(my_real_list), allocatable :: list, temp
  call move_alloc(list%next, temp)
  call move_alloc(temp, list)
end subroutine
```

as the executable parts of push and pop, respectively, but in fact the model for allocatable assignment in the standard specifies automatic deallocation only when an array shape, length type parameter, or dynamic type differs; that is not the case in these examples, so the compiler is expected to perform deep copying. (A standard-conforming program can only tell the difference when the type has any final subroutines or the list has the target attribute; so if the variables involved are not polymorphic and not targets, a compiler might produce more optimal code.)

6.12   Allocatable arrays vs. pointers

Why are allocatable arrays needed? Is all their functionality not available (and more) with pointers? The reason is that there are significant advantages for memory management and execution speed in using allocatable arrays when the added functionality of pointers is not needed.

- Code for an array pointer is likely to be less efficient because allowance has to be made for strides other than unity. For example, its target might be the section vector(1:n:2) or the section matrix(i,1:n) with non-unit strides, whereas most computers hold allocatable arrays in contiguous memory.

- If a defined operation involves a temporary variable of a derived type with a pointer component, the compiler will probably be unable to deallocate its target when storage for the variable is freed. Consider, for example, the statement

  ```
  a = b + c*d     ! a, b, c, and d are of the same derived type
  ```

  This will create a temporary for c*d, which is not needed once b + c*d has been calculated. The compiler is unlikely to be sure that no other pointer has the component or part of it as a target, so is unlikely to deallocate it.

- Intrinsic assignment is often unsuitable for a derived type with a pointer component because the assignment

```
a = b
```

will leave a and b sharing the same target for their pointer component. Therefore, a defined assignment that allocates a fresh target and copies the data will be used instead. However, this is very wasteful if the right-hand side is a temporary such as that of the assignment of the previous paragraph.

- Similar considerations apply to a function invocation within an expression. The compiler will be unlikely to be able to deallocate the pointer after the expression has been calculated.

- When a variable of derived type is deallocated, any ultimate allocatable component that is allocated is also deallocated. To avoid memory leakage with pointer components, the programmer would need to deallocate each one explicitly and be careful to order the deallocations correctly.

6.13 Summary

We have described how storage allocation for an object may be controlled in detail by a program.

Exercises

1. Using the type real_polynomial of Figure 6.4 in Section 6.11, write code to define a variable of that type with an allocatable component length of four and then to enlarge that allocatable array with two additional values.

2. Given the type

    ```
    type emfield
       real, allocatable :: strength(:,:)
    end type
    ```

    initialize a variable of type emfield so that its component has bounds (1:4,1:6) and value 1 everywhere. Enlarge this variable so that the component has bounds (0:5,0:8), keeping the values of the old elements and setting the values of the new elements to zero.

3. As Exercise 2, but with new bounds (1:6,1:9) and using the reshape intrinsic function.

4. Make an existing rank-2 integer array b, that has lower bounds of 1, two rows and two columns larger, with the old elements' values retained in the middle of the array.

7. Array features

7.1 Introduction

In an era when many computers have the hardware capability for efficient processing of array operands, it is self-evident that a numerically based language such as Fortran should have matching notational facilities. Such facilities provide not only notational convenience for the programmer, but also provide an opportunity to enhance optimization.

Arrays were introduced in Sections 2.10 to 2.13, their use in simple expressions and in assignments was explained in Sections 3.11 and 3.12, and they were used as procedure arguments in Section 5.7. These descriptions were deliberately restricted because Fortran contains a very full set of array features whose complete description would have unbalanced those chapters. The purpose of this chapter is to describe the array features in detail, but without anticipating the descriptions of the array intrinsic procedures of Chapter 9; the rich set of intrinsic procedures should be regarded as an integral part of the array features.

7.2 Zero-sized arrays

It might be thought that an array would always have at least one element. However, such a requirement would force programs to contain extra code to deal with certain natural situations. For example, the code in Figure 7.1 solves a lower-triangular set of linear equations. When i has the value n the sections have size zero, which is just what is required.

Figure 7.1 A do loop whose final iteration has a zero-sized array.

```
do i = 1,n
   x(i) = b(i) / a(i, i)
   b(i+1:n) = b(i+1:n) - a(i+1:n, i) * x(i)
end do
```

Fortran allows arrays to have zero size in all contexts. Whenever a lower bound exceeds the corresponding upper bound, the array has size zero.

There are few special rules for zero-sized arrays because they follow the usual rules, though some care may be needed in their interpretation. For example, two zero-sized arrays of the same rank may have different shapes. One might have shape (0,2) and the other (0,3) or (2,0).

Modern Fortran Explained, 3rd Edition. M. Metcalf, J. Reid, M. Cohen, and R. Bader. Oxford University Press (2024). © M. Metcalf, J. Reid, M. Cohen, and R. Bader (2024). DOI 10.1093/oso/9780198876571.001.0007

Such arrays of differing shape are not conformable and therefore may not be used together as the operands of a binary operation. However, an array is always conformable with a scalar, so the statement

zero-sized-array = scalar

is valid and the scalar is 'broadcast to all the array elements', making this a 'do nothing' statement.

A zero-sized array is regarded as being defined always, because it has no values that can be undefined.

7.3 Automatic objects

A procedure with dummy arguments that are arrays whose size varies from call to call may also need local arrays whose size varies. A simple example is the array `work` in the subroutine to interchange two arrays that is shown in Figure 7.2.

Figure 7.2 A procedure with an automatic array; `size` is described in Section 9.14.3.

```
subroutine swap(a, b)
    real, dimension(:), intent(inout) :: a, b
    real, dimension(size(a))          :: work ! automatic array
            ! size provides the size of an array
    work = a
    a = b
    b = work
end subroutine swap
```

An array whose extents vary in this way is called an **automatic array**, and is an example of an **automatic data object**. Such an object is not a dummy argument and its declaration contains one or more values that are not known at compile time; that is, not a constant expression (Section 8.4). An implementation is likely to bring them into existence when the procedure is called and destroy them on return, maintaining them on a stack.[1] The values must be defined by specification expressions (Section 8.17).

Another way that automatic objects arise is through varying character length, illustrated by the variable `double` in Figure 7.3.

A parameterized derived type (Section 13.2) may have a length type parameter that behaves very like character length, and a dummy argument of such a type may be an automatic object. We defer discussion of this to Section 13.2.2.

An array bound or the character length of an automatic object is fixed for the duration of each execution of the procedure and does not vary if the value of the specification expression varies or becomes undefined.

An automatic object must not be given the `save` attribute because this is obviously contrary to being automatic; and it must not be given an initial value, see Sections 8.5.1 and 8.5.2, because this gives it the `save` attribute.

---

[1] A stack is a memory management mechanism whereby fresh storage is established and old storage is discarded on a 'last in, first out' basis, often within contiguous memory.

Figure 7.3 A module containing a procedure with an automatic scalar.

```
program loren
   character (len = *), parameter :: a = 'just a simple test'
   print *, double(a)
contains
   function double(a)
      character (len = *), intent(in) :: a
      character (len = 2*len(a))      :: double
      double = a//a
   end function double
end program loren
```

7.4   Elemental operations and assignments

We saw in Section 3.11 that an intrinsic operator can be applied to conformable operands, to produce an array result whose element values are the values of the operation applied to the corresponding elements of the operands. Such an operation is called **elemental**.

It is not essential to use operator notation to obtain this effect. Many of the intrinsic procedures (Chapter 9) are elemental and have scalar dummy arguments that may be called with some or all array actual arguments provided all the array arguments have the same shape. Such a reference is called an **elemental reference**. For a function, the shape of the result is the shape of the array arguments. For example, we may find the square roots of all the elements of a real array a thus:

```
a = sqrt(a)
```

If any actual argument in an elemental subroutine invocation is array valued, all the actual arguments corresponding to dummy arguments with intent out or inout must be arrays. If a procedure that invokes an elemental procedure has an optional array-valued dummy argument that is absent, that dummy argument must not be used as an actual argument in the elemental invocation unless another array of the same rank is associated with a non-optional argument of the elemental procedure (to ensure that the rank does not vary from call to call).

Similarly, an intrinsic assignment may be used to assign a scalar to all the elements of an array, or to assign each element of an array to the corresponding element of an array of the same shape (Section 3.12). Such an assignment is also called **elemental**.

For a defined operator, a similar effect may be obtained with a generic interface to functions for each desired rank or pair of ranks. For example, the module in Figure 7.4 provides summation for scalars and rank-one arrays of intervals (Section 3.9). Alternatively, an elemental procedure (Section 7.9) can be defined for application to any rank. Similarly, elemental versions of defined assignments may be provided for the ranks required or an elemental procedure can be defined for any rank.

Figure 7.4 Interval addition for scalars and arrays of rank one.

```
module interval_addition
   type interval
      real :: lower, upper
   end type interval
   interface operator(+)
      module procedure add00, add11
   end interface
contains
   function add00 (a, b)
      type (interval)             :: add00
      type (interval), intent(in) :: a, b
      add00%lower = a%lower + b%lower  ! Production code would
      add00%upper = a%upper + b%upper  ! allow for roundoff.
   end function add00
   function add11 (a, b)
      type (interval), dimension(:), intent(in)        :: a
      type (interval), dimension(size(a))              :: add11
      type (interval), dimension(size(a)), intent(in) :: b
      add11%lower = a%lower + b%lower  ! Production code would
      add11%upper = a%upper + b%upper  ! allow for roundoff.
   end function add11
end module interval_addition
```

7.5   Array-valued functions

We mentioned in Section 5.11.1 that a function may have an array-valued result, and have used this language feature in Figure 7.4 where the interpretation is obvious.

The shape is specified within the function definition by the `dimension` attribute for the function name. Unless the function result is allocatable or a pointer, the bounds must be explicit expressions and they are evaluated on entry to the function. For another example, see the declaration of the function result in Figure 7.5, which uses the type

```
type matrix
   real :: element
end type matrix
```

of Section 3.11.

7.6   The `where` statement and construct

It is often desired to perform an array assignment only for certain elements, say those whose values are positive. The `where` statement provides this facility. A simple example is

```
where ( a > 1.0 ) a = 1.0/a    ! a is a real array
```

Figure 7.5 A function for matrix by vector multiplication; `size` is defined in Section 9.14.

```
function mult(a, b)
   type(matrix), dimension(:, :)          :: a
   type(matrix), dimension(size(a, 2)) :: b
   type(matrix), dimension(size(a, 1)) :: mult
   integer                                :: j, n

   mult = 0.0     ! A defined assignment from a real
                  ! scalar to a rank-one matrix.
   n = size(a, 1)
   do j = 1, size(a, 2)
      mult = mult + a(1:n, j) * b(j)
              ! Uses defined operations for addition of
              ! two rank-one matrices and multiplication
              ! of a rank-one matrix by a scalar matrix.
   end do
end function mult
```

which reciprocates those elements of a that are greater than 1.0 and leaves the rest unaltered. The general form is

>`where` (*logical-array-expr*) *array-assignment*

where *array-assignment* is

>*array-variable* = *expr*

The logical array expression *logical-array-expr* is known as the **mask** and must have the same shape as *array-variable*. It is evaluated first and then just those elements of *array-variable* that correspond to elements of *logical-array-expr* that have the value true are assigned values from *expr*. All other elements of *array-variable* are left unaltered. The assignment may be a defined assignment, provided that it is elemental (Section 7.9).

A single masking expression may be used for a sequence of array assignments all of the same shape. The simplest form of this construct is

>`where` (*logical-array-expr*)
> *[array-assignment]* ...
>`end where`

The masking expression *logical-array-expr* is first evaluated and then each array assignment is performed in turn, under the control of this mask. If any of these assignments affect entities in *logical-array-expr*, it is always the value obtained when the `where` statement is executed that is used as the mask.

The `where` construct may take the form

>`where` (*logical-array-expr*)
> *[array-assignment]* ...
>`elsewhere`
> *[array-assignment]* ...
>`end where`

Here, the assignments in the first block of assignments are performed in turn under the control of *logical-array-expr* and then the assignments in the second block are performed in turn under the control of .not. *logical-array-expr*. Again, if any of these assignments affect entities in *logical-array-expr*, it is always the value obtained when the where statement is executed that is used as the mask.

A simple example of a where construct is

```
where (pressure <= 1.0)
   pressure = pressure + inc_pressure
   temp = temp + 5.0
elsewhere
   raining = .true.
end where
```

where pressure, inc_pressure, temp, and raining are arrays of the same shape.

If a where statement or construct masks an elemental function reference, the function is called only for the wanted elements. For example,

```
where ( a > 0 ) a = log(a)
```

(log is defined in Section 9.4) would not lead to erroneous calls of log for negative arguments.

This masking applies to all elemental function references except any that are within an argument of a non-elemental function reference. The masking does not extend to array arguments of such a function. In general, such arguments have a different shape so that masking would not be possible. For example, in the case

```
where (a > 0) a = a/sum(log(a))
```

(sum is defined in Section 9.13) the logarithms of all of the elements of a are summed and the statement will fail if they are not all positive.

Masking does not extend to a non-elemental function reference or an array constructor. They are fully evaluated before the masking is applied.

It is permitted to mask not only the where statement of the where construct, but also any elsewhere statement that it contains. All the masking expressions must be of the same shape. A where construct may contain any number of masked elsewhere statements but at most one elsewhere statement without a mask, and if present this must be the final one. In addition, where constructs may be nested within one another; all the masking expressions of the nested construct must be of the same shape, and this must be the shape of all the array variables on the left-hand sides of the assignments within the construct.

The way this is interpreted is as if there were logical arrays control and pending of the same shape as the arrays of the where construct. The array control masks every assignment in the nested construct. Before executing the outermost where statement, they are assigned values thus

```
control = .true.
pending = .false.
```

A where (*logical-array-expr*) statement assigns their values thus

```
control = control .and. logical-array-expr
pending = pending .and. (control .and. .not. logical-array-expr)
```

An `elsewhere` statement assigns their values thus

```
control = pending
pending = .false.
```

An `elsewhere`(*logical-array-expr*) statement assigns their values thus

```
control = pending .and. logical-array-expr
pending = pending .and. not. logical-array-expr
```

An `endwhere` statement assigns their values to those they had before the corresponding `where` statement was executed (of significance only if the `where` construct is nested within another `where` construct).

A simple `where` statement such as that at the start of this section is permitted within a `where` construct and is interpreted as if it were the corresponding `where` construct containing one array assignment.

Finally, a `where` construct may be named in the same way as other constructs.

An example illustrating more complicated `where` constructs that are named is shown in Figure 7.6.

Figure 7.6 Nested `where` constructs, showing the masking.

```
assign_1: where (m_1)
                :      ! masked by m_1
          elsewhere (m_2)
                :      ! masked by .not.m_1 .and. m_2
assign_2:     where (m_4)
                : ! masked by .not.m_1 .and. m_2 .and. m_4
              elsewhere
                : ! masked by .not.m_1 .and. m_2 .and. .not.m_4
              end where assign_2
                :
          elsewhere (m_3) assign_1
                :      ! masked by .not.m_1 .and. .not.m_2 .and. m_3
          elsewhere assign_1
                :      ! masked by .not.m_1 .and. .not.m_2 .and. .not.m_3
          end where assign_1
```

All the statements of a `where` construct are executed one by one in sequence, including the `where` and `elsewhere` statements. The masking expressions in the `where` and `elsewhere` statements are evaluated when the statements are reached and subsequent changes to the values of these expressions do not affect the control of subsequent assignments.

7.7 Mask arrays

Logical arrays are needed for masking in `where` statements and constructs (Section 7.6), and they play a similar role in many of the array intrinsic functions (Chapter 9). Such arrays are often large, and there may be a worthwhile storage gain from using non-default logical types, if available. For example, some processors may use bytes to store elements of `logical(kind=1)` arrays, and bits to store elements of `logical(kind=0)` arrays. Unfortunately, there is no portable facility to specify such arrays, since there is no intrinsic function comparable to `selected_int_kind` and `selected_real_kind`.[2]

Logical arrays are formed implicitly in certain expressions, usually as compiler-generated temporary variables. In

```
where (a > 0.0) a = 2.0 * a
```

or

```
if (any(a > 0.0)) then
```

(`any` is described in Section 9.13.1) the expression `a > 0.0` is a logical array. In such a case, an optimizing compiler can be expected to choose a suitable kind type parameter for the temporary array.

7.8 Pure procedures

In the description of functions in Section 5.11 we noted that, although it is permissible to write functions with side-effects, this is regarded as undesirable. It is possible for the programmer to assert that a procedure has no side-effects by adding the `pure` keyword to the `subroutine` or `function` statement. A procedure that is not pure is **impure**. This may be declared explicitly with the `impure` keyword.

Declaring a procedure to be pure is an assertion that the procedure

- if a function, does not alter any dummy argument, unless it has the `value` attribute (Section 20.8);
- does not alter any part of a variable accessed by host or use association;
- contains no local variable with the `save` attribute;
- performs no operation on an external file (Chapters 10 and 12);
- contains no `stop` statement;
- cannot cause the execution of an image control statement; and
- does not reference an impure procedure.

To ensure that these requirements are met and that a compiler can easily check that this is so, there are the following further rules:

- any dummy argument that is a procedure, and any procedure referenced, is pure;
- the intent of a dummy argument is declared unless it is a procedure, a pointer, or has the `value` attribute (Section 20.8), and this intent must be `in` in the case of a function;

---

[2]This is remedied in Fortran 2023, see Section 23.3.4.

- any internal procedure is pure;
- the procedure does not reference an impure procedure through finalization (Section 15.11) of its function result or an intent out dummy argument;
- the procedure does not have an intent out dummy argument that is polymorphic or has a polymorphic allocatable ultimate component;
- for a function, the result is not polymorphic and allocatable, and does not have a polymorphic allocatable ultimate component; and
- a variable that is accessed by host or use association, is an intent in dummy argument, is a coindexed object, or any part of such a variable must not be used in any way that could alter its value, allocation, or pointer association status, or cause it to be the target of a pointer.

This last rule ensures that a local pointer cannot cause a side-effect.

The function in Figure 5.7 (Section 5.11) is suitable to be specified as pure:

```
pure function distance(p, q)
```

Unlike pure functions, pure subroutines may have dummy arguments that have intent out or inout, or the pointer attribute. The main reason now for allowing pure subroutines is to be able to use a defined assignment in a do concurrect statement (Section 7.16).[3] Their existence also gives the possibility of making subroutine calls from within pure functions.

All the intrinsic functions (Chapter 9) are pure, and can thus be referenced freely within pure procedures. Also, the elemental intrinsic subroutine mvbits (Section 9.10.5) is pure.

7.9   Elemental procedures

7.9.1   Pure elemental procedures

We have already met the notion of elemental intrinsic procedures (Section 7.4) – those with scalar dummy arguments that may be called with array actual arguments provided that the array arguments have the same shape (that is, provided all the arguments are conformable). For a function, the shape of the result is the shape of the array arguments. This feature also exists for non-intrinsic procedures. This requires the elemental prefix on the function or subroutine statement. For example, we could make the function add_intervals of Section 3.9 elemental, as shown in Figure 7.7. This is an aid to optimization on parallel processors.

Unless the impure prefix is used, an elemental procedure automatically has the pure attribute. In addition, any dummy argument must be a scalar variable that is not a coarray, is not allocatable, and is not a pointer; and a function result must be a scalar variable that is not allocatable, is not a pointer, and does not have a type parameter defined by an expression other than a constant expression. Each dummy argument must have specified intent unless it has the value attribute (Section 20.8).

The interface block for an external procedure must specify it as elemental if it is invoked elementally. This is because the compiler may use a different calling mechanism in order

---

[3]Pure procedures were originally introduced to support the forall feature (Section B.1.13), now superseded by do concurrent.

Figure 7.7 An elemental function.

```
elemental function add_intervals(a,b)
    type(interval)                :: add_intervals
    type(interval), intent(in) :: a, b
    add_intervals%lower = a%lower + b%lower ! Production code
    add_intervals%upper = a%upper + b%upper ! would allow for
end function add_intervals                  ! roundoff.
```

to accommodate the array case efficiently. It contrasts with the case of pure procedures, where more freedom is permitted. For consistency with Fortran 77, which allowed procedure arguments but lacked explicit interfaces, a non-intrinsic elemental procedure may not be used as an actual argument.

For an elemental subroutine, if any actual argument is array valued, all actual arguments corresponding to dummy arguments with intent inout or out must be arrays. For example, we can make the subroutine swap of Figure 7.2 (Section 7.3) perform its task on arrays of any shape or size, as shown in Figure 7.8. Calling swap with an array and a scalar argument is obviously erroneous and is not permitted.

Figure 7.8 Elemental version of the subroutine of Figure 7.2.

```
elemental subroutine swap(a, b)
    real, intent(inout)  :: a, b
    real                 :: work
    work = a
    a = b
    b = work
end subroutine swap
```

If a generic procedure reference (Section 5.18) is consistent with both an elemental and a non-elemental procedure, the non-elemental procedure is invoked. For example, we might write versions of add_intervals (Figure 7.7) for arrays of rank one and rely on the elemental function for other ranks. In general, one must expect the elemental version to execute more slowly for a specific rank than the corresponding non-elemental version.

A procedure is permitted to be both elemental and recursive.

7.9.2 Impure elemental procedures

Elemental procedures that are pure are an aid to parallel evaluation. They also permit a single function to replace separate functions for each permissible combination of ranks; for a procedure with two arguments, that is 46 separate procedures (16 cases where both arguments have the same rank, 15 where the first is scalar and the second an array, and 15 where the first is an array and the second scalar).

The impure prefix on the procedure heading allows one to get this effect when the function is not pure. It processes array argument elements one by one in array element order. An example is shown in Figure 7.9. This example is impure in three ways: it counts the number of overflows in the global variable `overflow_count`, it logs each overflow on the external unit `error_unit`, and it terminates execution with `stop` when too many errors have been encountered.

Figure 7.9 An impure elemental function. The `write` statement is described in Section 10.8.

```
module safe_arithmetic
  integer :: max_overflows = 1000
  integer :: overflow_count = 0
contains
  impure elemental integer function square(n)
    use iso_fortran_env, only:error_unit
    integer, intent(in)       :: n
    real(kind(0d0)), parameter :: sqrt_huge = &
                                   sqrt(real(huge(n), kind(0d0)))
    if (abs(n)>sqrt_huge) then
      write (error_unit,*) 'Overflow in square (', n, ')'
      overflow_count = overflow_count + 1
      if (overflow_count>max_overflows) stop '?Too many overflows'
      square = huge(n)
    else
      square = n**2
    end if
  end function
end module
```

Only the requirements relating to 'purity' (lack of side-effects) are lifted; the elemental requirements remain, that is:

- each dummy argument must be a scalar non-coarray dummy data object, must not have the `pointer` or `allocatable` attribute, and must either have specified intent or the `value` attribute;
- if the procedure is a function, its result must be scalar, must not have the `pointer` or `allocatable` attribute, and must not have a type parameter expression that depends on the value of a dummy argument, on the value of a deferred type parameter of a dummy argument, or on the bounds of a pointer or allocatable dummy array;
- in a reference to the procedure, all actual arguments must be conformable; and
- in a reference to the procedure, actual arguments corresponding to intent `out` and `inout` dummy arguments must either all be arrays or all be scalar.

7.10   Array elements

In Section 2.10 we restricted the description of array elements to simple cases. In general, an array element is a scalar of the form *part-ref [%part-ref] ...*

where *part-ref* is

part-name [(subscript-list)]

and the last *part-ref* has a *subscript-list*. The number of subscripts in each list must be equal to the rank of the array or array component, and each subscript must be a scalar integer expression whose value is within the bounds of the dimension of the array or array component. To illustrate this, take the type

```
type triplet
   real                :: u
   real, dimension(3)   :: du
   real, dimension(3,3) :: d2u
end type triplet
```

which was considered in Section 2.10.1. An array may be declared of this type:

```
type(triplet), dimension(10,20,30) :: tar
```

and

```
tar(n,2,n*n)          ! n of type integer
```

is an array element. It is a scalar of type `triplet` and

```
tar(n, 2, n*n)%du
```

is a real array with

```
tar(n, 2, n*n)%du(2)
```

as one of its elements.

If an array element is of type character, it may be followed by a substring reference:

(substring-range)

for example,

```
page (k*k) (i+1:j-5) ! i, j, k of type integer
```

By convention, such an object is called a substring rather than an array element.

Notice that it is the array *part-name* that the subscript list qualifies. It is not permitted to apply such a subscript list to an array designator unless the designator terminates with an array *part-name*. An array section, a function reference, or an array expression in parentheses must not be qualified by a subscript list.

7.11 Array subobjects

Array sections were introduced in Section 2.10.2 and provide a convenient way to access a regular subarray such as a row or a column of a rank-two array:

```
a(i, 1:n)    ! Elements 1 to n of row i
a(1:m, j)    ! Elements 1 to m of column j
```

For simplicity of description we did not explain that one or both bounds may be omitted when the corresponding bound of the array itself is wanted:

```
a(i, :)          ! The whole of row i
```

Another form of section subscript is a rank-one integer expression. All the elements of the expression must be defined with values that lie within the bounds of the parent array's subscript. For example, given an array v of extent 8,

```
v( [ 1, 7, 3, 2 ] )
```

is a section with elements v(1), v(7), v(3), and v(2), in this order. Such a subscript is called a **vector subscript**. If there are any repetitions in the values of the elements of a vector subscript, the section is called a **many–one section** because more than one element of the section is mapped onto a single array element. For example,

```
v( [ 1, 7, 3, 7 ] )
```

has elements 2 and 4 mapped onto v(7). A many–one section must not appear on the left of an assignment statement because there would be several possible values for a single element. For instance, the statement

```
v( [ 1, 7, 3, 7 ] ) = [ 1, 2, 3, 4 ]      ! Invalid
```

is not allowed because the values 2 and 4 cannot both be stored in v(7). The extent is zero if the vector subscript has zero size. Note that, with a sufficient repetition of subscript values, the section can actually be longer than the parent array.

When an array section with a vector subscript is an actual argument in a call of a non-elemental procedure, it is regarded as an expression and the corresponding dummy argument must not be defined or redefined and must not have intent out or inout. We expect compilers to make a copy as a temporary regular array on entry but to perform no copy back on return. Also, an array section with a vector subscript is not permitted to be a pointer target, since allowing them would seriously complicate the mechanism that compilers would otherwise have to establish for pointers. For similar reasons, such an array section is not permitted to be an internal file (Section 10.6).

In addition to the regular and irregular subscripting patterns just described, the intrinsic circular shift function cshift (Section 9.15.5) provides a mechanism that manipulates array sections in a 'wrap-round' fashion. This is useful in handling the boundaries of certain types of periodic grid problems, although it is subject to similar restrictions to those on vector subscripts. If an array v(5) has the value [1,2,3,4,5], then cshift(v, 2) has the value [3,4,5,1,2].

The general form of a subobject is

part-ref [%part-ref] ... [(substring-range)]

where *part-ref* now has the form

part-name [(section-subscript-list)]

and the number of section subscripts in each list must be equal to the rank of the array or array component. Each *section-subscript* is either a *subscript* (Section 7.10), a rank-one integer expression (vector subscript), or a *subscript-triplet* of the form

[lower] : [upper] [: stride]

where *lower*, *upper*, and *stride* are scalar integer expressions. If *lower* is omitted, the default value is the lower bound for this subscript of the array. If *upper* is omitted, the default value is the upper bound for this subscript of the array. If *stride* is omitted, the default value is one. The stride may be negative so that it is possible to take, for example, the elements of a row in reverse order by specifying a section such as

```
a(i, 10:1:-1)
```

The extent is zero if *stride* > 0 and *lower* $>$ *upper*, or if *stride* < 0 and *lower* $<$ *upper*. The value of *stride* must not be zero.

Unless the extent is zero, *lower* must be within the bounds of the corresponding array subscript and the *subscript-triplet* specifies the sequence of subscript values,

$$lower, \ lower + stride, \ lower + 2 \times stride, \ \ldots$$

going as far as possible without going beyond *upper* (above it when *stride* > 0 or below it when *stride* < 0). The length of the sequence for the *i*th *subscript-triplet* determines the *i*th extent of the array that is formed.

Normally, we expect the value of *upper* to be within the bounds of the corresponding array subscript, but this is not required. For example,

```
a(1, 2:11:2)
```

is allowed even if the upper bound of the second dimension of a is only 10.

The rank of a *part-ref* with a *section-subscript-list* is the number of vector subscripts and subscript triplets that it contains. So far in this section all the examples have been of rank one; by contrast, the ordinary array element

```
a(1,7)
```

is an example of a *part-ref* of rank zero, and the section

```
a(:,1:7)
```

is an example of a *part-ref* of rank two. The rank of a *part-ref* without a *section-subscript-list* is the rank of the object or component. A *part-ref* may be an array; for example,

```
tar%du(2)
```

for the array tar of Section 7.10 is an array section with elements tar(1,1,1)%du(2), tar(2,1,1)%du(2), tar(3,1,1)%du(2),... Being able to form sections in this way from arrays of derived type, as well as by selecting sets of elements, is a very useful feature of the language. A more prosaic example, given the specification

```
type(person), dimension(1:50) :: my_group
```

for the type person of Section 2.9, is the subobject my_group%id which is an integer array section of size 50.

Further, for the particular case of complex arrays, the re and im selectors can be applied to yield an array section comprising the real or imaginary part of each element of the array. For example,

```
complex :: x(n), y(n)
x%im = 2.0*y%im
```

Unfortunately, it is not permissible for more than one *part-ref* to be an array; for example, it is not permitted to write

```
tar%du    ! Invalid
```

for the array `tar` of Section 7.10. The reason for this is that if `tar%du` were considered to be an array, its element (1,2,3,4) would correspond to

```
tar(2,3,4)%du(1)
```

which would be too confusing a notation.

The *part-ref* with nonzero rank determines the rank and shape of the subobject. If any of its extents is zero, the subobject itself has size zero. It is called an array section if the final *part-ref* has a *section-subscript-list* or another *part-ref* has a nonzero rank.

A *substring-range* may be present only if the last *part-ref* is of type character and is either a scalar or has a *section-subscript-list*. By convention, the resulting object is called a section rather than a substring. It is formed from the unqualified section by taking the specified substring of each element. Note that, if c is a rank-one character array,

```
c(i:j)
```

is the section formed from elements i to j; if substrings of all the array elements are wanted, we may write the section

```
c(:)(k:l)
```

An array section that ends with a component name is also called a **structure component**. Note that if the component is scalar, the section cannot be qualified by a trailing subscript list or section subscript list. Thus, using the example of Section 7.10,

```
tar%u
```

is such an array section and

```
tar(1, 2, 3)%u
```

is a component of a valid element of `tar`. The form

```
tar%u(1, 2, 3)   ! not permitted
```

is not allowed.

Additionally, a *part-name* to the right of a *part-ref* with nonzero rank must not have the `allocatable` or `pointer` attribute. This is because such an object would represent an array whose elements were independently allocated and would require a very different implementation mechanism from that needed for an ordinary array. For example, consider the array

```
type(entry), dimension(n) :: rows  ! n of type integer
```

for the type `entry` defined in Figure 2.3. If we were allowed to write the object `rows%next`, it would be interpreted as another array of size n and type `entry`, but its elements are likely to be stored without any regular pattern (each having been separately given storage by an `allocate` statement) and indeed some will be null if any of the pointers are disassociated. Note that there is no problem over accessing individual pointers such as `rows(i)%next`.

7.12   Arrays of pointers

Although arrays of pointers as such are not allowed in Fortran, the equivalent effect can be achieved by creating a type containing a pointer component. This is useful when constructing a linked list that is more complicated than the chain described in Section 2.13.2. For instance, if a variable number of links are needed at each entry, the recursive type entry of Figure 2.3 might be expanded to the pair of types

```
type ptr
   type(entry), pointer :: point
end type ptr
type entry
   real              :: value
   integer           :: index
   type(ptr), pointer :: children(:)
end type entry
```

After appropriate allocations and pointer associations, it is then possible to refer to the index of child j of node as

```
node%children(j)%point%index
```

This extra level of indirection is necessary because the individual elements of children do not, themselves, have the pointer attribute – this is a property only of the whole array. For example, we can take two existing nodes, say a and b, each of which is a tree root, and make a big tree thus

```
tree%children(1)%point => a
tree%children(2)%point => b
```

which would not be possible with the original type entry.

7.13   Pointers as aliases

If an array section without vector subscripts, such as

```
table(m:n, p:q)
```

is wanted frequently while the integer variables m, n, p, and q do not change their values, it is convenient to be able to refer to the section as a named array such as

```
window
```

Such a facility is provided in Fortran by pointers and the pointer assignment statement. Here, window would be declared thus

```
real, dimension(:, :), pointer :: window
```

and associated with table, which must of course have the target or pointer attribute,[4] by the execution of the statement

---

[4]The associate construct (Section 15.6) provides a means of achieving this in a limited scope without the need for the target or pointer attribute.

```
window  => table(m:n, p:q)
```

If, later on, the size of `window` needs to be changed, all that is needed is another pointer assignment statement. Note, however, that the subscript bounds for `window` in this example are (1:n-m+1, 1:q-p+1) since they are as provided by the functions `lbound` and `ubound` (Section 9.14.3). To get the same bounds, we need to change the statement to

```
window(m:, p:) => table(m:n, p:q)
```

(see Section 7.14).

Pointer association provides a mechanism for subscripting or sectioning arrays such as

```
tar%u
```

where `tar` is an array and `u` is a scalar component, discussed in Section 7.10. Here we may perform the pointer association

```
taru => tar%u
```

if `taru` is a rank-three pointer of type real. Subscripting as in

```
taru(1, 2, 3)
```

is then permissible. Here the subscript bounds for `taru` will be those of `tar`.

7.14   Remapping bounds and rank in pointer assignments

There are two further facilities concerning the array pointer assignment statement. The first is that it is possible to set the desired lower bounds to any value by using the syntax

pointer-name (*lower-bound-list*) => *target*

where each *lower-bound* is an integer expression. The shape is that of the *target*. There is an example in Section 7.13.

The second facility for array pointer assignment is that the target of a multi-dimensional array pointer may be one-dimensional or simply contiguous (Section 7.17.2). The syntax is similar to that of the lower-bounds specification above, except that in this case one specifies each upper bound as well as each lower bound:

pointer-name (*remapping-list*) => *target*

where each *remapping* is

lower-bound : *upper-bound*

and each *lower-bound* and *upper-bound* is an integer expression. The elements of the array pointer, in array element order, are associated with the leading elements of the *target*. This can be used, for example, to provide a pointer to the diagonal of an array:

```
real, pointer :: base_array(:), matrix(:,:), diagonal(:)
allocate (base_array(n*n))
matrix(1:n, 1:n) => base_array
diagonal => base_array( : : n+1)
```

After execution of the pointer assignments, `diagonal` is now a pointer to the diagonal elements of `matrix`, and the three arrays, which have deferred shape, now have their shapes defined.

7.15   Array constructors

The syntax that we introduced in Section 2.10.3 for array constants may be used to construct more general rank-one arrays. The general form of an *array-constructor* is

   (/ *ac-spec* /)

or

   [*ac-spec*]

where *ac-spec* is

   [type-spec :: *] array-constructor-value-list*

or

   type-spec ::

The *type-spec* (if it appears) is the type name followed by the type parameter values in parentheses, if any. The form with square brackets is to make it easier to match parentheses; note that the two forms cannot be mixed, so (/...] and [.../) are not allowed.

   Each *array-constructor-value* is one of *expr* or *constructor-implied-do*, and a *constructor-implied-do* has the form

   (*array-constructor-value-list*, [integer *[(*[kind= *] kind-value *)] ::]
       variable = *expr*$_1$, *expr*$_2$ [, *expr*$_3$])

where *variable* is a named integer scalar variable and *expr*$_1$, *expr*$_2$, and *expr*$_3$ are scalar integer expressions. If integer *[(* [kind= *] kind-value*) *]* :: is present, it specifies the kind of *variable*. If *expr*$_3$ is absent, it is as if it were present with the value 1. Its interpretation is as if the *array-constructor-value-list* had been written

   $$\max(\,(expr_2 - expr_1 + expr_3) \div expr_3, 0\,)$$

times, with *variable* replaced by *expr*$_1$, *expr*$_1$ + *expr*$_3$, ..., as for the do construct (Section 4.4). A simple example is

   (/ (i,i=1,10) /)

which is equal to

   (/ 1, 2, 3, 4, 5, 6, 7, 8, 9, 10 /)

The array constructed is of rank one with its sequence of elements formed from the sequence of scalar expressions and elements of the array expressions in the *array-constructor-value-list*, including the expressions in the *constructor-implied-do* lists. The sequence may be empty, in which case a zero-sized array is constructed.

   Note that the syntax permits nesting of one *constructor-implied-do* inside another, as in the example

   (/ ((i,i=1,3), j=1,3) /)

which is equal to

   (/ 1, 2, 3, 1, 2, 3, 1, 2, 3 /)

and the nesting of structure constructors within array constructors (and vice versa); for instance, for the type in Section 7.5,

```
(/ (matrix(0.0), i = 1, limit) /)
```

The scope of the *variable* is the *constructor-implied-do*. Other statements, or even other parts of the array constructor, may refer to another variable having the same name. The value of the other variable is unaffected by execution of the array constructor and is available outside the *constructor-implied-do*.

An array of rank greater than one may be constructed from an array constructor by using the intrinsic function reshape (Section 9.15.3). For example,

```
reshape( source = [ 1,2,3,4,5,6 ], shape = [ 2,3 ] )
```

has the value

1   3   5
2   4   6

If the array constructor does not begin with *type-spec* ::, its type and type parameters are those of the first *expr*, and each *expr* must have the same type and type parameters. If every *expr*, *expr₁*, *expr₂*, and *expr₃* is a constant expression (Section 8.4), the array constructor is a constant expression.

If an array constructor begins with *type-spec* ::, its type and type parameters are those specified by the *type-spec*. In this case, the array constructor values may have any type and type parameters (including character length) that are assignment-compatible with the specified type and type parameters, and the values are converted to that type by the rules of intrinsic assignment.

Here are some examples:

```
[ character(len=33) :: 'the good', 'the bad', 'and', &
                       'the appearance-challenged' ]
[ complex(kind(0d0)) :: 1, (0,1), 3.14159265358979323846264338327d0 ]

[ matrix(kind=kind(0.0), n=10, m=20) :: ] ! zero-sized array

[ (a(i,i), integer(int64) :: i=1,n) ]
```

In the final case, the type specified applies only to the do variable of that implied-do loop.

7.16   The do concurrent construct

A further form of the do construct, the do concurrent construct, is provided to help improve performance by enabling parallel execution of the loop iterations. The basic idea is that by using this construct the programmer asserts that there are no interdependencies between loop iterations. The effect is similar to that of various compiler-specific directives such as !dec$ivdep; such directives have been available for a long time, but often have slightly different meanings on different compilers.

Use of do concurrent has a long list of requirements that can be grouped into 'limitations' on what may appear within the construct and 'guarantees' by the programmer that the computation has certain properties (essentially, no dependencies) that enable parallelization. Note that in this context parallelization does not necessarily require that multiple processors will be used: other optimizations that improve single-threaded performance are also enabled by these properties, including vectorization, pipelining, and other possibilities for overlapping the execution of instructions from more than one iteration on a single processor.

If no change is made to the value or status of a variable within the construct or if a change is made only in one iteration and it is not referenced in any others, parallel execution can occur with the variable being accessed in the normal way. Alternatively, if each iteration references it only after having defined it and the final value is not required, parallel execution can occur with each iteration using a separate local variable. By using do concurrent, the programmer is asserting that each variable that is accessed is done so in one of these ways and can be implemented as a 'shared' or 'local' variable, respectively.

The form of the do concurrent statement is

    do [,] concurrent (*concurrent-header*) [*locality-spec*]...

where

- *concurrent-header* is

    [integer [(*kind*)] ::] *index-spec-list* [, *mask-expr*]

    where *index-spec-list* is a list of index specifications of the form

    index-variable-name = *initial-value* : *final-value* [: *step-value*]

    and the type declaration (if present) specifies the kind of each *index-variable-name*, and

- *locality-spec* is one of:

    ```
    local  (variable-name-list)
    local_init  (variable-name-list)
    shared  (variable-name-list)
    default  (none)
    ```

Each *index-variable-name* is an integer variable. Each variable named in the *locality-spec*s of the statement must appear only once, must exist in the scope of the statement, and must not be the same as any *index-variable-name* of the statement.

Here are two simple examples:

```
do concurrent (i=1:n, j=1:m) shared(a, alpha, b)
   a(i, j) = a(i, j) + alpha*b(i, j)
end do
```

and

```
do concurrent (integer(int64)::i=1:n, j=1:m, i/=j) shared(x)
   x(i,j) = 1/abs(i-j)
end do
```

where int64 is the named constant from the intrinsic module iso_fortran_env (Section 9.24.4).

Because changes to do indices are never allowed in a do construct, a separate variable may be used for each do index in each iteration, as for a local variable, but on exit the outside variable will have been updated as in serial execution. The outside variable must be scalar and have `integer` type. If the type declaration is omitted, the `do` index is an integer with the kind of the outside variable. Each *initial-value*, *final-value*, and *step-value* is a scalar integer expression.

The optional *mask-expr* is a scalar expression of type `logical`; if it appears, only those iterations that satisfy the condition are executed. For example,

```
do concurrent (i=1:n,  j=1:m,  i/=j)
   ⋮
end do
```

has exactly the same meaning as

```
do concurrent (i=1:n,  j=1:m)
  if (i/=j) then
     ⋮
  end if
end do
```

Any procedure referenced in the *mask-expr* must be pure (Section 7.8).

The `local` and `local_init` clauses declare that each named variable is treated as a *construct entity* of the `do concurrent`, and that the outside variable with the same name is inaccessible in the construct. The construct entity has the same type, type parameters, and rank as the outside variable, and has the `asynchronous`, `contiguous`, `pointer`, `target`, or `volatile` (Appendix A.15) attribute if the outside variable has the attribute. Even if the outside variable has the `bind`, `intent`, `protected`, `save`, or `value` attribute, the construct entity does not have the attribute. The outside variable is not permitted to be allocatable, intent in, optional, a coarray, an assumed-size array, a non-pointer polymorphic dummy argument, of finalizable type, or be a variable that is not permitted to appear in a context that could change its value (for example, a `protected` variable outside its module).

At the beginning of each iteration, a variable that is `local` is undefined and has undefined association status if a pointer, whereas a `local_init` variable begins each iteration with the value or pointer association status of the outside variable.

The `shared` clause declares that each named variable is not a construct entity, but is the outside variable. If it is defined or becomes undefined during any iteration, it must not be referenced, defined, or become undefined during any other iteration. If it is allocated, deallocated, nullified, or pointer-assigned during an iteration, it must not have its allocation or association status, dynamic type, array bounds, shape, or a deferred type parameter value inquired about in any other iteration. A non-contiguous array with `shared` locality must not be supplied as an actual argument corresponding to a contiguous intent `inout` dummy argument.

A variable that does not appear in a `local`, `local_init`, or `shared` clause has *unspecified locality*. In this case,

- if the variable is referenced in an iteration it must either be previously defined in the same iteration, or its value must not be affected by any other iteration;
- if the variable is a non-contiguous array that is supplied as an actual argument corresponding to a contiguous intent `inout` dummy argument in an iteration, it must

either be previously defined in that iteration or must not be defined in any other iteration;

- if the variable is a pointer that is referenced in an iteration it must either be previously pointer associated in the same iteration, or must not have its pointer association changed by any other iteration;
- if the variable is an allocatable object that is allocated or deallocated by an iteration it must not be used[5] by any other iteration, unless every iteration that uses the object first allocates it and finally deallocates it.

Furthermore, when execution of the construct has completed,

- any variable whose value is affected by more than one iteration becomes undefined on termination of the loop; and
- any pointer whose association status is changed by more than one iteration has an association status of undefined.

The `default (none)` *locality-spec* requires the *locality* of each variable used in the loop to be specified in another *locality-spec*; we recommend this because it is very helpful to the compiler. If a variable has unspecified locality, the compiler has to decide whether to treat it as shared or local. While this is straightforward for simple cases, it can be impossible to decide at compile time if there is a conditional assignment to the variable. In such cases there is no choice but for sequential execution.

A simple example of the use of `local` is shown in Figure 7.10.

Figure 7.10 Simple example of use of `local` in do concurrent.

```
real a(:), b(:), x
   :
do concurrent (i=1:size(a)) local (x) shared (a, b)
   if (a(i)>0) then
      x = sqrt (a(i))
      a(i) = a(i) - x**2
   end if
   b(i) = b(i) - a(i)
end do
   :
```

The following items are all prohibited within a do concurrent construct (and except for the last two the compiler is required to be able detect them):

- a statement that would terminate the do concurrent construct: that is, a return statement (Section 5.8), a branch (Appendix A.12) with a label that is outside the construct, an exit statement that would exit from the do concurrent construct, or a cycle statement that names an outer do construct;

---

[5]That is, referenced, defined, allocated, deallocated, or have any dynamic property inquired about.

- an image control statement;
- a reference to a procedure that is not pure;
- a statement that might result in the deallocation of a polymorphic entity;
- a reference to one of the procedures `ieee_get_flag`, `ieee_set_halting_mode`, or `ieee_get_halting_mode` from the intrinsic module `ieee_exceptions` (Section 19.6); and
- an input/output statement with an `advance=` specifier (Section 10.11);
- writing data to a file in one iteration and reading any of them from the file in another iteration.

The above are all prohibited because they are either impossible or extremely difficult to use without breaking the 'no interdependencies between iterations' rule.

If records are written to a sequential file (Chapter 10) by more than one iteration of the loop, the ordering between the records written by different iterations is indeterminate. That is, the records written by one iteration might appear before the records written by the other, after the records written by the other, or be interspersed.

Note that any ordinary do construct that satisfies the limitations and which obviously has the required properties can be parallelized, so use of do concurrent is not necessary for parallel execution. In fact, a compiler that parallelizes do concurrent is likely to treat it as a request that it should parallelize that construct; if the loop iteration count is very small, this could result in worse performance than an ordinary do loop due to the overhead of initiating parallel threads of execution. Thus, even when the programmer-provided guarantees are trivially derived from the loop body itself, do concurrent is still useful for

- indicating to the compiler that this is likely to have a high enough iteration count to make parallelization worthwhile;
- using the compiler to enforce the prohibitions (e.g., no calls to impure procedures);
- documenting the parallelizability for code reading and maintenance; and
- as a crutch to compilers whose analysis capabilities are limited.

7.17   Contiguous arrays

7.17.1   The `contiguous` attribute

It is expected that most arrays will be held in contiguous memory, but there are exceptions such as these array sections:

```
vector(::2)      ! all the odd-numbered elements
vector(10:1:-1) ! reverse order
cxarray%re       ! the real parts of a complex array
```

(where the third line uses the syntax of Section 2.9). A pointer or assumed-shape array that is not contiguous can come about by association with an array section that is not contiguous, and the compiler has to allow for this. The contiguous attribute is an attribute for pointer and assumed-shape dummy arrays. For an array pointer, it restricts its target to being contiguous.

For an assumed-shape array, it specifies that if the corresponding actual argument is not contiguous, copy-in copy-out is used make the dummy argument contiguous.

Knowing that an array is contiguous in this sense can improve performance by reducing cache misses and simplifying array traversals and array element address calculations. Whether this improvement is significant depends on the fraction of time spent performing suitable operations; in some programs this time is substantial, but in many cases it is insignificant.

Traditionally, the Fortran standard has shied away from specifying whether arrays are contiguous in the sense of occupying sequential memory locations with no intervening unoccupied spaces. In the past this tradition has enabled high-performance multi-processor implementations of the language, but the `contiguous` attribute is a move towards more specific hardware limitations. Although contiguous arrays are described only in terms of language restrictions and not in terms of the memory hardware, the interaction between these and interoperability with the C language (Chapter 20) means that these arrays will almost certainly be stored in contiguous memory locations.

Any of the following arrays are considered to be contiguous by the standard:

- an array with the `contiguous` attribute;
- a whole array (named array or array component without further qualification) that is not a pointer or assumed-shape;
- an assumed-shape array that is argument associated with an array that is contiguous;
- an array allocated by an `allocate` statement;
- a pointer associated with a contiguous target; or
- a nonzero-sized array section provided that
  - its base object is contiguous;
  - it does not have a vector subscript;
  - the elements of the section, in array element order, are elements of the base object that are consecutive in array element order;
  - if the array is of type character and a substring selector appears, the selector specifies all of the characters of the string;
  - it is not a component of an array; and
  - it is not the real or imaginary part of an array of type complex.

A subobject (of an array) is definitely not contiguous if all of these conditions apply:

- it (the subobject) has two or more elements;
- its elements in array element order are not consecutive in the elements of the original array;
- it is not a zero-length character array; and
- it is not of a derived type with no ultimate components other than zero-sized arrays and zero-length character strings.

Whether an array that is in neither list is contiguous or not is compiler-specific.

The `contiguous` attribute can be specified with the `contiguous` keyword on a type declaration statement, for example

```
subroutine s(x)
  real, contiguous :: x(:,:)
  real, pointer, contiguous :: column(:)
```

It can also be specified by the `contiguous` statement, which has the form

   `contiguous` *[::] object-name-list*

Contiguity can be tested with the inquiry function `is_contiguous` (Section 9.14.2).

Arrays in C are always contiguous, so referencing the intrinsic function `c_loc` (Section 20.3) is permitted for any target that is contiguous (at execution time). The example in Figure 7.11 uses `is_contiguous` to check that it is being asked to process a contiguous object, and produces an error message if it is not. It also makes use of the `c_sizeof` function to calculate the size of x in bytes (see Section 20.7).

Figure 7.11 Using `is_contiguous` before using `c_loc`.

```
subroutine process(x)
  use iso_c_binding
  real(c_float), target :: x(:)
  interface
    subroutine c_routine(a, nbytes)
      use iso_c_binding
      type(c_ptr), value       :: a
      integer(c_size_t), value :: nbytes
    end subroutine
  end interface
  :
  if (is_contiguous(x)) then
    call c_routine(c_loc(x), c_sizeof(x))
  else
    stop 'x needs to be contiguous'
  end if
end subroutine
```

There is also the concept of **simply contiguous**; that is, not only is the object contiguous, but it can be seen to be obviously so at compilation time. Unlike 'being contiguous', this is completely standardized. This is further discussed in the next section.

When dealing with `contiguous` assumed-shape arrays and array pointers, it is important to keep in mind the various run-time requirements and restrictions. For assumed-shape arrays, the `contiguous` attribute makes no further requirements on the program: if the actual argument is not contiguous, a local copy is made on entry to the procedure, and any changes to its value are copied back to the actual argument on exit. Depending on the number and manner of the references to the array in the procedure, the cost of copying can be higher than the performance savings given by the `contiguous` attribute. For example, in

```
complex function f(v1, v2, v3)
  real, contiguous, intent(in) :: v1(:), v2(:), v3(:)
  f = cmplx(sum(v1*v2*v3))**(-size(v1))
end function
```

since the arrays are only accessed once, if any actual argument is discontiguous this will almost certainly perform much worse than if the contiguous attribute were not present.

For array pointers, the contiguous attribute has a run-time requirement that it be associated only with a contiguous target (via pointer assignment). However, it is the programmer's responsibility to check this, or to 'know' that the pointer will never become associated with a discontiguous section. (Such knowledge is prone to becoming false in the course of program maintenance, so checking on each pointer assignment is recommended.) Similar comments apply to the use of the c_loc function on an array that might not be contiguous. If these requirements are violated, the program will almost certainly produce incorrect answers with no indication of the failure.

7.17.2 Simply contiguous array designators

A **simply contiguous** array designator is, in principle, a designator that not only describes an array (or array section) that is contiguous, but one that can easily be seen at compilation time to be contiguous. Whether a designator is simply contiguous does not depend on the value of any variable.

A simply contiguous array can be used in the following ways:

- as the target of a rank-remapping pointer assignment (that is, associating a pointer with a target of a different rank, see Section 7.14);
- as an actual argument corresponding to a dummy argument that is not an assumed-shape array or which is an assumed-shape array with the contiguous attribute, when both have either the asynchronous or volatile attribute;
- as an actual argument corresponding to a dummy pointer with the contiguous attribute (this also requires that the actual argument have the pointer or target attribute).

The example in Figure 7.12 'flattens' the matrix a into a simple vector, and then uses that to associate another pointer with the diagonal of the matrix.

Figure 7.12 Diagonal of contiguous matrix.

```
real, target  :: a(n, m)
real, pointer :: a_flattened(:), a_diagonal(:)
a_flattened(1:n*m) => a
a_diagonal        => a_flattened(::n+1)
```

Another example of the use of simply contiguous to enforce contiguity of an actual argument is explained in Section 7.17.3.

Also, when a simply contiguous array with the target attribute (and not the value attribute, Section 20.8) is used as the actual argument corresponding to a dummy argument that has the `target` attribute and is an assumed-shape array with the `contiguous` attribute or is an explicit-shape array,

- a pointer associated with the actual argument becomes associated with the dummy argument on invocation of the procedure; and
- when execution of the procedure completes, pointers in other scopes that were associated with the dummy argument are associated with the actual argument.

However, we do not recommend using this complicated fact, as it is difficult to understand and program maintenance is quite likely to break one of the essential conditions for its applicability.

An array designator is simply contiguous if and only if it is

- a whole array that has the `contiguous` attribute;
- a whole array that is not an assumed-shape array or array pointer; or
- a section of a simply contiguous array that
  - is not the real or imaginary part of a complex array (see Section 2.9);
  - does not have a substring selector;
  - is not a component of an array; and
  - either does not have a *section-subscript-list*, or has a *section-subscript-list* which specifies a simply contiguous section.

A *section-subscript-list* specifies a simply contiguous section if and only if

- it does not have a vector subscript;
- all but the last *subscript-triplet* is a colon;
- the last *subscript-triplet* does not have a stride; and
- no *subscript-triplet* is preceded by a *section-subscript* that is a subscript.

An array variable is simply contiguous if and only if it is a simply contiguous array designator or a reference to a function that returns a pointer with the `contiguous` attribute.

7.17.3   Automatic pointer targetting

An actual argument with the `target` attribute is permitted to correspond to a dummy pointer with the intent `in` attribute. This is known as automatic targetting and is illustrated by Figure 7.13. Here, the module `solver` places data in the pointer array `solver_field` and this is given the target `field`. It saves the user from having to create a local pointer, pointing it at `args`, and passing the local pointer to `set_params`. Furthermore, it enforces contiguity requirements; if a dummy pointer array has the `contiguous` attribute, the actual argument must be simply contiguous (see Section 7.17.2). This means that the user can be sure that no unintended copying, by a copy-in copy-out argument-passing mechanism, takes place when the module calls procedures to perform its work.

Figure 7.13 Automatic targetting. The `protected` attribute (Section 8.6.2) limits changing the pointer status of `solver_field` to within the module.

```
module solver
  real, pointer, contiguous, protected :: solver_field(:)
contains
  subroutine set_field(field)
    real, pointer, contiguous, intent(in) :: field(:)
    solver_field => field
  end subroutine
  ⋮
end module
⋮
  use solver
  real, allocatable, target :: field(:)
  ⋮
  allocate (field(n))
  call set_field(field)
  ⋮
end subroutine
```

7.18 Assumed rank

The concept of **assumed rank** is available, making the writing of procedures that can handle any rank easier. It is particularly useful in combination with C functions that have been written to handle arguments of any rank (Section 21.3.6).

A dummy argument that is not a coarray and does not have the `value` attribute (Section 20.8) may be declared to have assumed rank with the syntax (`..`). For example, the procedure

```
subroutine scale(a) bind(c)
  real a(..)
  ⋮
end subroutine scale
```

may be called with an array of any rank or even a scalar as an actual argument. The dummy argument assumes the rank and shape of the actual argument.

The elements of an assumed-rank array are accessible only within code in a `select rank` construct of the form

> *[name:]* select rank (*[associate-name =>] selector*)
>   *[rank-case-stmt [name]*
>   *block]* ...
>   end select *[name]*

where *selector* is the name of an assumed-rank object and each *rank-case-stmt* is one of

```
rank ( expr )
rank (*)
rank default
```

Each *expr* must be a scalar integer constant expression and have a distinct non-negative value. There must be at most one rank (*) statement and at most one rank default statement. The value in a rank (*expr*) statement is permitted to exceed the maximum possible array rank, in which case its block can never be executed. This is to enhance portability of programs between processors with different maximum array rank.

Execution of a select rank construct involves a single block. If the selector is not assumed size (Section 20.5) and a rank (*expr*) statement matches its rank, this block is executed; if the selector is assumed size and a rank (*) statement appears, this block is executed; otherwise, if a rank default statement appears, this block is executed. For example, the function in Figure 7.14 computes a function of an object depending on its rank.

Figure 7.14 Simple example of assumed rank.

```
real(kind(0d0)) function f(x)
    real(kind(0d0)), intent(in) :: x(..)
    select rank (x)
    rank (*)
        error stop 'Function not defined on assumed-size arrays'
    rank (0)
        f = abs(x)
    rank (1)
        f = sum(abs(x))
    rank (2)
        f = sqrt(sum(x**2))
    rank (3)
        f = sum(abs(x)**3)**(1/3.0d0)
    rank (4)
        f = sum(x**4)**(0.25d0)
    rank default
        error stop 'Function not supported for rank>4 (unstable)'
    end select
end function
```

As for other constructs, the leading and trailing statements must either both be unnamed or both bear the same name; a *rank-case-stmt* within it may be named only if the select rank statement is named and bears the same name.

If *associate-name* is present, it is used in the block for the name of the selector; otherwise, the same name is used.

Within the *block* of a rank (*expr*) statement, the selector has the rank specified by *expr* and the bounds returned by the intrinsic functions lbound and ubound. Within the *block* of a rank (*) statement, the selector is assumed size with rank one and lower bound one. Within the *block* of a rank default statement, the selector is of the same assumed rank.

Because an assumed-rank dummy argument may be associated with an assumed-size dummy argument, the same restrictions apply as for assumed size when it has intent out, see Section 20.5.

The intrinsic inquiry functions of Section 9.14, is_contiguous, lbound, rank, shape, size, and ubound, are available for an assumed-rank object.

An assumed-rank object must not appear in any other context, including in a variable designator. For example, an assumed-rank complex variable arc may be passed to a complex assumed-rank dummy, but its real or imaginary parts arc%re and arc%im are not permitted anywhere, not even as an actual argument to a real assumed-rank dummy.

Because a scalar object has rank zero, when an assumed-rank object arx is associated with a scalar, shape(arx) will return a zero-sized array, and size(arx) will return the value 1. Note that because the dim argument to the size intrinsic is required to satisfy $1 \leq \text{dim} \leq \text{rank(array)}$, it must not be used on an assumed-rank object when it is associated with a scalar.

When an assumed-rank object arx is associated with an assumed-size array, because it has no final extent, shape(arx) will return an array with a final extent value of -1. Similarly, size(arx, rank(arx)) will return the value -1. However, care should be taken with size(arx) with no dim argument; this could return a negative value, but due to the possibility of integer overflow, could return any value. For example, when associated with an assumed-size array declared as a(65536,65536,65536,0), a reference to size(arx) is likely either to return the value zero, or produce a run-time error.

Similar considerations apply to the intrinsic functions lbound and ubound. When associated with a scalar, the dim form must not be used and the non-dim form returns a zero-sized array. When associated with an assumed-size array, ubound treats it as having a final extent of -1 by returning a value for the final dimension that is two less than the lower bound.

For the purposes of generic resolution, an assumed-rank dummy argument is considered not to be distinguishable from any other rank, and therefore

```
interface g
   subroutine scalar(x)
      real x
   end subroutine
   subroutine sassumed(x)
      real x(..)
   end subroutine
end interface
```

is not a valid generic interface.

7.19 Summary

We have explained that arrays may have zero size and that no special rules are needed for them. Storage for an array may be allocated automatically on entry to a procedure and automatically deallocated on return. Functions may be array valued either through the mechanism of an elemental reference that performs the same calculation for each array element, or through the truly array-valued function. Elemental procedures may be pure or impure. Array assignments may be masked through the use of the `where` statement and construct. Structure components may be arrays if the parent is an array or the component is an array, but not both. A subarray may either be formulated directly as an array section, or indirectly by using pointer assignment to associate it with a pointer. An array may be constructed from a sequence of expressions. A logical array may be used as a mask. Performance may be enhanced by the use of `do concurrent` and by the use of arrays declared as `contiguous`. Assumed-rank arrays allow procedures to be written for any rank.

The intrinsic functions are an important part of the array features and will be described in Chapter 9.

We conclude this chapter by drawing attention to a complete program, *linear.f90* accessible through the bullet for this book on `http://www.oup.com/fortran2023` that illustrates the use of array expressions, array assignments, allocatable arrays, automatic arrays, and array sections. The module `linear` contains a subroutine for solving a set of linear equations, and this is called from a main program that prompts the user for the problem and then solves it. The function `size` is described in Section 9.14.3 and the function `maxloc` is described in Section 9.16. The edit descriptors used in the `write` statements in the main program are described in Chapter 11.

Exercises

1. Given the array declaration

   ```
   real, dimension(50,20) :: a
   ```

   write array sections representing

   i) the first row of a;

   ii) the last column of a;

   iii) every second element in each row and column;

   iv) as for iii) in reverse order in both dimensions;

   v) a zero-sized array.

2. Write a `where` statement to double the value of all the positive elements of an array z.

3. Write an array declaration for an array j which is to be completely defined by the statement

   ```
   j = (/ (3, 5, i=1,5), 5,5,5, (i, i = 5,3,-1 ) /)
   ```

4. Classify the following arrays:

   ```
   subroutine example(n, a, b)
      real, dimension(n, 10) :: w
      real                   :: a(:), b(0:)
      real, pointer          :: d(:, :)
   ```

5. Write a declaration and a pointer assignment statement suitable to reference as an array all the third elements of component du in the elements of the array tar having all three subscript values even (Section 7.10).

6. Given the array declarations
   ```
   integer, dimension(100, 100), target :: l, m, n
   integer, dimension(:, :), pointer   :: ll, mm, nn
   ```
   rewrite the statements
   ```
   l(j:k+1, j-1:k) = l(j:k+1, j-1:k) + l(j:k+1, j-1:k)
   l(j:k+1, j-1:k) = m(j:k+1, j-1:k) + n(j:k+1, j-1:k) + n(j:k+1, j:k+1)
   ```
   as they could appear following execution of the statements
   ```
   ll => l(j:k+1, j-1:k)
   mm => m(j:k+1, j-1:k)
   nn => n(j:k+1, j-1:k)
   ```

7. Complete Exercise 1 of Chapter 4 using array syntax instead of do constructs.

8. Write a module containing a data structure suitable for holding the entries of a sparse matrix linked by rows with pointers to the row starts. Include a subroutine for inserting an entry specified by a value together with a row index and a column index. Ensure that the entries are linked in order of increasing column indices. Also include a subroutine for printing the matrix entries row by row.

9. Write a module that contains the example in Figure 7.5 (Section 7.5) as a module procedure and supports the defined operations and assignments that it contains.

8. Specification statements

8.1 Introduction

In the preceding chapters we have learnt the elements of the Fortran language, how they may be combined into expressions and assignments, how we may control the logic flow of a program, how to divide a program into manageable parts, and have considered how arrays may be processed. We have seen that this knowledge is sufficient to write programs, when combined with rudimentary `read` and `print` statements and with the `end` statement.

In Chapters 2 to 7 we met some specification statements when declaring the type and other properties of data objects, but to ease the reader's task we did not always explain all the available options. In this chapter we fill this gap. To begin with, however, it is necessary to recall the place of specification statements in a programming language. A program is processed by a computer in stages. In the first stage, compilation, the source code (text) of the program is read by a program known as a **compiler** which analyses it, and generates files containing **object code**. Each program unit of the complete program is usually processed separately. The object code is a translation of the source code into a form that can be understood by the computer hardware, and contains the precise instructions as to what operations the computer is to perform. Using these files, an executable program is constructed. The final stage consists of the execution, whereby the coded instructions are performed and the results of the computations made available.

During the first stage, the compiler requires information about the entities involved. This information is provided at the beginning of each program unit or subprogram by specification statements. The description of most of these is the subject of this chapter. The specification statements associated with procedure interfaces, including interface blocks and the `interface` statement, were explained in Chapter 5. The `intrinsic` statement is explained in Section 9.1.4.

8.2 Implicit declarations

8.2.1 Implicit typing

Many programming languages require that all typed entities have their types specified explicitly. Any data entity that is encountered in an executable statement without its type having been declared will cause the compiler to indicate an error. Fortran, however, has implicit typing: an entity is assigned a type according to the initial letter of its name unless

Modern Fortran Explained, 3rd Edition. M. Metcalf, J. Reid, M. Cohen, and R. Bader. Oxford University Press (2024). © M. Metcalf, J. Reid, M. Cohen, and R. Bader (2024). DOI 10.1093/oso/9780198876571.001.0008

it is explicitly typed by appearing in a type declaration statement or being accessed by use or host association. The default is that an entity whose name begins with one of the letters i, j, ..., n is of type default integer, and an entity whose name begins with one of the letters a, b, ..., h, or o, p, ..., z is of type default real.[1]

Implicit typing can lead to program errors; for instance, if a variable name is misspelt, the misspelt name will give rise to a separate variable which, if used, can lead to unforeseen consequences. For this reason, we recommend that implicit typing be avoided. The statement

```
implicit none
```

avoids implicit typing in a scoping unit. It may be preceded within the scoping unit only by use (and format, Section 10.2) statements. We recommend that each implicit none statement be at the start of the specifications, immediately following any use statements. A scoping unit without an implicit statement has the same implicit typing (or lack of it) as its host. For example, a single implicit none statement in a module avoids implicit typing in the module and all the scoping units it contains.

8.2.2 Requiring explicit procedure and type declarations

Fortran also allows a name to be implicitly declared as that of an external procedure if it is used as a procedure without being accessed from a module, declared in an interface block, or given the external attribute (Section A.13). This is as undesirable as implicit typing. For example, mistyping the name of a module procedure will not be detected by the compiler. An implicit procedure declaration can be avoided with either of the statements

```
implicit none (type, external)
implicit none (external)
```

We recommend the first form because this also disallows implicit typing. There is a fourth form of the statement,

```
implicit none (type)
```

which has exactly the same interpretation as implicit none.

As before, these implicit none statements affect the scoping unit and any contained scoping units. Note that there is no mechanism for overriding

```
implicit none (external)
```

in a contained scoping unit.

8.3 Named constants

Inside a program, we often need to define a constant or set of constants. For instance, in a program requiring repeated use of the speed of light, we might use a real variable c that is given its value by the statement

---

[1]See Section A.9 for means of specifying other mappings between the letters and types, using an implicit statement. This can be used even in a contained scoping unit to override an implicit none for one or more initial letters.

```
c = 2.99792458
```

A danger in this practice is that the value of c may be overwritten inadvertently, for instance because another programmer reuses c as a variable to contain a different quantity, failing to notice that the name is already in use.

It might also be that the program contains specifications such as

```
real    :: x(10), y(10), z(10)
integer :: mesh(10, 10), ipoint(100)
```

where all the dimensions are 10 or 10^2. Such specifications may be used extensively, and 10 may even appear as an explicit constant, say as a parameter in a do-construct which processes these arrays:

```
do i = 1, 10
```

Later, it may be realized that the value 20 rather than 10 is required, and the new value must be substituted everywhere the old one occurs, an error-prone undertaking.

Yet another case was met in Section 2.6, where named constants were needed for kind type parameter values.

In order to deal with all of these situations, Fortran contains what are known as named constants. These may never appear where a variable is needed, such as on the left-hand side of an assignment statement, but may be used in expressions in any way in which a literal constant may be used. A type declaration statement may be used to specify such a constant:

```
real, parameter :: c = 2.99792458
```

The value is protected, as c is now the name of a constant and may not be used as a variable name in the same scoping unit. Similarly, we may write

```
integer, parameter :: length = 10
real               :: x(length), y(length), z(length)
integer            :: mesh(length, length), ipoint(length**2)
   :
do i = 1, length
```

which has the clear advantage that in order to change the value of 10 to 20 only a single line need be modified, and the new value is then correctly propagated.

A named constant may be an array, as in the case

```
real, dimension(3), parameter :: field = [ 0.0, 10.0, 20.0 ]
```

However, it is not necessary to declare the shape in advance: the shape may be taken from the value. This is called an **implied-shape array**, and is specified by an asterisk as upper bound, for example,

```
character, parameter :: vowels(*) = [ 'a', 'e', 'i', 'o', 'u' ]
```

For an array of rank greater than one, the reshape function described in Section 9.15.3 must be applied. If the array has implied shape, an asterisk must be specified for each upper bound. For example

```
integer, parameter :: powers(0:*,*) = &
reshape( [ 0, 1, 2, 3, 0, 1, 4, 9, 0, 1, 8, 27 ], [ 4, 3 ] )
```

declares `powers` to have the bounds `(0:3, 1:3)`.

Similarly, the length of a scalar named constant of type `character` may be specified as an asterisk and taken directly from its value, which obviates the need to count the length of a character string, making modifications to its definition much easier. An example of this is

```
character(len=*), parameter :: string = 'No need to count'
```

Unfortunately, counting is needed when a character array is defined using an array constructor without a type specifier, since all the elements must have the same length:

```
character(len=7), parameter, dimension(3) ::                &
                c=['Cohen  ', 'Metcalf', 'Reid   ']
```

would not be correct without the two blanks in `'Cohen  '` and the three in `'Reid   '`. This need can be circumvented by using a *type-spec*, as shown in the final two paragraphs of Section 7.15; but that has the possibility of truncation if the wrong length is specified in the *type-spec*.

A named constant may be of derived type, as in the case

```
type(posn), parameter :: a = posn(1.0,2.0,0)
```

for the type

```
type posn
   real    :: x, y
   integer :: z
end type posn
```

Note that a subobject of a constant need not necessarily have a constant value. For example, if `i` is an integer variable and `field` is the array constant defined earlier in this section, `field(i)` may have the value 0.0, 10.0, or 20.0. Note also that a constant may not be a pointer, allocatable object, dummy argument, or function result, because these are always variables. However, it may be of a derived type with an allocatable component that is unallocated or a pointer component that is disassociated. Clearly, because such a component is part of a constant, it is not permitted to be allocated or pointer assigned.

The `parameter` attribute is an important means whereby constants may be protected from overwriting, and programs modified in a safe way. It should be used for these purposes on every possible occasion.

8.4   Constant expressions

In an example in the previous section, the expression `length**2` appeared in one of the array bound specifications. This is a particular example of a **constant expression**, which is an expression whose value cannot change during execution and which can be verified at compile time to satisfy this rule. The standard specifies a long list of possibilities for primaries in constant expressions, see below, in order that compilers can perform this verification. It is so long that if the programmer is satisfied that an expression satisfies the rule, it almost certainly does.

In the definition of a named constant we may use any constant expression, and the constant becomes defined with the value of the expression according to the rules of intrinsic assignment. This is illustrated by the example

```
integer, parameter :: length=10, long=selected_real_kind(12)
real, parameter    :: lsq = length**2
```

Note from this example that it is possible in one statement to define several named constants, in this case two, separated by commas.

A **constant expression** is an expression in which each operation is intrinsic and each primary is

i) a constant or a subobject of a constant;

ii) an array constructor in which every expression is a constant expression;

iii) a structure constructor in which each allocatable component is a reference to the intrinsic function `null`, each pointer component is an initialization target (Section 8.5.4) or a reference to the intrinsic function `null`, and each other component is a constant expression;

iv) a specification inquiry (Section 8.17) where each designator or function argument is a constant expression or a variable whose properties inquired about are not assumed, deferred, or defined by an expression that is not a constant expression;

v) a reference to an elemental or transformational intrinsic function (Chapter 9) other than `command_argument_count`, `null`, `num_images`, `this_image`, or `transfer`, provided each argument is a constant expression;

vi) a reference to the intrinsic function `null` that does not have an argument with a type parameter that is assumed or is defined by an expression that is not a constant expression;

vii) a reference to the intrinsic function `transfer` where each argument is a constant expression and each ultimate pointer component of the `source` argument is disassociated;

viii) a reference to a function of one of the intrinsic modules `ieee_arithmetic` and `ieee_exceptions` (Chapter 19), where each argument is a constant expression;

ix) within a derived-type definition, a previously declared kind type parameter of the type being defined;

x) a *do-var* within a *data-implied-do* loop in a data statement (Section 8.5.2);

xi) a *variable* within a *constructor-implied-do* in an array constructor (Section 7.15) where each loop parameter $expr_i$ is a constant expression; or

xii) a constant expression enclosed in parentheses;

where each subscript, section subscript, substring bound, or type parameter value is a constant expression.

In the initialization of a named constant or variable, it is permitted to use a property that does not depend on its value, such as an array bound or type parameter. For example,

```
integer :: b = bit_size(b)
real, parameter :: rreps = sqrt(sqrt(epsilon(rreps)))
integer :: iota(10) = [ ( i, i = 1, size(iota,1) ) ]
character(len=6), parameter :: stars6 = repeat('*',len(stars6))
```

are now valid. This only applies to the initialization, and when the property was determined by the declaration; that is, a property determined by the declaration cannot be used within the declaration itself, nor may a property declared by the initialization be used within that expression. For example, the following remain prohibited:

```
integer :: x(10, size(x, 1))
character(*),parameter :: c = repeat('c', len(c))
```

On the other hand, if in a module, a generic intrinsic function name must not be referenced if the generic name has a specific non-intrinsic version defined later in the module. Furthermore, an intrinsic function must not be referenced by its generic name if a specific non-intrinsic version is defined later.

Any named constant used in a constant expression must either be accessed from the host, be accessed from a module, be declared in a preceding statement, or be declared to the left of its use in the same statement. An example using a constant expression including a named constant that is defined in the same statement is

```
integer, parameter :: apple = 3, pear = apple**2
```

An example using one of the mathematical intrinsic functions (sin, cos, etc.) is

```
real :: root2 = sqrt(2.0)
```

8.5 Initial values for variables

8.5.1 Initialization in type declaration statements

A non-pointer non-allocatable variable may be assigned an initial value in a type declaration statement simply by following the name of the variable by an equals sign and a constant expression, as in the examples

```
real              :: a = 0.0
real, dimension(3) :: b = [ 0.0, 1.2, 4.5 ]
```

The initial value is defined by the value of the corresponding expression according to the rules of intrinsic assignment. The variable automatically acquires the save attribute. It must not be a dummy argument, an automatic object, or a function result.

8.5.2 The data statement

An alternative way to specify an initial value for a non-pointer non-allocatable variable is by the data statement. It has the general form

```
data object-list /value-list/ [ [,] object-list /value-list/ ] ...
```

where *object-list* is a list of variables and *data-implied-do* loops, and *value-list* is a list of scalar constants, structure constructors, and 'boz' constants (Section 2.6.7). A simple example is

```
real     :: a, b, c
integer :: i, j, k
data      a,b,c/1.,2.,3./, i,j,k/1,2,z'b'/
```

in which the variable a acquires the initial value 1.0, b acquires the value 2.0, ..., k acquires the value 11. A 'boz' constant must correspond to an integer, and this is given the value with the bit sequence of the 'boz' constant after truncation on the left or padding with zeros on the left to the bit length of the integer.

A *data-implied-do* has the form

(*dlist*, [*integer-type-spec* ::] *variable* = *expr*₁, *expr*₂ [, *expr*₃])

where *dlist* is a list of array elements, scalar structure components, and *data-implied-do* loops; *integer-type-spec* is integer [([kind=] *kind-expr*)]; and *expr*₁, *expr*₂, and *expr*₃ are scalar integer expressions. The variable must be of type integer, with its kind specified by *integer-type-spec* if it appears, and otherwise is the type and kind it would have in the surrounding context. If *expr*₃ is absent, it is as if it were present with the value 1. Its interpretation is as if the *dlist* had been written

$$\max(\,(expr_2 - expr_1 + expr_3) \div expr_3, 0\,)$$

times, with *variable* replaced by *expr*₁, *expr*₁ + *expr*₃, ..., as for the do construct (Section 4.4).

If any part of a variable is initialized by a data statement, the variable automatically acquires the save attribute. The variable must not be a dummy argument, an automatic object, or a function result.

After any array or array section in *object-list* has been expanded into a sequence of scalar elements in array element order, there must be as many constants in each *value-list* as scalar elements in the corresponding *object-list*. Each scalar element is assigned the corresponding scalar constant.

Constants which repeat may be written once and combined with a scalar integer **repeat count** which may be a literal constant, named constant, or subobject of a constant, for example

```
data i,j,k/3*0/
```

The value of the repeat count must be positive or zero. As an example, consider the statement

```
data r(1:length)/length*0./
```

where r is a real array and length is a named constant which might take the value zero.

Arrays may be initialized in three different ways: as a whole, by element, or by a *data-implied-do* loop. These three ways are shown below for an array declared by

```
real :: a(5, 5)
```

Firstly, for the whole array, the statement

```
data a/25*1.0/
```

sets each element of a to 1.0.

Secondly, individual elements and sections of a may be initialized, as in

```
data a(1,1), a(3,1), a(1,2), a(3,3) /2*1.0, 2*2.0/
data a(2:5,4) /4*1.0/
```

in each of which only the four specified elements and the section are initialized. Each array subscript must be a constant expression, as must any character substring subscript.

Thirdly, when the elements to be selected fall into a pattern that can be represented by do-loop indices, it is possible to use a *data-implied-do* statement thus:

```
data ((a(i,j), i=1,5,2), j=1,5) /15*0./
```

A variable in an *expr* must be a *do-var* of an outer *data-implied-do*:

```
integer            :: j, k
integer, parameter :: lgth=5, lgth2=((lgth+1)/2)**2
real               :: a(lgth,lgth)
data ((a(j,k), k=1,j), j=1,lgth,2) / lgth2 * 1.0 /
```

This example sets to 1.0 the first element of the first row of a, the first three elements of the third row, and all the elements of the last row, as shown in Figure 8.1.

Figure 8.1 Result of the implied-do loop in the data statement.

1.0	.	.	.	.
.	.	.	.	.
1.0	1.0	1.0	.	.
.	.	.	.	.
1.0	1.0	1.0	1.0	1.0

A further example, where the kind and type of the do variable are specified, is

```
data ((a(i,j), integer(int64) :: i=1,5,2), j=1,5) /15*0./
```

In this case, the type declaration applies only to the variable i.

The only variables permitted in subscript expressions in data statements are do indices of the same or an outer-level loop, and all operations must be intrinsic. The requirements on these expressions are identical to those on other constant expressions, allowing, for example,

```
real :: a(10,7,3)
data ((a(i,i,j),i=1,min(size(a,1),size(a,2))),j=1,size(a,3))/21*1.0/
```

Unless it has default initialization (Section 8.5.5), an object of derived type may appear in a data statement. In this case, the corresponding value must be a structure constructor having a constant expression for each component. Using the derived-type definition of posn in Section 8.3, we can write

```
type(posn) :: position1, position2
data position1 /posn(2., 3., 0)/, position2%z /4/
```

In the examples given so far, the types and type parameters of the constants in a *value-list* have always been the same as the type of the variables in the *object-list*. This need not be the case, but they must be compatible for intrinsic assignment since the entity is initialized following the rules for intrinsic assignment. It is thus possible to write code such as

```
real, parameter :: b, q
integer, parameter :: i
data q/1/, i/3.1/, b/(0.,1.)/
```

Each variable must either have been typed in a previous type declaration statement in the scoping unit, or its type is that associated with the first letter of its name according to the implicit typing rules of the scoping unit. In the case of implicit typing, the appearance of the name of the variable in a subsequent type declaration statement in the scoping unit must confirm the type and type parameters. Similarly, any array variable must have previously been declared as such.

No variable or part of a variable may be initialized more than once in a scoping unit.

We recommend using the type declaration statement rather than the data statement, but the data statement must be employed when only part of a variable is to be initialized.

8.5.3   Pointer initialization as disassociated

Means are available to avoid the initial status of a pointer being undefined. Pointers may be given the initial status of disassociated in a type declaration statement such as

```
real, pointer, dimension(:) :: vector => null()
```

a data statement

```
real, pointer, dimension(:) :: vector
data vector/ null() /
```

or, for procedure pointers (Section 14.2), a procedure statement

```
procedure(real), pointer :: pp => null()
```

This, of course, implies the save attribute, which applies to the pointer association status. The pointer must not be a dummy argument or function result. Here, or if the save attribute is undesirable (for a local variable in a recursive procedure, for example), the variable may be explicitly nullified early in the subprogram.

Our recommendation is that all pointers be initialized to reduce the risk of bizarre effects from the accidental use of undefined pointers. This is an aid too in writing code that avoids memory leaks.

The function null is an intrinsic function (Section 9.17), whose simple form null(), as used in the above example, is almost always suitable since the attributes are immediately apparent from the context. For example, given the type entry of Figure 2.3, the structure constructor

```
entry (0.0, 0, null())
```

is available. Also, pointer assignment to null() has the same effect as the nullify statement. For example, the statement

```
vector => null()
```

is equivalent to

```
nullify(vector)
```

The form `null(mold)` is needed for an actual argument that corresponds to a dummy argument with an assumed length type parameter (Sections 5.19 and 13.2.2), or for an argument to a generic procedure if the type, type parameters, or rank are needed to resolve the reference (Section 5.18).

8.5.4 Pointer initialization as associated

The initial association status of a pointer can instead be defined to be associated with a target, as long as that target has the `save` attribute, is not coindexed, and does not have the `allocatable` attribute. For example,

```
real, target  :: x(10,10) = 0
real, pointer :: p(:,:) => x
```

Note that the `data` statement cannot be used for this purpose; however, a procedure pointer can also be initialized to associated in a `procedure` statement, for example,

```
procedure(real), pointer :: pp => myfun
```

where `myfun` is not a dummy procedure.

Furthermore, a pointer can be associated with a part of such a target, including an array section (but not one with a vector subscript). Any subscript or substring position in the target specification must be a constant expression. For example,

```
real, pointer :: column_one(:) => x(:,1)
```

8.5.5 Default initialization of components

It is possible to specify that any object of a derived type be given a default initial value for a component. The value must be specified when the component is declared as part of the derived-type definition (Section 2.9). If the component is neither allocatable nor a pointer, this is done in a type declaration statement in the usual way (Section 8.5), with the equals sign followed by a constant expression, and the rules of intrinsic assignment apply (including specifying a scalar value for all the elements of an array component). For a pointer component, initialization is as described in Sections 8.5.3 and 8.5.4. If the component is allocatable, no initialization is allowed.

Initialization does not have to apply to all components of a given derived type. An example is

```
type entry
   real                 :: value = 2.0
   integer              :: index
   type(entry), pointer :: next => null()
end type entry
```

Given an array declaration such as

```
type(entry), dimension(100) :: matrix
```

subobjects such as `matrix(3)%value` will have the initial value 2.0, and the reference `associated(matrix(3)%next)` will return the value false.

For an object of a nested derived type, the initializations associated with components at all levels are recognized. For example, given the specifications

```
type node
    integer     :: counter
    type(entry) :: element
end type node
type (node) :: n
```

the component `n%element%value` will have the initial value 2.0.

Unlike **explicit initialization** in a type declaration or `data` statement, default initialization does not imply that the objects have the `save` attribute. However, all data objects declared in a module do have the `save` attribute.

Objects may still be explicitly initialized in a type declaration statement, as in

```
type(entry), dimension(100) :: matrix=entry(huge(0.0),huge(0),null())
```

in which case the default initialization is ignored. Similarly, default initialization may be overridden in a nested derived-type definition such as

```
type node
    integer     :: counter
    type(entry) :: element=entry(0.0, 0 , null())
end type node
```

However, no part of a non-pointer object with default initialization is permitted in a `data` statement (Section 8.5.2).

As well as applying to the initial values of static data, default initialization also applies to any data that is dynamically created during program execution. This includes allocation with the `allocate` statement. For example, the statement

```
allocate (matrix(1)%next)
```

creates a partially initialized object of type `entry`. It also applies to unsaved local variables (including automatic objects), function results, and dummy arguments with intent out.

It applies even if the derived-type definition is `private` (Section 8.6.1) or the components are `private`.

8.6   Accessibility

8.6.1   The `public` and `private` attributes

Modules (Section 5.5) permit specifications to be 'packaged' into a form that allows them to be accessed elsewhere in the program. So far, we have assumed that all the entities in the module are to be accessible, that is, have the `public` attribute, but sometimes it is desirable to limit the access. For example, several procedures in a module may need access to a work array containing the results of calculations that they have performed. If access is limited to only the procedures of the module, there is no possibility of an accidental corruption of these data by another procedure and design changes can be made within the module without affecting the rest of the program. In cases where entities are not to be accessible outside their own module, they may be given the `private` attribute.

These two attributes may be specified with the `public` and `private` keywords on type declaration statements in the module, as in

```
real, public    :: x, y, z
integer, private :: u, v, w
```

or in `public` and `private` statements, as in

```
public  :: x, y, z, operator(.add.)
private :: u, v, w, assignment(=), operator(*)
```

which have the general forms

> `public` *[[::] access-id-list]*
> `private` *[[::] access-id-list]*

where *access-id* is a name or a *generic-spec* (Section 5.18.1).

Note that if a procedure has such a generic identifier, the accessibility of its specific name is independent of the accessibility of its generic identifier. One may be `public` while the other is `private`, which means that it is accessible only by its specific name or only by its generic identifier.

If a `public` or `private` statement has no list of entities, it confirms or resets the default value. Thus, the statement

```
public
```

confirms `public` as the default value, and the statement

```
private
```

sets the default value for the module to `private` accessibility. For example,

```
private
public :: means
```

gives the entity `means` the `public` attribute whilst all others are `private`. There may be at most one accessibility statement without a list in a scoping unit.

The entities that may be specified by name in `public` or `private` lists are named variables, procedures (including generic procedures), derived types, named constants, modules, and namelist groups. Thus, to make a generic procedure name accessible but the corresponding specific names inaccessible, we might write the code of Figure 8.2.

A type that is accessed from a module may be given the `private` attribute in the accessing module. If an entity of this type has the `public` attribute, a subsequent `use` statement for it may be accompanied by a `use` statement for the type from the original module.

Entities of private type are not themselves required to be private; this applies equally to procedures with arguments that have private type. This means that a module writer can

Figure 8.2 Making a generic procedure name accessible and the specific names inaccessible.

```
module example
   private specific_int, specific_real
   interface generic_name
      module procedure specific_int, specific_real
   end interface
contains
   subroutine specific_int(i)
      ⋮
   subroutine specific_real(a)
      ⋮
end module example
```

provide very limited access to values or variables without thereby giving the user the power to create new variables of the type. For example, the widely used LAPACK library requires character arguments such as uplo, a character variable that must be given the value 'L' or 'U' according to whether the matrix is upper or lower triangular. The value is checked at run time and an error return occurs if it is invalid. This could be replaced by values lower and upper of private type. This would be clearer and the check would be made at compile time.

When a module contains use statements, the entities accessed are treated as entities in the module. They may be given the private or public attribute explicitly or through the default rule in effect in the module. Thus, given the two modules in Figure 8.3 and a third program unit containing a use statement for two, the variable i is accessible there only if it also contains a use statement for one or if i is made public explicitly in two.

Figure 8.3 Making private an entity accessed from a module.

```
module one
   integer :: i
end module one
module two
   use one
   private
   ⋮
end module two
```

If a module a_mod uses module b_mod, the default accessibility for entities it accesses from b_mod is the overall default accessibility for entities in a_mod. This is inconvenient when a_mod is default public but does not want to export very much if anything from b_mod, or when a_mod is default private but wants to export everything or nearly everything from b_mod. However, having to specify the non-default accessibility for most or all of b_mod can be tedious and error-prone. To overcome this, the default accessibility for entities accessed from a module can be controlled separately from the default accessibility of entities declared or defined locally. This is done by using the module name in a `public` or `private` statement; for example,

```
private b_mod
```

makes entities accessed by `use b_mod` have a default accessibility of `private` in `a_mod`, regardless of the overall default accessibility in `a_mod`. An explicit `public` or `private` specification for an entity used from `b_mod` is still permitted, and overrides or confirms the default.

A module name must not appear more than once in all the `public` and `private` statements in the using module.

The use of the `private` statement for components of derived types in the context of defining an entity's access within a module will be described in Section 8.14.

The `public` and `private` attributes may appear only in the specifications of a module.

8.6.2   The `protected` attribute

It is sometimes desirable to allow the user of a module to be able to reference the value of a module variable without allowing it to be changed. Such control is provided by the `protected` attribute. This attribute does not affect the visibility of the variable, which must still be `public` to be visible, but confers the same protection against modification that intent in does for dummy arguments. It may appear only in the specifications of a module.

The `protected` attribute may be specified with the `protected` keyword in a type declaration statement. For example, in

```
module m
   public
   real, protected    :: v
   integer, protected :: i
```

both v and i have the `protected` attribute. The attribute may also be specified separately, in a `protected` statement, just as for other attributes (see Section 8.8).

Variables with this attribute may be modified only within the defining module. Outside the module they are not allowed to appear in a context in which they would be altered, such as on the left-hand side of an assignment statement. For example, in the code of Figure 8.4, the `protected` attribute allows users of `thermometer` to read the temperature in either Fahrenheit or Celsius, but the variables can only be changed via the provided subroutines which ensure that both values agree.

8.7   Pointer functions denoting variables

When a **pointer function** returns an associated pointer, that pointer is always associated with a variable that has the `target` attribute, either by pointer assignment or by allocation. Such a reference to a pointer function may also be used in contexts that require a variable, in particular

---

Figure 8.4 Using `protected` to ensure the consistency of Fahrenheit and Celsius values.

```
module thermometer
   real, protected :: temperature_celsius    = 0
   real, protected :: temperature_fahrenheit = 32
contains
   subroutine set_celsius(new_celsius_value)
      real, intent(in) :: new_celsius_value
      temperature_celsius = new_celsius_value
      temperature_fahrenheit = temperature_celsius*(9.0/5.0) + 32
   end subroutine set_celsius
   subroutine set_fahrenheit(new_fahrenheit_value)
      real, intent(in) :: new_fahrenheit_value
      temperature_fahrenheit = new_fahrenheit_value
      temperature_celsius = (temperature_fahrenheit - 32)*(5.0/9.0)
   end subroutine set_fahrenheit
end module thermometer
```

- as an actual argument for an intent `inout` or `out` dummy argument;
- on the left-hand side of an assignment statement.

In this respect, a pointer function reference can be used exactly as if it were the variable that is the target of the pointer result.

These are sometimes known as **accessor functions**; by abstracting the location of the variable, they enable objects with special features such as sparse storage, instrumented accesses, and so on to be used as if they were normal arrays. They also allow changing the underlying implementation mechanisms without needing to change the code using the objects. An example of this feature is shown in Figure 8.5.

8.8 The `pointer`, `target`, and `allocatable` statements

For the sake of regularity in the language, there are statements for specifying the `allocatable`, `pointer`, and `target` attributes of entities. They take the forms:

```
allocatable [::] object-decl-list
pointer [::] pointer-decl-list
target [::] object-decl-list
```

where *pointer-decl* is an *object-decl* or *proc-name* and each *object-decl* has the form

object-name [(array-spec)] [[coarray-spec]]

Each *array-spec* provides rank and bounds information for an array, and each *coarray-spec* provides corank and cobounds information for a coarray. They must satisfy the same rules as apply to providing this information in a type declaration statement for an array or coarray with the `allocatable`, `pointer`, or `target` attribute, respectively.

Figure 8.5 An accessor function. The function `findloc` is defined in Section 9.16.

```
module indexed_store
    real, private, pointer      :: values(:) => null()
    integer, private, pointer :: keys(:) => null()
    integer, private          :: maxvals = 0
contains
    function storage(key)
        integer, intent(in) :: key
        real, pointer :: storage
        integer :: loc(1)
        if (.not.associated(values)) allocate (values(100), keys(100))
        loc = findloc(keys(:maxvals), key)
        if (loc(1)>0) then
            storage => values(loc(1))
        else
            maxvals = maxvals + 1
            :    (Code to enlarge arrays if necessary elided.)
            keys(maxvals) = key
            storage => values(maxvals)
        end if
    end function
end module

program main
    use indexed_store
    storage(13) = 100
    print *, storage(13)
end program main
```

A *proc-name* in a `pointer` statement specifies that the procedure name is a procedure pointer.

If the array information for an object is not supplied, this may be because the object is scalar or it may be because this information is supplied in another statement. Similarly, if the coarray information is not supplied, this may be because the object is not a coarray or it may be because this information is supplied in another statement. Neither the attribute, the array information, nor the coarray information may be supplied twice for an entity.

Here is an example of the use of this feature:

```
real        :: a, son, y
allocatable :: a(:,:)
pointer     :: son
target      :: a, y(10)
```

We believe that it is much clearer to specify these attributes for objects in type declaration statements, and therefore do not use these forms for objects. However, a `pointer` statement is needed to specify a procedure pointer (see Section 14.2).

8.9   The `intent` and `optional` statements

The `intent` attribute (Section 5.10) for a dummy argument that is not a dummy procedure (but may be a dummy procedure pointer) may be specified in a type declaration statement, a `procedure` statement, or in an `intent` statement of the form

   `intent` (*inout*) [::] *dummy-argument-name-list*

where *inout* is in, out, or inout. Examples are

```
subroutine solve (a, b, c, x, y, z)
    real         :: a, b, c, x, y, z
    intent(in)   :: a, b, c
    intent(out)  :: x, y, z
```

The `optional` attribute (Section 5.14) for a dummy argument may be specified in a type declaration statement or in an `optional` statement of the form

   `optional` [::] *dummy-argument-name-list*

An example is

```
optional :: a, b, c
```

The `optional` attribute is the only attribute which may be specified for a dummy argument that is a procedure.

   Note that the `intent` and `optional` attributes may be specified only for dummy arguments. As for the statements of Section 8.8, we believe that it is much clearer to specify these attributes for objects in the type declaration statements, and therefore do not use these forms for objects. However, a separate `optional` statement is needed to give a procedure pointer the optional attribute.

8.10   The `save` attribute

The `save` attribute was explained in Section 5.9. It specifies that the value of a local variable is retained between calls of a subprogram, for example to count the number of times the subprogram is entered.

   All variables in modules (Section 5.5) automatically have the `save` attribute.

   The `save` attribute must not be specified for a dummy argument, a function result, or an automatic object (Section 7.3). It may be specified for a pointer, in which case the pointer association status is saved. It may be specified for an allocatable array, in which case the allocation status and value are saved. A saved variable in a subprogram that is referenced recursively is shared by all executing instances of the subprogram.

   An alternative to specifying the `save` attribute on a type declaration statement is the `save` statement:

   `save` [[::] *variable-name-list*]

A `save` statement with no list is equivalent to a list containing all possible names, and in this case the scoping unit must contain no other `save` statements and no `save` attributes in type declaration statements. Our recommendation is against this form of `save`. If a programmer tries to give the `save` attribute explicitly to an automatic object, a diagnostic will result. On the other hand, he or she might think that `save` without a list would do this too, and not get the behaviour intended. Also, there is a loss of efficiency associated with `save` on some processors, so it is best to restrict it to those objects for which it is really needed.

The `save` statement or `save` attribute may appear in the declaration statements in a main program but has no effect.

8.11 Asynchronous actions

8.11.1 Introduction

The `asynchronous` attribute for a variable indicates that it may be subject to asynchronous input/output or asynchronous communication.

In **asynchronous input/output** an input/output statement that affects the value of an asynchronous variable continues to execute while subsequent statements execute up to a completing statement (usually a `wait` statement). The completing statement delays further execution until the input/output has finished. This case is discussed fully in Section 10.15. The variable must not be referenced or defined while an asynchronous input statement executes and must not be defined while an asynchronous output statement executes.

Asynchronous communication is similar but occurs through the action of procedures defined by means other than Fortran that behave similarly to asynchronous input/output procedures. It is initiated by execution of one procedure and completed by execution of another. It is either input communication or output communication. An asynchronous variable must not be referenced or defined while involved in asynchronous input communication and must not be defined while involved in asynchronous output communication. The main application is for non-blocking calls of `MPI_Irecv` and `MPI_Isend`. A call of `MPI_Irecv` is very like asynchronous input – data is put in the buffer array while execution continues; a call of `MPI_Isend` is very like asynchronous output – data is copied from the buffer array while execution continues. For both, a call of `MPI_Wait` can play the role of the completing statement. We illustrate this in Figure 8.6.

Figure 8.6 Asynchronous communication.

```
real, asynchronous :: buf(100, 100) ! Input buffer
integer :: req ! Request handle
: ! Code that involves buf and defines req
call MPI_Irecv(buf, ..., req, ...)
: ! Code that does not involve buf or req
call MPI_Wait(req, ...)
: ! Code that involves buf.
```

The standard does not limit asynchronous communication to these MPI functions. Instead, it talks in general of procedures that initiate input or output asynchronous communication or complete it. Whether a procedure has such a property is processor dependent.

8.11.2   The `asynchronous` attribute

A named variable may be declared with the `asynchronous` attribute:

```
integer, asynchronous :: int_array(10)
```

or given it by the `asynchronous` statement

```
asynchronous :: int_array, another
```

This statement may be used to give the attribute to a variable that is accessed by use or host association.

Whether an object has the `asynchronous` attribute may vary between scoping units. If a variable is accessed by use or host association, it may gain the attribute, but it never loses it. For dummy and corresponding actual arguments, there is no requirement for agreement in respect of the `asynchronous` attribute. This provides useful flexibility, but needs to be used with care. If the programmer knows that all asynchronous action will be within the called procedure, there is no need for the actual argument to have the `asynchronous` attribute. Similarly, there is no need for a dummy argument to have the `asynchronous` attribute if the programmer knows that no asynchronous operation for it can be incomplete when the procedure is called.

All subobjects of a variable with the `asynchronous` attribute have the attribute.

There are restrictions that avoid copy-in copy-out for an actual argument when the corresponding dummy argument has the `asynchronous` attribute but does not have the `value` attribute:[2] the actual argument must not be an array section with a vector subscript; if the actual argument is an array section or an assumed-shape array, the dummy argument must be an assumed-shape array; and if the actual argument is an array pointer, the dummy argument must be an array pointer or assumed-shape array.

8.12   The `block` construct

The `block` construct is a scoping unit of the form

```
[ name: ] block
            block
      end block [ name ]
```

It is an executable construct that allows the declaration of entities whose scope is the construct. Such entities may be variables, types, constants, or even procedures. Each is known as a **construct entity**. Any entity of the host scoping unit with the same name is hidden by the declaration.

For example, in Figure 8.7 the variables `alpha`, `temp`, and `j` are local to the block, and have no effect on any variables outside the block that might have the same names. Used

---

[2]It is copy-out that gives a problem and this does not happen if the dummy argument has the `value` attribute.

Figure 8.7 A block within a do construct.

```
do i=1, m
   block
      real alpha, temp(n)
      integer j
      ⋮
      temp(j) = alpha*a(j, i) + b(j)
      ⋮
   end block
end do
```

judiciously, this can make code easier to understand (there is no need to look through the whole subprogram for later accesses to alpha, for instance) and since the compiler also knows that these are local to each iteration, this can aid optimization. To ensure that adding a block construct to existing code has no effect on the semantics of the existing code, the scope of labels includes any contained block constructs, that is, labels have inclusive scope (Section 5.15).

Another example is

```
block
   use convolution_module
   intrinsic norm2
   ⋮
   x = convolute(y)*norm2(z)
   ⋮
end block
```

Here, the entities brought in by the use statement are visible only within the block, and the declaration of the norm2 intrinsic avoids clashing with any norm2 that might exist outside the block. These techniques can be useful in large subprograms, or during code maintenance when it is desired to access a module or procedure without risking disturbance to the rest of the subprogram.

Not all declarations are permitted in a block construct. The intent, optional, and value attributes are not available (because a block has no dummy arguments).[3] The namelist statement (Section 8.19) is prohibited because of potential ambiguity or confusion. Finally, a save statement that specifies entities in the block is permitted, but a save without a list is prohibited, again because it would be ambiguous as to just exactly what would be saved.

Like other constructs, the block construct may be given a construct name, and that construct name may be used in exit statements to exit from the construct (see Section 4.5). Similarly, block constructs may be nested the same way that other constructs are nested. An example of this is shown in Figure 8.8.

---

[3] Also, the implicit statement (Section A.9) is prohibited because it would be confusing to change the implicit typing rules in the middle of a subprogram.

Figure 8.8 Nesting `block` constructs. For `epsilon`, see Section 9.9.2.

```
find_solution: block
  real :: work(n)
    ⋮
  loop: do i=1, n
    block
        real :: residual
          ⋮
        if (residual < epsilon(x)) exit find_solution
      end block
    end do loop
    ⋮
end block find_solution
```

The `block` construct is only of limited use in normal programming, but is really useful when program-generation techniques such as macros are being used, to avoid conflicts with entities elsewhere in a subprogram. (Macros are not part of Fortran, but various macro processors are widely used with Fortran.)

8.13   The use statement

In Section 5.5 we introduced the `use` statement in its simplest form

> use *module-name*

which provides access to all the public named data objects, derived types, interface blocks, procedures, generic identifiers, and namelist groups (Section 8.19) in the module named. Any `use` statements must precede other specification statements in a scoping unit. Apart from `volatile` (Appendix A.15), the only attribute of an accessed entity that may be specified afresh is `public` or `private` (and this only in a module), but the entity may be included in one or more namelist groups.

If access is needed to two or more modules that have been written independently, the same name might be in use in more than one module. This is the main reason for permitting accessed entities to be renamed by the `use` statement. Renaming is also available to resolve a name clash between a local entity and an entity accessed from a module, though our preference is to use a text editor or other tool to change the local name. With renaming, the `use` statement has the form

> use *module-name*, *rename-list*

where each *rename* has the form

> *local-id* => *use-id*

where *use-id* is a public name or *generic-spec* (for a user-defined operator) in the module, and *local-id* is the identifier to use for it locally.

As an example,

```
use stats_lib, sprod => prod
use maths_lib
```

makes all the public entities in both `stats_lib` and `maths_lib` accessible. If `maths_lib` contains an entity called `prod`, it is accessible by its own name while the entity `prod` of `stats_lib` is accessible as `sprod`.

Renaming is not needed if there is a name clash between two entities that are not required. A name clash is permitted if there is no reference to the name in the scoping unit.

A name clash is also permissible for a generic name that is required. Here, all generic interfaces accessed by the name are treated as a single concatenated interface block. This is true too for defined operators, and also for defined assignment (for which no renaming facility is available). In all these cases, any two procedures having the same generic identifier must differ as explained in Section 5.18.2. We imagine that this will usually be exactly what is needed. For example, we might access modules for interval arithmetic and matrix arithmetic, both needing the functions sqrt, sin, etc., the operators +, -, etc., and assignment, but for different types.

Just as with variable and procedure names, user-defined operators may also be renamed in a use statement. For example,

```
use fred, operator(.nurke.) => operator(.banana.)
```

renames the `.banana.` operator located in module `fred` so that it may be referenced by using `.nurke.` as an operator.

However, this only applies to user-defined operators. Intrinsic operators cannot be renamed, so all of the following are invalid:

```
use fred, only: operator(.equal.) => operator(==)     ! Invalid
use fred, only: operator(/=) => operator(.notequal.)  ! Invalid
use fred, only: operator(*) => assignment(=)          ! Invalid
```

For cases where only a subset of the names of a module is needed, the `only` option is available, having the form

```
use module-name, only: [only-list]
```

where each *only* has the form

   access-id

or

   [local-id =>] use-id

where each *access-id* is a public entity in the module, and is either a name or a *generic-spec* (Section 5.18.1). This provides access to an entity in a module only if the entity is public and is specified as a *use-id* or *access-id*. Where a *use-id* is preceded by a *local-id*, the entity is known locally by the *local-id*. An example of such a statement is

```
use stats_lib, only : sprod => prod, mult
```

which provides access to `prod` by the local name `sprod` and to `mult` by its own name.

We would recommend that only one use statement for a given module be placed in a scoping unit, but more are allowed. If there is a use statement without an `only` qualifier, all public entities in the module are accessible and the *rename-list*s and *only-list*s are interpreted as if concatenated into a single *rename-list* (with the form *use-id* in an *only-list* being treated as the rename *use-id => use-id*). If all the statements have the `only` qualification, only those entities named in one or more of the *only-lists* are accessible, that is, all the *only-lists* are interpreted as if concatenated into a single *only-list*.

An `only` list will be rather clumsy if almost all of a module is wanted. The effect of an 'except' clause can be obtained by renaming unwanted entities. For example, if a large program (such as one written in Fortran 77) contains many external procedures, a good practice is to collect interface blocks for them all into a module that is referenced in each program unit for complete mutual checking. In an external procedure, we might then write

```
use all_interfaces, except_this_one => special
```

to avoid having two explicit interfaces for itself (where `all_interfaces` is the module name and `special` is the procedure name).

An entity may be accessed by more than one local name. This is illustrated in Figure 8.9, where module b accesses s of module a by the local name bs; if a subprogram such as c accesses both a and b, it will access s by both its original name and by the name bs. Figure 8.9 also illustrates that an entity may be accessed by the same name by more than one route (see variable t).

Figure 8.9 Accessing a variable by more than one local name.

```
module a
   real :: s, t
   ⋮
end module a
module b
   use a, bs => s
   ⋮
end module b
subroutine c
   use a
   use b
   ⋮
end subroutine c
```

A more direct way for an entity to be accessed by more than one local name is for it to appear more than once as a *use-name*. This is not a practice that we recommend.

Of course, all the local names of entities accessed from modules must differ from each other and from names of local entities. If a local entity is accidentally given the same name as an accessible entity from a module, this will be noticed at compile time if the local entity is declared explicitly (since no accessed entity may be given a type locally). However, if the local entity is intended to be implicitly typed (Section 8.2.1) and appears in no specification statements, each appearance of the name will be taken, incorrectly, as a reference to the accessed variable. To avoid this, we recommend, as always, the conscientious use of explicit typing in a scoping unit containing one or more use statements. For greater safety, the only option may be employed in a use statement to ensure that all accesses are intentional.

8.14   Derived-type definitions

When derived types were introduced in Section 2.9, some simple example definitions were given, but the full generality was not included. An example illustrating more features is

```
type, public :: padlock
   private
   integer, pointer :: key(:)
   logical          :: state
end type padlock
```

The general form (apart from redundant features, see Appendix A.8) is

type *[[, type-attr]* ... :: *] type-name [(type-param-name-list)]*
   [type-param-def-stmt] ...
   [private *]*
   [component-def-stmt] ...
   [type-bound-procedure-part]
end type *[type-name]*

where each *type-attr* is one of

abstract	(Section 15.10),
access	(described later in this section),
bind(c)	(Section 20.4), or
extends *(parent-type-name)*	(Section 15.2).

Derived type parameters (*type-param-name-list* and *type-param-def-stmt*) are described in Section 13.2. The *type-bound-procedure-part* is described in Section 15.8.

Each *component-def-stmt* is either a procedure statement (Section 14.2.3) or declares one or more data components and has the form

type *[[, component-attr]* ... :: *] component-decl-list*

where *type* specifies the type and type parameters (see Section 8.16), each *component-attr* is *access*, allocatable, codimension[*cobounds-list*], contiguous, dimension(*bounds-list*), or pointer, and each *component-decl* is

*component-name [(bounds-list)] [*char-len] [[cobounds-list]] [comp-init]*

The meaning of *char-len* is explained at the end of Section 8.16, and *comp-int* represents component initialization, as explained in Section 8.5.5. If the *type* is a derived type and neither the allocatable nor the pointer attribute is specified, the type must be previously defined in the host scoping unit or accessible there by use or host association. If the allocatable or pointer attribute is specified, the type may also be the one being defined (for example, the type entry of Figure 2.3), or one defined elsewhere in the scoping unit.

A *type-name* must not be the same as the name of any intrinsic type or a derived type accessed from a module.

The bounds of an array component are declared by a *bounds-list*, where each *bounds* is just

    ⋮

for an allocatable or a pointer component (see example in Section 7.12) or

   [lower-bound: *] upper-bound*

for a component that is neither allocatable nor a pointer and *lower-bound* and *upper-bound* are specification expressions (Section 8.17) whose values do not depend on those of variables. Similarly, the character length of a component of type character must be a specification expression whose value does not depend on that of a variable. If there is a *bounds-list* attached to the *component-name*, this defines the bounds. If a dimension attribute is present in the statement, its *bounds-list* applies to any component in the statement without its own *bounds-list*. Similarly, a codimension attribute in the statement applies to any component in the statement without its own *cobounds-list*.

    Only if the host scoping unit is a module may the private statement, or an *access* attribute on the type statement, or a *component-def-stmt* appear. The *access* attribute on a type statement may be public or private and specifies the accessibility of the type. If it is private, then the type name, the structure constructor for the type, any entity of the type, and any procedure with a dummy argument or function result of the type are all inaccessible outside the host module. The accessibility may also be specified in a private or public statement in the host. In the absence of both of these, the type takes the default accessibility of the host module.

    If a private statement appears, any component whose *component-def-stmt* does not have a public attribute is private. A component that is private (whether by a private statement or private attribute) is inaccessible in any scoping unit accessing the host module; this means it cannot be selected as a component, and a structure constructor cannot provide an explicit value for it.

    The accessibility of a type and of each component are completely independent, so all combinations are possible: a public component in a public type, a public component in a private type, a private component in a public type, and a private component in a private type. Even when the type and all of its components are private, entities of that type can be accessible: that is, public variables, constants, functions, and components of other types can be of the private type.

    Having mixed component accessibility can be useful in some situations. For example, in

```
module mytype_module
   type mytype
      private
      character(20), public :: debug_tag = ''
      ⋮      ! private components omitted
   end type mytype
   ⋮
end module mytype_module
```

although some of the components of mytype are private (protecting the integrity of those components), the debug_tag field is public so it can be used directly outside of the module mytype_module. If any component of a derived type is private, the structure constructor can be used outside the module in which it is defined only if that component has default initialization, which allows the value for that component to be omitted.

In practice the full generality of accessibility is often not needed, most cases being effectively addressed by one of three simpler levels of access:

i) all public, where the type and all its components are accessible, and the components of any object of the type are accessible wherever the object is accessible;

ii) a public type with all private components, where the type is accessible but all of its components are hidden (commonly referred to as an opaque type);

iii) all private, where both the type and its components are used only within the host module, and are hidden to an accessing procedure.

An opaque type (case ii) has, where appropriate, the advantage of enabling changes to be made to the type without in any way affecting the code in the accessing procedure. A fully private type (case iii) offers this advantage and has the additional merit of not cluttering the name space of the accessing procedure. The use of private accessibility for the components or for the whole type is thus recommended whenever possible.

We note that, even if two derived-type definitions are identical in every respect except their names, then entities of those two types are not equivalent and are regarded as being of different types. Even if the names, too, are identical, the types are different (unless they have the sequence attribute, a feature that we do not recommend and whose description is left to Appendix A.8). If a type is needed in more than one program unit, the definition should be placed in a module and accessed by a use statement wherever it is needed. Having a single definition is far less prone to errors.

8.15   The type declaration statement

We have already met many simple examples of the declarations of named entities by integer, real, complex, logical, character, and type(*type-name*) statements. The general form is

type [[, type-attr] ... ::] entity-list

where *type* specifies the type and type parameters (Section 8.16), *type-attr* is one of the following:

allocatable	dimension (*bounds-list*)	parameter	save
asynchronous	external	pointer	target
bind(c...)	intent (*inout*)	private	value
codimension[*cobounds-list*]	intrinsic	protected	volatile
contiguous	optional	public	

and each *entity* is one of

object-name [(bounds-list)] [[cobounds-list]] [*char-len] [= constant-expr]
function-name [*char-len]
pointer-name [(bounds-list)] [*char-len] [=> null-init]

where *null-init* is a reference to the intrinsic function null with no arguments. The meaning of *char-len* is explained at the end of Section 8.16; a *bounds-list* specifies the rank and possibly bounds of array-valued entities. If = *constant-expr* or => *null-init* appears, any array bound or character length that is an expression must be a constant expression.

No attribute may appear more than once in a given type declaration statement. The double colon :: need not appear in the simple case without any *attributes* and without any initialization (= *constant-expr* or => *null-init*); for example

```
real a, b, c(10)
```

If the statement specifies the parameter attribute, = *constant-expr* must appear.

If the pointer attribute is specified, the allocatable, intrinsic, and target attributes must not be specified. If the parameter attribute is specified, the allocatable, external, intent, intrinsic, save, optional, and target attributes must not be specified. If either the external or the intrinsic attribute is specified, the target attribute must not be specified.

If an object is specified with the intent or parameter attribute, this is shared by all its subobjects. The pointer attribute is not shared in this manner, but note that a derived-type component may itself be a pointer. However, the target attribute is shared by all its subobjects, except for any that are pointer components.

The bind, intrinsic, parameter, and save attributes must not be specified for a dummy argument or function result.

The intent, optional, and value attributes may be specified only for dummy arguments.

For a function name, specifying the external attribute is an alternative to the external statement (Section 5.12) for declaring the function to be external, and specifying the intrinsic attribute is an alternative to the intrinsic statement (Section 9.1.4) for declaring the function to be intrinsic. These two attributes are mutually exclusive.

Each of the attributes may also be specified in statements (such as save) that list entities having the attribute. This leads to the possibility of an attribute being specified explicitly more than once for a given entity, but this is not permitted. Our recommendation is to avoid such statements because it is much clearer to have all the attributes for an entity collected in one place.

8.16   Type and type parameter specification

We have used *type* to represent one of the following

```
integer    [ ( [ kind= ] kind-value ) ]
real       [ ( [ kind= ] kind-value ) ]
complex    [ ( [ kind= ] kind-value ) ]
character  [ ( actual-parameter-list ) ]
logical    [ ( [ kind= ] kind-value ) ]
type       ( type-name [ actual-parameter-list ] )
```

in the function statement (Section 5.20), the component definition statement (Section 8.14), and the type declaration statement (Section 8.15). A *kind-value* must be a constant expression (Section 8.4) and must have a value that is valid on the processor being used.

For `character`, each *actual-parameter* has the form

[`len=`] *len-value*   or   [`kind=`] *kind-value*

and provides a value for one of the parameters. The default for character length is 1. It is permissible to omit `kind=` from a kind *actual-parameter* only when `len=` is omitted and *len-value* is both present and comes first, just as for an actual argument list (Section 5.14). Neither parameter may be specified more than once. The *actual-parameter-list* for a derived type is described in Section 13.2.

For a scalar named constant or for a dummy argument of a subprogram, a *len-value* may be specified as an asterisk, in which case the value is assumed from that of the constant itself or the associated actual argument. In both cases, the `len` intrinsic function (Section 9.7.1) is available if the actual length is required directly, for instance as a do-construct iteration count. A combined example is

```
function line(char_arg)
character(len=*)                 :: char_arg
character(len=len(char_arg))  :: line
character(len=*), parameter   :: char_const = 'page'
   if ( len(char_arg) < len(char_const) ) then
   :
```

A *len-value* that is neither an asterisk nor a colon (Section 6.3) must be a specification expression (Section 8.17) and must be a constant expression if = *constant-expr* or => *null-init* appears. Negative values declare character entities to be of zero length.

In addition, it is possible to specify the character length for individual entities in a type declaration statement or component declaration statement using the syntax *entity*char-len*, where *char-len* is either (*len-value*) or *len*, where *len* is a scalar integer literal constant that does not have a kind type parameter specified for it. An illustration of this form is

```
character(len=8) :: word(4), point*1, text(20)*4
```

where `word`, `point`, and `text` have character lengths 8, 1, and 4, respectively. Similarly, the alternative form may be used for individual components in a component definition statement.

8.17 Specification expressions

A scalar integer expression known as a **specification expression** may be used to specify the array bounds and character lengths (examples in Section 7.3) of data objects in a subprogram, and of function results. Such an expression need not be constant but may depend only on data values that are always available on entry to the subprogram. Just as for constant expressions (Section 8.4), the standard specifies a long list of rules in order to ensure that this is so and that the execution of a specification expression does not have undesirable side effects.

Non-intrinsic functions are permitted to be called. They are required to be pure to ensure that they cannot have side effects on other objects being declared in the same specification sequence. They are required not to be internal, which ensures that they cannot inquire, via host association, about other objects being declared. Recursion is disallowed, to avoid the creation of a new instance of a procedure while the construction of one is in progress.

We begin by defining a **specification inquiry**, which is a reference to

- an intrinsic inquiry function (Section 9.1.3), other than `present`;
- a type parameter inquiry (Section 13.1);
- an inquiry function from the intrinsic module `ieee_arithmetic` or `ieee_exceptions`;
- the function `c_sizeof` from the intrinsic module `iso_c_binding`; or
- the function `compiler_version` or `compiler_options` from the intrinsic module `iso_fortran_env`.

Next, a function is a **specification function** if it is a pure function, is not a standard intrinsic function, is not an internal function, and does not have a dummy procedure argument.

Now we need to define a **restricted expression**. A restricted expression is an expression in which each operation is intrinsic or defined by a specification function and each primary is

- a constant or subobject of a constant;
- all or part of a dummy argument that is not optional or of intent `out`;
- all or part of an object that is made accessible by use or host association;[4]
- in a `block` construct, all or part of a local variable either of the procedure containing the construct or of an outer `block` construct containing the construct;
- an array constructor where each *expr* is a restricted expression;
- a structure constructor where each component is a restricted expression;
- a specification inquiry where each designator or function argument is

    i) a restricted expression or

    ii) a variable other than an optional argument whose properties inquired about are not dependent on the upper bound of the last dimension of an assumed-size array, deferred, or defined by an expression that is not a restricted expression;

- a specification inquiry where each designator or function argument is a reference to the intrinsic function `present`;
- a reference to any other standard intrinsic function where each argument is a restricted expression;
- a reference to a transformational function from the intrinsic module `ieee_arithmetic`, `ieee_exceptions`, or `iso_c_binding` where each argument is a restricted expression;
- a reference to a specification function where each argument is a restricted expression;
- in a definition of a derived type, a type parameter of the type being defined;
- a *variable* within an array constructor where each *expr* of the corresponding *constructor-implied-do* is a restricted expression; or
- a restricted expression enclosed in parentheses,

where each subscript, section subscript, substring starting or ending point, and type parameter value is a restricted expression.

---

[4]Or is in a common block (Section B.1.7).

Evaluation of a specification expression in a subprogram must not cause a procedure defined by the subprogram to be invoked.

No variable referenced is allowed to have its type and type parameters specified later in the same sequence of specification statements, unless they are those implied by the implicit typing rules.

If a specification expression includes a specification inquiry that depends on a type parameter or an array bound, the type parameter or array bound must be specified earlier.[5] It may be to the left in the same statement, but not within the same entity declaration. If a specification expression includes a reference to the value of an element of an array, the array shall be completely specified in prior declarations.[6]

A generic entity referenced in a specification expression in a specification sequence must have no specific procedures defined in the scoping unit, or its host scoping unit, subsequent to the specification expression.

An array whose bounds are declared using specification expressions is called an **explicit-shape array**.

A variety of possibilities are shown in Figure 8.10.

Figure 8.10 A variety of declarations in a subprogram.

```
subroutine sample(arr, n, string)
  use definitions  ! Contains the real a and the integer datasetsize
  integer, intent(in)              :: n
  real, dimension(n), intent(out)   :: arr    ! Explicit-shape array
  character(len=*), intent(in)      :: string ! Assumed length
  real, dimension(datasetsize+5)    :: x      ! Automatic array
  character(len=n+len(string))      :: cc     ! Automatic object
  integer, parameter :: pa2 =  selected_real_kind(2*precision(a))
  real(kind=pa2)     :: z      ! Precision of z is at least twice
                               ! the precision of a
```

The bounds and character lengths are not affected by any redefinitions or undefinitions of variables in the expressions during execution of the procedure.

As the interfaces of specification functions must be explicit yet they cannot be internal functions,[7] such functions are probably most conveniently written as module procedures.

This feature is a great convenience for specification expressions that cannot be written as simple expressions. Here is an example:

---

[5]This avoids such a case as
```
    character (len=len(a)) ::  fun
    character (len=len(fun)) ::  a
```
[6]This avoids such a case as
```
    integer, parameter, dimension (j(1):j(1)+1) ::  i = (/0,1/)
    integer, parameter, dimension (i(1):i(1)+1) ::  j = (/1,2/)
```
[7]This prevents them inquiring, via host association, about objects being specified in the set of statements in which the specification function itself is referenced.

```
function solve (a, ...
   use matrix_ops
   type(matrix), intent(in) :: a
   real                     :: work(wsize(a))
```

where `matrix` is a type defined in the module `matrix_ops` and intended to hold a sparse matrix and its LU factorization:

```
type matrix
   integer :: n               ! Matrix order.
   integer :: nz              ! Number of nonzero entries.
   logical :: new = .true.    ! Whether this is a new, unfactorized
                              ! matrix.
      ⋮
end type matrix
```

and `wsize` is a module procedure that calculates the required size of the array `work`:

```
pure integer function wsize(a)
   type(matrix), intent(in) :: a
   wsize = 2*a%n + 2
   if(a%new) wsize = a%nz + wsize
end function wsize
```

Note that a dummy argument of an elemental subprogram is permitted to be used in a specification expression for a local variable. Here is a partial example:

```
elemental real function f(a, b, order)
   real, intent (in)    :: a, b
   integer, intent (in) :: order
   real                 :: temp(order)
      ⋮
```

In this elemental function, the local variable `temp` is an array whose size depends on the `order` argument.

8.18   Structure constructors

Like procedures, structure constructors can have keyword arguments and optional arguments; moreover, a generic procedure name can be the same as the structure constructor name (which is the same as the type name), with any specific procedures in the generic set taking precedence over the structure constructor if there is any ambiguity. This can be used effectively to produce extra 'constructors' for the type, as shown in Figure 8.11.

Keyword arguments permit a parent (or ancestor) component to be specified in a structure constructor, as long as it does not result in a component being given a value more than once.

If a component of a type has default initialization or is an allocatable component, its value may be omitted in the structure constructor as if it were an optional argument. For example, in

Figure 8.11 Keywords in a structure constructor and a function as a structure constructor.

```
module mycomplex_module
   type mycomplex
      real :: argument, modulus
   end type
   interface mycomplex
      module procedure complex_to_mycomplex, two_reals_to_mycomplex
   end interface
   ⋮
contains
   type(mycomplex) function complex_to_mycomplex(c)
      complex, intent(in) :: c
      ⋮
   end function complex_to_mycomplex
   type(mycomplex) function two_reals_to_mycomplex(x, y)
      real, intent(in)           :: x
      real, intent(in), optional :: y
      ⋮
   end function two_reals_to_mycomplex
   ⋮
end module mycomplex_module
⋮
use mycomplex_module
type(mycomplex) :: a, b, c
⋮
a = mycomplex(argument=5.6, modulus=1.0) ! The structure constructor
c = mycomplex(x=0.0, y=1.0)              ! A function reference
```

```
type real_list_element
   real                                :: value
   type(real_list_element), pointer :: next => null()
end type real_list_element
⋮
type(real_list_element) :: x = real_list_element(3.5)
```

the omitted value for the `next` component means that it takes on its default initialization value
– that is, a null pointer. For an allocatable component, it is equivalent to specifying `null()`
for that component value.

8.19   The `namelist` statement

It is sometimes convenient to gather a set of variables into a single group in order to facilitate
input/output operations on the group as a whole. The actual use of such groups is explained

in Section 10.10. The method by which a group is declared is via the `namelist` statement, which in its simple form has the syntax

    `namelist` *namelist-spec*

where *namelist-spec* is

    /namelist-group-name/ variable-name-list

The *namelist-group-name* is the name given to the group for subsequent use in the input/output statements. An example is

```
real :: carpet, tv, brushes(10)
namelist /household_items/ carpet, tv, brushes
```

It is possible to declare several namelist groups in one statement, with the syntax

    `namelist` *namelist-spec [[,] namelist-spec]* ...

as in the example

```
namelist /list1/ a, b, c /list2/ x, y, z
```

It is possible to continue a list within the same scoping unit by repeating the namelist name on more than one statement. Thus,

```
namelist /list/ a, b, c
namelist /list/ d, e, f
```

has the same effect as a single statement containing all the variable names in the same order. A namelist group object may appear more than once in a namelist group and may belong to more than one namelist group.

If the type, kind type parameters, or rank of a namelist variable is specified in a specification statement in the same scoping unit, the specification statement must either appear before the `namelist` statement, or be a type declaration statement that confirms the implicit typing rule in force in the scoping unit for the initial letter of the variable. Also, if the namelist group has the `public` attribute, no variable in the list may have the `private` attribute or have private components.[8]

8.20   Summary

In this chapter most of the specification statements of Fortran have been described. The following concepts have been introduced: implicit typing and its attendant dangers, named constants, constant expressions, data initialization, control of the accessibility of entities in modules, saving data between procedure calls, volatility, selective access of entities in a module, renaming entities accessed from a module, specification expressions that may be used when specifying data objects and function results, and the formation of variables into namelist groups. We have also explained alternative ways of specifying attributes.

We conclude this chapter with a complete program, Figure 8.12, that uses a module to sort US-style addresses (name, street, town, and state with a numerical zip code) in order of zip code. It illustrates the interplay between many of the features described so far, but note that it is not production code since the sort subroutine is not very efficient and the full range of US addresses is not handled. Suitable test data are given in Figure 8.13.

---

[8] A variable that is an assumed-size array (Section 20.5) is prohibited.

Figure 8.12 A module to sort postal addresses and a program that uses it; `maxloc` is described in Section 9.16. The `read` and `write` statements here are explained in Section 10.7, Section 10.8, and Chapter 11.

```fortran
module sort              ! To sort postal addresses by zip code.
   implicit none
   private
   public :: selection_sort
   integer, parameter :: string_length = 30
   type, public :: address
      character(len = string_length) :: name, street, town, &
                                        state*2
      integer                        :: zip_code
   end type address
contains
   recursive subroutine selection_sort (array_arg)
      type (address), dimension (:), intent (inout)        &
                                        :: array_arg
      integer                          :: current_size
      integer                          :: big
      current_size = size (array_arg)
      if (current_size > 0) then
         big = maxloc (array_arg(:)%zip_code, dim=1)
         call swap (big, current_size)
         call selection_sort (array_arg(1: current_size - 1))
      end if
   contains
      subroutine swap (i, j)
         integer, intent (in) :: i, j
         type (address)       :: temp
         temp = array_arg(i)
         array_arg(i) = array_arg(j)
         array_arg(j) = temp
      end subroutine swap
   end subroutine selection_sort
 end module sort
 program zippy
   use sort
 use iso_Fortran_env
   implicit none
   integer, parameter                 :: array_size = 100
   type (address), dimension (array_size) :: data_array
   integer                            :: i, ios, n
   do i = 1, array_size
      read  (*, '(/a/a/a/a2,i8)', iostat=ios) data_array(i)
 if (ios == iostat_end ) exit
      write (*, '(/a/a/a/a2,i8)')          data_array(i)
   end do
   n = i - 1
   call selection_sort (data_array(1: n))
   write (*, '(//a)') 'after sorting:'
   do i = 1, n
      write (*, '(/a/a/a/a2,i8)') data_array(i)
   end do
 end program zippy
```

Figure 8.13 Sample data for Figure 8.12.

```
Prof. James Bush,
206 Church St. SE,
Minneapolis,
MN 55455

J. E. Dougal,
Rice University,
Houston,
TX 77251

Jack Finch,
104 Ayres Hall,
Knoxville,
TN 37996
```

Exercises

1. Write suitable type statements for the following quantities:

   i) an array to hold the number of counts in each of the 100 bins of a histogram numbered from 1 to 100;

   ii) an array to hold the temperature to two decimal places at points, on a sheet of iron, equally spaced at 1 cm intervals on a rectangular grid 20 cm square, with points in each corner (the melting point of iron is 1530 °C);

   iii) an array to describe the state of 20 on/off switches;

   iv) an array to contain the information destined for a printed page of 44 lines, each of 70 letters or digits.

2. Explain the difference between the following pair of declarations:

```
real :: i = 3.1
```

and

```
real, parameter :: i = 3.1
```

What is the value of i in each case?

3. Write type declaration statements which initialize:

   i) all the elements of an integer array of length 100 to the value zero;

   ii) all the odd elements of the same array to 0 and the even elements to 1;

   iii) the elements of a real 10×10 square array to 1.0;

   iv) a character string to the digits '0' to '9'.

4.   i) Write a type declaration statement that declares and initializes a variable of derived type
      person (Section 2.9).

      ii) Either

            a. write a type declaration statement that declares and initializes a variable of type entry
               (Section 2.13.2); or

            b. write a type declaration statement for such a variable and a data statement to initialize
               its non-pointer components.

5. Which of the following are constant expressions?

      i) kind(x), for x of type real

      ii) selected_real_kind(6, 20)

      iii) 1.7**2

      iv) 1.7**2.0

      v) (1.7, 2.3)**(-2)

      vi) (/ (7*i, i=1, 10) /)

      vii) person("Reid", 25*2.0, 22**2)

      viii) entry(1.7, 1, null_pointer)

9. Intrinsic procedures and modules

9.1 Introduction

9.1.1 Extent and performance of the intrinsic procedures

In a language that has a clear orientation towards scientific applications, there is an obvious requirement for the most frequently required mathematical functions to be provided as part of the language itself, rather than expecting each user to code them afresh. When provided with the compiler, they are normally coded to be very efficient and will have been well tested over the complete range of values that they accept. It is difficult to compete with the high standard of code provided by the vendors.

The efficiency of the intrinsic procedures when handling arrays is particularly marked because a single call may cause a large number of individual operations to be performed, during the execution of which advantage may be taken of the specific nature of the hardware.

Another feature of a substantial number of the intrinsic procedures is that they extend the power of the language by providing access to facilities that are not otherwise available. Examples are inquiry functions for the presence of an optional argument, the parts of a floating-point number, and the length of a character string.

There are over 190 intrinsic procedures in all, a particularly rich set. They fall into distinct groups, each of which we describe in turn. Some processors may offer additional intrinsic procedures. Note that a program containing references to such procedures is portable only to other processors that provide those same procedures. In fact, such a program does not conform to the standard.

All the intrinsic procedures are generic.

9.1.2 Keyword calls

The procedures may be called with keyword actual arguments, using the dummy argument names as keywords. This facility is not very useful for those with a single non-optional argument, but is useful for those with several optional arguments. For example,

```
call date_and_time (date=d)
```

returns the date in the scalar character variable d. The rules for positional and keyword argument lists were explained in Section 5.14. In this chapter, the dummy arguments that are optional are indicated with square brackets. We have taken some 'poetic licence' with this

Modern Fortran Explained, 3rd Edition. M. Metcalf, J. Reid, M. Cohen, and R. Bader. Oxford University Press (2024). © M. Metcalf, J. Reid, M. Cohen, and R. Bader (2024). DOI 10.1093/oso/9780198876571.001.0009

notation, which might suggest to the reader that the positional form is permitted following an absent argument (this is not the case).

9.1.3 Categories of intrinsic procedures

There are six categories of intrinsic procedures.

   i) **Elemental procedures** (Section 7.4).

   ii) **Inquiry functions** return properties of their principal arguments that do not depend on their values; indeed, for variables, their values may be undefined.

   iii) **Transformational functions** are functions that are neither elemental nor inquiry; they usually have array arguments and an array result whose elements depend on many of the elements of the arguments.

   iv) **Non-elemental subroutines**.

   v) **Atomic subroutines** (Section 17.23).

   vi) **Collective subroutines** (Section 17.22).

All the functions are pure (Section 7.8). Of the subroutines, only two are pure: the subroutine mvbits, and the subroutine move_alloc unless it is applied to coarrays.

In this chapter we present the intrinsic procedures in the first four categories. Atomic and collective subroutines are presented in Chapter 17.

9.1.4 The `intrinsic` statement

A name may be specified to be that of an intrinsic procedure in an `intrinsic` statement, which has the general form

    `intrinsic` *[::] intrinsic-name-list*

where *intrinsic-name-list* is a list of intrinsic procedure names. A name must not appear more than once in the `intrinsic` statements of a scoping unit but may appear as a generic name on an interface block if an intrinsic procedure is being extended, see Section 5.18.1. It is possible to include such a statement in every scoping unit that contains references to intrinsic procedures, in order to make the use clear to the reader. We particularly recommend this practice when referencing intrinsic procedures that are not defined by the standard, for then a clear diagnostic message should be produced if the program is ported to a processor that does not support the extra intrinsic procedures.

9.1.5 Argument intents

Since all the functions are pure, their arguments all have intent in. For the subroutines, the intents vary from case to case (see the descriptions given later in the chapter).

9.2   Inquiry functions for any type

The following are inquiry functions whose arguments may be of any type.

allocated (array) or **allocated (scalar)** has type default logical. It returns, when the allocatable array `array` or the allocatable scalar `scalar` is currently allocated, the value true; otherwise it returns the value false.

associated (pointer [,target]) has type default logical. When `target` is absent, it returns the value true if the pointer `pointer` is associated with a target and false otherwise. The pointer association status of `pointer` must not be undefined. If `target` is present, it must be allowable as a target of `pointer`. The value is true if `pointer` is associated with `target`, and false otherwise.

In the array case, true is returned only if the shapes are identical and corresponding array elements, in array element order, are associated with each other. If the character length or array size is zero, false is returned. A different bound, as in the case of `associated(p,a)` following the pointer assignment `p => a(:)` when `lbound(a) = 0`, is insufficient to cause false to be returned.

In the scalar case, true is returned only if `pointer` is associated with `target`. If the character length is zero, false is returned.

The argument `target` may itself be a pointer, in which case its target is compared with the target of `pointer`; the pointer association status of `target` must not be undefined and if either `pointer` or `target` is disassociated, the result is false.

If `target` is an internal procedure or a pointer associated with an internal procedure, the result is true only if `pointer` and `target` also have the same host instance.

present (a) has type default logical. It may be called in a subprogram that has an optional dummy argument `a` or has access to such a dummy argument. It returns the value true if the corresponding actual argument is present in the current call to it, and false otherwise. The exact meaning of 'present' is explained in the final two paragraphs of Section 5.14.

There is an inquiry function whose argument may be of any intrinsic type:

kind (x) has type default integer and value equal to the kind type parameter value of x.

9.3   Elemental numeric functions

9.3.1   Introduction

There are 17 elemental functions for performing simple numerical tasks, many of which perform type conversions for some or all permitted types of arguments.

9.3.2 Elemental functions that may convert

If kind is present in the following elemental functions, it must be a scalar integer constant expression and provide a kind type parameter that is supported on the processor.

abs (a) returns the absolute value of an argument of type integer, real, or complex. The result is of type integer if a is of type integer and otherwise it is real. It has the same kind type parameter as a.

aimag (z) returns the imaginary part of the complex value z. The type of the result is real and its kind type parameter is that of z.

aint (a [, kind]) truncates a real value a towards zero to produce a real that is a whole number. The value of the kind type parameter of the result is the value of the argument kind if it is present, or that of a otherwise.

anint (a [, kind]) returns a real whose value is the nearest whole number to the real value a. If two are equally near, that of greater magnitude is returned. The value of the kind type parameter of the result is the value of the argument kind, if it is present, or that of a otherwise.

ceiling (a [, kind]) returns the least integer greater than or equal to the real value a. If kind is present, the value of the kind type parameter of the result is the value of kind, otherwise it is that of the default integer type.

cmplx (x [, y] [, kind]) or **cmplx (x [, kind])** converts (x, y) or x to complex type with the value of the kind type parameter of the result being the value of kind if it is present or of default complex otherwise. The argument x may be of type integer or real, or it may be a 'boz' constant. The argument x may be of type complex if y is absent. If y is absent and x is not of type complex, the result is as if y were present with the value zero. If x is of type complex, the result is as if x were present with the value real(x, kind) and y were present with the value aimag(x, kind). The value of cmplx(x, y, kind) has real part real(x, kind) and imaginary part real(y, kind).

floor (a [, kind]) returns the greatest integer less than or equal to its real argument. If kind is present, the value of the kind type parameter of the result is the value of kind, otherwise it is that of the default integer type.

int (a [, kind]) converts to integer type with the value of the kind type parameter being the value of the argument kind, if it is present, or that of the default integer otherwise. The argument a may be

- integer, in which case int(a) = a;
- real, in which case the value is truncated towards zero;
- complex, in which case the real part is truncated towards zero; or
- a 'boz' constant, in which case the result has the bit sequence of the specified integer after truncation on the left or padding with zeros on the left to the bit length of the result; if the leading bit is 1, the value is processor dependent.

nint (a [,kind]) returns the integer value that is nearest to the real a, or whichever such value has greater magnitude if two are equally near. If kind is present, the value of the kind type parameter of the result is the value of kind, otherwise it is that of the default integer type.

real (a [,kind]) converts to real type with the value of the kind type parameter being that of kind if it is present. If kind is absent, the kind type parameter is that of a if a is of type complex and default real otherwise. The argument a may be

- integer or real, in which case the result is a processor-dependent approximation to a;
- complex, in which case the result is a processor-dependent approximation to the real part of a; or
- a 'boz' constant, in which case the result has the bit sequence of a after truncation on the left or padding with zeros on the left to the bit length of the result.

9.3.3   Elemental functions that do not convert

The following are elemental functions whose result is of type and kind type parameter that are those of the first or only argument.

conjg (z) returns the conjugate of the complex value z.

dim (x, y) returns max(x-y, 0.) for arguments that are both integer or both real. The arguments must have the same kind.

max (a1, a2 [,a3, ...]) returns the maximum of two or more values, which must all be integer, all be real, or all be character (character comparison is explained in Section 3.5). The arguments must all have the same kind.

min (a1, a2 [,a3, ...]) returns the minimum of two or more values, which must all be integer, all be real, or all be character (character comparison is explained in Section 3.5). The arguments must all have the same kind.

mod (a, p) returns the remainder of a modulo p, that is, a-int(a/p)*p. The arguments must have the same kind. The value of p must not be zero; a and p must be both integer or both real.

modulo (a, p) returns a modulo p when a and p are both integer or both real, that is, a-floor(a/p)*p in the real case, and a-floor(a÷p)*p in the integer case, where ÷ represents ordinary mathematical division. The arguments must have the same kind. The value of p must not be zero. The difference between mod and modulo, is illustrated in this table:

a	−8	−8	+8	+8
p	−5	+5	−5	+5
mod(a, p)	−3	−3	+3	+3
modulo(a, p)	−3	+2	−2	+3

sign (a, b) returns the absolute value of a times the sign of b. The arguments a and b must be integer or real, and can be of different kinds. The kind of the result is that of argument a, which provides the magnitude. If b is real with the value zero and the processor can distinguish between a negative and a positive real zero, the result has the sign of b (see also Section 9.9.1). Otherwise, if b is zero, its sign is taken as positive.

9.4 Elemental mathematical functions

The following are elemental functions that evaluate elementary mathematical functions. The type and kind type parameter of the result are those of the argument x, which is usually the only argument.

acos (x) returns the arc cosine (inverse cosine) function value for real or complex values x. If x is real it must be such that $|x| \leq 1$; the result is expressed in radians in the range $0 \leq \text{acos}(x) \leq \pi$. If x is complex, the real part is expressed in radians in the range $0 \leq \text{real}(\text{acos}(x)) \leq \pi$.

acosh (x) returns the inverse hyperbolic cosine for real or complex values of x, that is, y such that cosh(y) would be approximately equal to x. If the result is complex, the imaginary part is expressed in radians in the range $0 \leq \text{aimag}(\text{acosh}(x)) \leq \pi$.

asin (x) returns the arc sine (inverse sine) function value for real or complex values x. If x is real it must be such that $|x| \leq 1$; the result is expressed in radians in the range $-\pi/2 \leq \text{asin}(x) \leq \pi/2$. If x is complex, the real part is expressed in radians in the range $-\pi/2 \leq \text{real}(\text{asin}(x)) \leq \pi/2$.

asinh (x) returns the inverse hyperbolic sine for real or complex values of x. If the result is complex, the imaginary part is expressed in radians in the range $-\pi/2 \leq \text{aimag}(\text{asinh}(x)) \leq \pi/2$.

atan (x) returns the arc tangent (inverse tangent) function value for real or complex x, whose real part is expressed in radians in the range $-\pi/2 \leq \text{real}(\text{atan}(x)) \leq \pi/2$.

atan (y, x) returns the arc tangent (inverse tangent) function value for pairs of reals, x and y, of the same type and type parameter. They must not both have the value zero. The result is the principal value of the argument of the complex number (x, y), expressed in radians in the range $-\pi \leq \text{atan}(y, x) \leq \pi$. If the processor can distinguish between positive and negative real zero (e.g., if the arithmetic is IEEE), an approximation to $-\pi$ is returned if $x < 0$ and y is a negative zero. If x is zero, the absolute value of the result is approximately $\pi/2$.

atan2 (y, x) is the same as atan (y, x).

atanh (x) returns the inverse hyperbolic tangent for real or complex values of x. If the result is complex, the imaginary part is expressed in radians and it satisfies the inequality $-\pi/2 \leq \text{aimag}(\text{atanh}(x)) \leq \pi/2$.

bessel_j0 (x) returns the Bessel function of the first kind and order zero; x must be of type real.

bessel_j1 (x) returns the Bessel function of the first kind and order one for a real argument.

bessel_jn (n, x) returns the Bessel function of the first kind and order n; x must be of type real, and n must be of type integer with a non-negative value. See Section 9.5 for `bessel_jn (n1, n2, x)`.

bessel_y0 (x) returns the Bessel function of the second kind and order zero for a real argument.

bessel_y1 (x) returns the Bessel function of the second kind and order one for a real argument whose value is greater than zero.

bessel_yn (n, x) returns the Bessel function of the second kind and order n; x must be of type real, and n must be of type integer with a non-negative value. See Section 9.5 for `bessel_yn (n1, n2, x)`.

cos (x) returns the cosine function value for an argument of type real or complex. The result, or its real part if complex, is treated as a value in radians.

cosh (x) returns the hyperbolic cosine function value for a real or complex argument x. If x is complex, the imaginary part of the result is treated as a value in radians.

erf (x) returns the value of the error function of x, $\frac{2}{\sqrt{\pi}} \int_0^x e^{-t^2} \, dt$, for x of type real.

erfc (x) returns the complement of the error function, $1 - \text{erf}(x)$, for x of type real. This has the mathematical form $\frac{2}{\sqrt{\pi}} \int_x^\infty e^{-t^2} \, dt$.

erfc_scaled (x) returns the exponentially scaled complementary error function, `exp(x**2)*erfc(x)`, for x of type real. Note that as x increases, its error function very rapidly approaches one, and its complement quite quickly approaches zero. In IEEE double precision, `erf(x)` is equal to 1 for $x > 6$ and `erfc(x)` underflows for $x > 26.7$. Thus `erfc_scaled` may be more useful than `erfc` when x is not small.

exp (x) returns the exponential function value for a real or complex argument x. If x is complex, the imaginary part of the result is expressed in radians.

gamma (x) returns the value of the gamma function at x; x must be of type real with a value that is not zero or a negative whole number.

hypot (x, y) returns the Euclidean distance, $\sqrt{x^2 + y^2}$, calculated without undue overflow or underflow. The arguments x and y must be of type real with the same kind type parameter, and the result is also real of that kind.[1]

---

[1]This intrinsic function means that there are three intrinsic functions that calculate Euclidean distances, which seems a trifle unnecessary for such a simple thing. (The other two functions are `abs(cmplx(x, y))`, which has been available for this purpose since Fortran 90, and `norm2([x, y])`.)

log (x) returns the natural logarithm function for a real or complex argument x. In the real case, x must be positive. In the complex case, x must not be zero, and the imaginary part w of the result lies in the range $-\pi \leq w \leq \pi$. If the processor can distinguish between positive and negative real zero (e.g., if the arithmetic is IEEE), an approximation to $-\pi$ is returned if the real part of x is less than zero and the imaginary part is a negative zero.

log_gamma (x) returns the natural logarithm of the absolute value of the gamma function, `log(abs(gamma(x)))`; x must be of type real with a value that is not zero or a negative whole number.

log10 (x) returns the common (base 10) logarithm of a real argument whose value must be positive.

sin (x) returns the sine function value for a real or complex argument. The result, or its real part if complex, is treated as a value in radians.

sinh (x) returns the hyperbolic sine function value for a real or complex argument. If x is complex, the imaginary part of the result is treated as a value in radians.

sqrt (x) returns the square root function value for a real or complex argument x. If x is real, its value must not be negative. In the complex case, the result is the principal value with the real part greater than or equal to zero; if the real part of the result is zero, the imaginary part has the same sign as the imaginary part of x.

tan (x) returns the tangent function value for a real or complex argument. The result, or its real part if complex, is treated as a value in radians.

tanh (x) returns the hyperbolic tangent function value for a real or complex argument. If x is complex, the imaginary part of the result is treated as a value in radians.

9.5 Transformational functions for Bessel functions

Two transformational functions return rank-one arrays of multiple Bessel function values:

bessel_jn (n1, n2, x) first kind and orders n1 to n2;

bessel_yn (n1, n2, x) second kind and orders n1 to n2.

In this case n1 and n2 must be of type integer with non-negative values, and all three arguments must be scalar. If n2 < n1, the result has zero size.

It is potentially more efficient to calculate successive Bessel function values together rather than separately, so if these are required the transformational forms should be used instead of multiple calls to the elemental ones.

9.6   Elemental character and logical functions

9.6.1   Character–integer conversions

The following are elemental functions for conversions from a single character to an integer, and vice versa.

achar (i [, kind]) is of type character with length one and returns the character in the position in the ASCII collating sequence that is specified by the integer i. The value of i must be in the range $0 \leq i \leq 127$, otherwise the result is processor dependent. The optional kind argument specifies the kind of the result, which is of default kind if it is absent. For instance, if the processor had an extra character kind 37 for EBCDIC, achar(iachar('h'),37) would return the EBCDIC lower-case 'h' character.

char (i [, kind]) is of type character and length one, with kind type parameter value that of the value of kind if present, or default otherwise. It returns the character in position i in the processor collating sequence associated with this kind parameter. The value of i must be in the range $0 \leq i \leq n-1$, where n is the number of characters in this collating sequence. If kind is present, it must be a scalar integer constant expression.

iachar (c [, kind]) is of type integer and returns the position in the ASCII collating sequence of the character c if it is in the ASCII character set and a processor-dependent value otherwise. The optional kind argument must be a scalar integer constant expression. If present, it specifies the kind of the result, which is otherwise default.

ichar (c [, kind]) is of type integer and returns the position of the character c in the processor collating sequence associated with the kind parameter of c. The optional kind argument must be a scalar integer constant expression. If present, it specifies the kind of the result, which is otherwise default.

9.6.2   Lexical comparison functions

The following elemental functions accept arguments that are of type character and are both of default kind or both of ASCII kind, make a lexical comparison based on the ASCII collating sequence, and return a default logical result. If the strings have different lengths, the shorter one is padded on the right with blanks.

lge (string_a, string_b) returns the value true if string_a follows string_b in the ASCII collating sequence or is equal to it, and the value false otherwise.

lgt (string_a, string_b) returns the value true if string_a follows string_b in the ASCII collating sequence, and the value false otherwise.

lle (string_a, string_b) returns the value true if string_b follows string_a in the ASCII collating sequence or is equal to it, and the value false otherwise.

llt (string_a, string_b) returns the value true if string_b follows string_a in the ASCII collating sequence, and false otherwise.

9.6.3 String-handling elemental functions

The following are elemental functions that manipulate strings. The arguments `string`, `substring`, and `set` are always of type character, and where two are present have the same kind type parameter. The kind type parameter value of the result is that of `string`.

adjustl (string) adjusts left to return a string of the same length by removing all leading blanks and inserting the same number of trailing blanks.

adjustr (string) adjusts right to return a string of the same length by removing all trailing blanks and inserting the same number of leading blanks.

index (string, substring [,back] [,kind]) returns an integer holding the starting position of `substring` as a substring of `string`, or zero if it does not occur as a substring. If `back` is absent or present with value false, the starting position of the first such substring is returned; the value 1 is returned if `substring` has zero length. If `back` is present with value true, the starting position of the last such substring is returned; the value `len(string)+1` is returned if `substring` has zero length. If `kind` is present, it specifies the kind of the result, which is otherwise default; it must be a scalar integer constant expression.

len_trim (string [,kind]) returns an integer whose value is the length of `string` without trailing blank characters. If `kind` is present, it specifies the kind of the result, which is otherwise default; it must be a scalar integer constant expression.

scan (string, set [,back] [,kind]) returns an integer whose value is the position of a character of `string` that is in `set`, or zero if there is no such character. If the logical `back` is absent or present with value false, the position of the leftmost such character is returned. If `back` is present with value true, the position of the rightmost such character is returned. If `kind` is present, it specifies the kind of the result, which is otherwise default; it must be a scalar integer constant expression.

verify (string, set [,back] [,kind]) returns the integer value 0 if each character in `string` appears in `set`, or the position of a character of `string` that is not in `set`. If the logical `back` is absent or present with value false, the position of the leftmost such character is returned. If `back` is present with value true, the position of the rightmost such character is returned. If `kind` is present, it specifies the kind of the result, which is otherwise default; it must be a scalar integer constant expression.

9.6.4 Logical conversion

The following elemental function converts from a logical value with one kind type parameter to another.

logical (l [,kind]) returns a logical value equal to the value of the logical `l`. The value of the kind type parameter of the result is the value of `kind` if it is present or that of default logical otherwise. If `kind` is present, it must be a scalar integer constant expression.

9.7   Non-elemental string-handling functions

9.7.1   String-handling inquiry function

len (string [,kind]) for `string` of type character is an inquiry function that returns a scalar integer holding the number of characters in `string` if it is scalar, or in an element of `string` if it is array valued. The value of `string` need not be defined. It is permitted to be a disassociated pointer or unallocated allocatable unless it has a deferred type parameter. If `kind` is present, it specifies the kind of the result, which is otherwise default; it must be a scalar integer constant expression.

9.7.2   String-handling transformational functions

There are two functions that cannot be elemental because the length type parameter of the result depends on the value of an argument.

repeat (string, ncopies) for `string` of type character forms the string consisting of the concatenation of `ncopies` copies of `string`, where `ncopies` is of type integer and its value must not be negative. Both arguments must be scalar.

trim (string) for `string` of type character returns `string` with all trailing blanks removed. The argument `string` must be scalar.

9.8   Character inquiry function

The intrinsic inquiry function `new_line(a)` returns the character that can be used to cause record termination (this is the equivalent of the C language `'\n'` character).

new_line (a) returns the newline character used for formatted stream output. The argument a must be of type character. The result is of type `character` with the same kind type parameter value as a. In the unlikely event that there is no suitable character for newline in that character set, a blank is returned.

9.9   Numeric inquiry and manipulation functions

9.9.1   Models for integer and real data

The numeric inquiry and manipulation functions are defined in terms of a model set of integers and a model set of reals for each kind of integer and real data type implemented. For each kind of integer, it is the set

$$i = s \times \sum_{k=1}^{q} w_k \times r^{k-1},$$

where s is ± 1, q is a positive integer, r is an integer exceeding 1 (usually 2), and each w_k is an integer in the range $0 \le w_k < r$. For each kind of real, it is the set

$$x = 0 \qquad \text{and} \qquad x = s \times b^e \times \sum_{k=1}^{p} f_k \times b^{-k},$$

where s is ± 1, p and b are integers exceeding 1, e is an integer in a range $e_{\min} \le e \le e_{\max}$, and each f_k is an integer in the range $0 \le f_k < b$ except that f_1 is also nonzero.

Values of the parameters in these models are chosen for the processor so as best to fit the hardware with the proviso that all model numbers are representable. Note that it is quite likely that there are some machine numbers that lie outside the model. For example, many computers represent the integer $-r^q$, and the IEEE standard for floating-point arithmetic (Section 19.2) contains reals with $f_1 = 0$ (called subnormal numbers), register numbers with increased precision and range, and special values known as NaNs (Not a Number) for exceptions.

In Section 2.6.1, we noted that the value of a signed zero is regarded as being the same as that of an unsigned zero. However, many processors distinguish at the hardware level between a negative real zero value and a positive real zero value, and the IEEE standard makes use of this where possible. For example, when the exact result of an operation is nonzero but the rounding produces a zero, the sign is retained.

In Fortran, the two zeros are treated identically in all relational operations and as input arguments to all intrinsic functions except sign. If the processor can distinguish between positive and negative zeros, the function sign (Section 9.3.3) takes the sign of the second argument into account even if its value is zero. For example, the value of sign(2.0, -0.0) is -2.0. Also, such a processor is required to represent negative zero numbers with a minus sign.

9.9.2 Numeric inquiry functions

There are nine inquiry functions that return values from the models associated with their arguments. Each has a single argument that may be scalar or array valued and each returns a scalar result. The value of the argument need not be defined.

digits (x) for real or integer x, returns the default integer whose value is the number of significant digits in the model that includes x, that is, p or q.

epsilon (x) for real x, returns a real result with the same type parameter as x that is almost negligible compared with the value 1.0 in the model that includes x, that is, b^{1-p}.

huge (x) for real or integer x, returns the largest value in the model that includes x. It has the type and type parameter of x. The value is $(1 - b^{-p})b^{e_{\max}}$ for real x, or $r^q - 1$ for integer x.

maxexponent (x) for real x, returns the default integer $e_{\max}$, the maximum exponent in the model that includes x.

minexponent (x) for real x, returns the default integer e_{min}, the minimum exponent in the model that includes x.

precision (x) for real or complex x, returns a default integer holding the equivalent decimal precision in the model representing real numbers with the same type parameter value as x. The value is $int((p-1)*log10(b))+k$, where k is 1 if b is an integral power of 10 and 0 otherwise.

radix (x) for real or integer x, returns the default integer that is the base in the model that includes x, that is, b or r.

range (x) for integer, real, or complex x, returns a default integer holding the equivalent decimal exponent range in the models representing integer or real numbers with the same type parameter value as x. The value is $int(log10(huge))$ for integers and $int(min(log10(huge), -log10(tiny)))$ for reals, where *huge* and *tiny* are the largest and smallest positive numbers in the models.

tiny (x) for real x, returns the smallest positive number $b^{e_{min}-1}$ in the model that includes x. It has the type and type parameter of x.

9.9.3   Elemental functions to manipulate reals

There are seven elemental functions whose first or only argument is of type real and that return values related to the components of the model values associated with the actual value of the argument. For the functions exponent, fraction, and set_exponent, if the value of x lies outside the range of model numbers, its e value is determined as if the model had no exponent limits.

exponent (x) returns the default integer whose value is the exponent part e of x when represented as a model number. If x = 0, the result has value zero. If x is an IEEE infinity or NaN, the result has the value huge (0).

fraction (x) returns a real with the same type parameter as x whose value is the fractional part of x when represented as a model number, that is, xb^{-e}. If x is an IEEE NaN, the result is that NaN. If x is an IEEE infinity, the result is an IEEE NaN.

nearest (x, s) returns a real with the same type parameter as x whose value is the nearest different machine number in the direction given by the sign of the real s. The value of s must not be zero.

rrspacing (x) returns a real with the same type parameter as x whose value is the reciprocal of the relative spacing of model numbers near x. If y is a model number nearest to x the result is $|yb^{-e}|b^p$. If x is an IEEE NaN, the result is that NaN. If x is an IEEE infinity, the result is an IEEE NaN.

scale (x, i) returns a real with the same type parameter as x, whose value is xb^i, where b is the base in the model for x, and i is of type integer, provided this result is representable; if not, the result is processor dependent.

set_exponent (x, i) returns a real with the same type parameter as x, whose fractional part is the fractional part of the model representation of x and whose exponent part is i, that is, xb^{i-e}. If x has the value zero, the result has the same value. If x is an IEEE NaN, the result is that NaN. If x is an IEEE infinity, the result is an IEEE NaN.

spacing (x) returns a real with the same type parameter as x whose value is the absolute spacing of model numbers near x, that is, b^{e-p}, for the parameters of the model number nearest to x. If x = 0, the result is tiny(x). If x is an IEEE NaN, the result is that NaN. If x is an IEEE infinity, the result is an IEEE NaN.

9.9.4 Transformational functions for kind values

There are three functions that return the least kind type parameter value that will meet a given requirement. They have scalar arguments and results, so are classified as transformational.

selected_char_kind (name) returns the kind value for the character set whose name is given by the character string name, or -1 if it is not supported (or if the name is not recognized). In particular, if name is

default the result is the kind of the default character type (equal to kind('A'));

ascii the result is the kind of the ASCII character type;

iso_10646 the result is the kind of the ISO/IEC 10646 UCS-4 character type.

Other character set names are processor dependent. The character set name is not case sensitive (lower case is treated as upper case), and any trailing blanks are ignored.

Note that the only character set which is guaranteed to be supported is the default character set; a processor is not required to support ASCII or ISO 10646.

selected_int_kind (r) returns the default integer scalar that is the kind type parameter value for an integer data type able to represent all integer values n in the range $-10^r < n < 10^r$, where r is a scalar integer. If more than one is available, a kind with least decimal exponent range is chosen (and least kind value if several have least decimal exponent range). If no corresponding kind is available, the result is -1.

selected_real_kind ([p] [,r] [,radix]) returns the default integer scalar that is the kind type parameter value for a real data type with decimal precision (as returned by the function precision) at least p, decimal exponent range (as returned by the function range) at least r, and radix (as returned by the function radix) radix. If more than one is available, a kind with the least decimal precision is chosen (and least kind value if several have least decimal precision). All three arguments are scalar integers; at least one of them must be present. If radix is absent, there is no requirement on the radix selected. If no corresponding kind value is available, the result is -1 if sufficient precision is unavailable, -2 if sufficient exponent range is unavailable, -3 if both are unavailable, -4 if the radix is available with the precision or the range but not both, and -5 if the radix is not available at all.

9.9.5 Checking for unsafe conversions

The elemental intrinsic function `out_of_range` tests whether the real or integer value x can be safely converted to a real or integer type and kind of `mold`.

`out_of_range (x, mold [, round])` where x is of type real or integer, `mold` is of type real or integer, and `round` is of type logical returns the default logical that has the value true if and only if

- the value of x is an IEEE infinity or NaN, and `mold` is of type integer or is of type real and of a kind that does not support such a value;
- `mold` is of type integer, `round` is absent or present with the value false, and the integer with largest magnitude that lies between zero and x inclusive is not representable by objects with the type and kind of `mold`;
- `mold` is of type integer, `round` is present with the value true, and the integer nearest x, or the integer of greater magnitude if two integers are equally near to x, is not representable by objects with the type and kind of `mold`; or
- `mold` is of type real and the result of rounding the value of x to the extended model for the kind of `mold` has magnitude larger than that of the largest finite number with the same sign as x that is representable by objects with the type and kind of `mold`.

The arguments `mold` and `round` are required to be scalars. The argument `round` may be present only if x is of type real and `mold` is of type integer.

9.10 Bit manipulation procedures

9.10.1 Model for bit data

There are intrinsic procedures for manipulating bits held within integers. They are based on a model in which an integer holds b bits w_k, $k = 0, 1, \ldots, b - 1$, in a sequence from right to left, based on the non-negative value

$$\sum_{k=0}^{b-1} w_k \times 2^k.$$

This model is valid only in the context of these intrinsics. It is identical to the model for integers in Section 9.9.1 when $r = 2$, $s = 1$, and $w_{b-1} = 0$, but otherwise the models do not correspond, and the value expressed as an integer may vary from processor to processor.

9.10.2 Inquiry function

`bit_size (i)` returns the number of bits in the model for bits within an integer of the same type parameter as i. The result is a scalar integer having the same type parameter as i.

9.10.3   Basic elemental functions

btest (i, pos) returns the default logical value true if bit `pos` of the integer `i` has value 1 and false otherwise; `pos` must be an integer with value in the range $0 \le \text{pos} < \text{bit\_size(i)}$.

iand (i, j) returns the logical and of all the bits in `i` and corresponding bits in `j`, according to the truth table

i	1	1	0	0
j	1	0	1	0
iand(i, j)	1	0	0	0

The arguments `i` and `j` must be integers of the same kind or be 'boz' constants. At least one of `i` and `j` must be an integer, and a 'boz' constant is converted to that type as if by the `int` intrinsic; the result is of type integer with the same kind.

ibclr (i, pos) returns an integer, with the same type parameter as `i`, and value equal to that of `i` except that bit `pos` is cleared to 0. The argument `pos` must be an integer with value in the range $0 \le \text{pos} < \text{bit\_size(i)}$.

ibits (i, pos, len) returns an integer, with the same type parameter as `i`, and value equal to the `len` bits of `i` starting at bit `pos` right adjusted and all other bits zero. The arguments `pos` and `len` must be integers with non-negative values such that $\text{pos+len} \le \text{bit\_size(i)}$.

ibset (i, pos) returns an integer, with the same type parameter as `i`, and value equal to that of `i` except that bit `pos` is set to 1. The argument `pos` must be an integer with value in the range $0 \le \text{pos} < \text{bit\_size(i)}$.

ieor (i, j) returns the logical exclusive or of all the bits in `i` and corresponding bits in `j`, according to the truth table

i	1	1	0	0
j	1	0	1	0
ieor(i, j)	0	1	1	0

The arguments `i` and `j` must be integers of the same kind or be 'boz' constants. At least one of `i` and `j` must be an integer, and a 'boz' constant is converted to that type as if by the `int` intrinsic; the result is of type integer with the same kind.

ior (i, j) returns the logical inclusive or of all the bits in `i` and corresponding bits in `j`, according to the truth table

i	1	1	0	0
j	1	0	1	0
ior(i, j)	1	1	1	0

The arguments `i` and `j` must be integers of the same kind or be 'boz' constants. At least one of `i` and `j` must be an integer, and a 'boz' constant is converted to that type as if by the `int` intrinsic; the result is of type integer with the same kind.

not (i) returns an integer with the same type parameter as the integer i and value the logical complement of all the bits in i, according to the truth table

i	1	0
not(i)	0	1

9.10.4   Shift operations

There are seven elemental functions for bit shifting: five described here, and two 'double-width' functions described in Section 9.10.7.

ishft (i, shift) returns an integer, with the same type parameter as i, and value equal to that of i except that the bits are shifted shift places to the left (-shift places to the right if shift is negative). Zeros are shifted in from the other end. The argument shift must be an integer with value satisfying the inequality $|shift| \leq bit\_size(i)$.

ishftc (i, shift [,size]) returns an integer, with the same type parameter as i, and value equal to that of i except that the size rightmost bits (or all the bits if size is absent) are shifted circularly shift places to the left (-shift places to the right if shift is negative). The argument shift must be an integer with absolute value not exceeding the value of size (or bit_size(i) if size is absent).

shifta (i, shift) returns an integer, with the same type parameter as i, holding the bits of i shifted right by shift bits, but instead of shifting in zero bits from the left, the leftmost bit is replicated. The argument shift must be an integer with value satisfying the inequality $0 \leq shift \leq bit\_size(i)$.

shiftl (i, shift) returns the bits of i shifted left, equivalent to ishft(i, shift). The argument shift must be an integer with value satisfying the inequality $0 \leq shift \leq bit\_size(i)$.

shiftr (i, shift) returns the bits of i shifted right, equivalent to ishft(i, -shift). The argument shift must be an integer with value satisfying the inequality $0 \leq shift \leq bit\_size(i)$.

The advantages of shiftl and shiftr over ishft are:

- the shift direction is implied by the name, so one doesn't have to remember that a positive shift value means 'shift left' and a negative shift value means 'shift right';
- if the shift amount is variable, the code generated for shifting is theoretically more efficient (in practice, unless a lot of other things are being done to the values, the performance is going to be limited by the main memory bandwidth anyway, not the shift function).

9.10.5   Elemental subroutine

call mvbits (from, frompos, len, to, topos) copies the sequence of bits in from that starts at position frompos and has length len to to, starting at position topos. The other bits of to are not altered. The arguments from, frompos, len, and topos are all integers with intent in, and they must have values that satisfy the inequalities frompos ≥ 0, topos ≥ 0, len ≥ 0, frompos+len $\leq$ bit_size(from), and topos+len $\leq$ bit_size(to). The argument to is an integer with intent inout; it must have the same kind type parameter as from. The same variable may be specified for from and to.

9.10.6 Bitwise (unsigned) comparison

Four elemental functions are provided for performing bitwise comparisons, returning a default logical result. Bitwise comparisons treat integer values as **unsigned** integers; that is, the most significant bit is not treated as a sign bit but as having the value of 2^{b-1}, where b is the number of bits in the integer. Each argument is an integer or a 'boz' constant.

bge (i, j) returns the value true if i is bitwise greater than or equal to j, and the value false otherwise.

bgt (i, j) returns the value true if i is bitwise greater than j, and the value false otherwise.

ble (i, j) returns the value true if i is bitwise less than or equal to j, and the value false otherwise.

blt (i, j) returns the value true if i is bitwise less than j, and the value false otherwise.

The arguments i and j must either be of type integer or be 'boz' constants (Section 2.6.7); if of type integer, they need not have the same kind type parameter. For example, on a two's-complement processor with integer(int8) holding eight-bit integers, -1_int8 has the bit pattern z'ff', and this has the value 255 when treated as unsigned, so bge(-1_int8, 255) is true and blt(-1_int8, 255) is false.

9.10.7 Double-width shifting

Two unusual elemental functions provide double-width shifting. These functions concatenate i and j and shift the combined value left or right by shift; the result is the left half for a left shift and the right half for a right shift.

dshiftl (i, j, shift) returns the left half of a double-width left shift.

dshiftr (i, j, shift) returns the right half of double-width right shift.

One of the arguments i and j must be of type integer. The other is either of type integer with the same kind or is a 'boz' literal constant and is converted to the kind as if by the int function. The result is integer and of this kind. The shift argument must be an integer, but can be of any kind. For example, if integer(int8) holds eight-bit integers, dshiftl(21_int8, 64_int8, 2), has the value 85_int8.

In general, these functions are harder to understand and will perform worse than simply using ordinary shifts on integers of double the width, so they should be used only if the exact functionality is really what is required.

9.10.8   Bitwise reductions

Three transformational functions that perform bitwise reductions to reduce an integer array by one rank or to a scalar, iall, iany, and iparity, are described in Section 9.13.1.

9.10.9   Counting bits

Several elemental functions are provided for counting bits within an integer.

leadz (i) returns the number of leading (on the left) zero bits in i.

popcnt (i) returns the number of nonzero bits in i.

poppar (i) returns the value 1 if popcnt (i) is odd and 0 otherwise.

trailz (i) returns the number of trailing (on the right) zero bits in i.

   The argument i may be any kind of integer, and the result is a default integer.
   Note that the values of popcnt, poppar, and trailz depend only on the value of the argument, whereas the value of leadz depends also on the kind of the argument. For example, if integer(int8) holds eight-bit integers and integer(int16) holds 16-bit integers, leadz(64_int8) has the value 1, while leadz(64_int16) has the value 9; the values of popcnt(64_k), poppar(64_k), and trailz(64_k) are 1, 1, and 6, respectively, no matter what the kind value k is.

9.10.10   Producing bitmasks

Other elemental functions facilitate producing simple bitmasks:

maskl (i [,kind]) returns an integer with the leftmost i bits set and the rest zero.

maskr (i [,kind]) returns an integer with the rightmost i bits set and the rest zero.

The result type is integer with the specified kind (or default integer if no kind is specified). The argument i must be of type integer of any kind (the kind of i has no effect on the result), and with value in the range $0 \leq i \leq b$, where b is the bit size of the result.
   For example, if integer(int8) holds eight-bit integers, maskl(3,int8) is equal to int(b'11100000',int8) and maskr(3,int8) is equal to 7_int8.

9.10.11   Merging bits

An elemental function merges bits from separate integers.

merge_bits (i, j, mask) returns the bits of i and j merged under the control of mask. The arguments i, j, and mask must be integers of the same kind or be 'boz' constants. At least one of i and j must be an integer, and a 'boz' constant is converted to that type as if by the int intrinsic; the result is of type integer with the same kind.

This function is modelled on the `merge` intrinsic, treating 1 and 0 bits as true and false, respectively. The value of the result is determined by taking the bit positions where `mask` is 1 from `i`, and the bit positions where `mask` is 0 from `j`; this is equal to `ior(iand(i, mask), iand(j, not(mask)))`.

9.11 Transfer function

The transformational function `transfer` allows data of one type to be transferred to another without the physical representation being altered. This would be useful, for example, in writing a generic data storage and retrieval system. The system itself could be written for one type, default integer say, and other types handled by transfers to and from that type, for example:

```
integer          :: store
character(len=4) :: word         ! To be stored and retrieved
  ⋮
store = transfer(word, store)    ! Before storage
  ⋮
word  = transfer(store, word)    ! After retrieval
```

transfer (source, mold [, size]) returns a result of type and type parameters those of `mold`. When `size` is absent, the result is scalar if `mold` is scalar, and it is of rank one and size just sufficient to hold all of `source` if `mold` is array valued. When `size` is present, the result is of rank one and size `size`. If the physical representation of the result is as long as or longer than that of `source`, the result contains `source` as its leading part and the value of the rest is processor dependent; otherwise the result is the leading part of `source`. As the rank of the result can depend on whether or not `size` is specified, the corresponding actual argument must not itself be an optional dummy argument.

Note that if, say, `carray` is specified as `character(1)`, `dimension(n)` and `string` is `character(n)`, then both of the assignments `carray = transfer(string,carray)` and `string = transfer(carray,string)` are valid.

9.12 Vector and matrix multiplication functions

There are two transformational functions that perform vector and matrix multiplications. They each have two arguments that are both of numeric type (integer, real, or complex) or both of logical type. The result is of the same type and type parameter as for the multiply or and operation between two such scalars. The functions `sum` and `any`, used in the definitions, are defined in Section 9.13.1.

dot_product (vector_a, vector_b) requires two arguments each of rank one and the same size. If `vector_a` is of type integer or type real, the function returns `sum(vector_a*vector_b)`; if `vector_a` is of type complex, it returns `sum(conjg(vector_a)*vector_b)`; and if `vector_a` is of type logical, it returns `any(vector_a .and. vector_b)`.

matmul (matrix_a, matrix_b) performs matrix multiplication. For numeric arguments, three cases are possible:

   i) `matrix_a` has shape (n, m) and `matrix_b` has shape (m, k). The result has shape (n, k), and element (i, j) has the value `sum(matrix_a(i, :)*matrix_b(:, j))`.

  ii) `matrix_a` has shape (m) and `matrix_b` has shape (m, k). The result has shape (k) and element (j) has the value `sum(matrix_a*matrix_b(:, j))`.

 iii) `matrix_a` has shape (n, m) and `matrix_b` has shape (m). The result has shape (n) and element (i) has the value `sum(matrix_a(i, :)*matrix_b)`.

For logical arguments, the shapes are as for numeric arguments and the values are determined by replacing 'sum' and '*' in the above expressions by 'any' and '.and.'.

9.13   Transformational functions that reduce arrays

9.13.1   Single-argument case

In their simplest form, the following transformational functions have a single array argument and return a scalar result. All except `count` have a result of the same type and kind as the argument. The mask array `mask`, used as an argument in `any`, `all`, `count`, `parity`, and optionally in others, is also described in Section 7.7.

all (mask)   returns the value true if all elements of the logical array `mask` are true or `mask` has size zero, and otherwise returns the value false.

any (mask)   returns the value true if any of the elements of the logical array `mask` is true, and returns the value false if no elements are true or if `mask` has size zero.

count (mask)   returns the default integer value that is the number of elements of the logical array `mask` that have the value true.

iall (array)   returns an integer value in which each bit is 1 if all the corresponding bits of the elements of the integer array `array` are 1, and is 0 otherwise.

iany (array)   returns an integer value in which each bit is 1 if any of the corresponding bits of the elements of the integer array `array` are 1, and is 0 otherwise.

iparity (array)   returns an integer value in which each bit is 1 if the number of corresponding bits of the elements of the integer array `array` is odd, and is 0 otherwise.

maxval (array)   returns the maximum value of an element of an integer, real, or character array (character comparison is explained in Section 3.5). If `array` is of type integer or real and has size zero, the function returns the negative value of largest magnitude supported by the processor. If `array` is of type logical and has size zero, every character of the result is `char(0,kind(array))`.

minval (array) returns the minimum value of an element of an integer, real, or character array (character comparison is explained in Section 3.5). If array is of type integer or real and has size zero, the function returns the largest positive value supported by the processor. If array is of type logical and has size zero, every character of the result is char$(n-1,$ kind(array)) where n is the number of characters in the collating sequence.

norm2 (x) returns the L_2 norm of a real array x, that is, the square root of the sum of the squares of the elements. It returns the value zero if array has size zero. The standard recommends, but does not require, that norm2 be calculated without undue overflow or underflow.

parity (mask) returns the value true if an odd number of the elements of the logical array mask are true, and false otherwise.

product (array) returns the product of the elements of an integer, real, or complex array. It returns the value one if array has size zero.

sum (array) returns the sum of the elements of an integer, real, or complex array. It returns the value zero if array has size zero.

9.13.2 Additional arguments dim and kind

Each of the functions of Section 9.13.1 except count has an alternative form with an additional second argument dim, as in

sum (array, dim)

where dim is a scalar integer. The operation is applied to all rank-one sections that span right through dimension dim to produce an array of rank reduced by one and extents equal to the extents in the other dimensions, or a scalar if the original rank is one. For example, if a is a real array of shape (4,5,6), sum(a,dim=2) is a real array of shape (4,6) and element (i, j) has value sum(a(i,:,j)).

The function count is a little different. It has the form

count (mask [,dim] [,kind])

which allows control of the kind of the result. The optional argument dim plays the same role as in the other functions. Because the rank of the result depends on whether dim is specified (unless the original is rank one), the corresponding actual argument is not permitted itself to be an optional dummy argument, a disassociated pointer, or an unallocated allocatable object.[2] If the argument kind of count is present, it must be a scalar integer constant expression and specifies the kind of the result, which is otherwise of default kind.

---

[2]This restriction is not needed for the other functions because for them dim is not an optional argument, thus a pointer must be associated, an allocatable must be allocated, and an optional argument must be present.

9.13.3   Optional argument `mask`

For both forms, the functions `iall`, `iany`, `iparity`, `maxval`, `minval`, `product`, and `sum` have an optional final argument `mask` as in

sum (array [,mask]) or **sum (array, dim [,mask])**

If `mask` is present, it must be conformable with the first argument and the operation is applied to the elements corresponding to true elements of `mask`; for example, `sum(a, mask = a>0)` sums the positive elements of the array `a`. The argument `mask` affects only the value of the function and does not affect the evaluation of arguments that are array expressions. Indeed, any element not selected by the mask need not be defined at the time the function is invoked.

9.13.4   Generalized array reduction

The transformational intrinsic function `reduce` performs user-defined array reductions.

reduce (array, operation [,mask] [,identity] [,ordered]) or

reduce (array, operation, dim [,mask] [,identity] [,ordered])

>   **array**   is an array of any type.

>   **operation**   is a pure function that provides the binary operation for reducing the array. It is recommended that the operation be mathematically associative.[3] The function must have two scalar arguments and a scalar non-coarray result, all non-polymorphic with the same declared type and type parameters as `array`. The arguments must not have the `allocatable`, `optional`, or `pointer` attributes; if one argument has the `asynchronous`, `target`, or `volatile` attribute, the other must have the same attribute.

>   **dim**   is an integer scalar with value $1 \leq$ dim $\leq n$, where n is the rank of `array`.

>   **mask**   is logical and conformable with `array`.

>   **identity**   is a scalar of the type and type parameters of `array`.

>   **ordered**   is a logical scalar.

The first form of this function returns a scalar of the declared type and type parameters of `array`. The result value is produced by taking the sequence of elements of `array` in array element order, and repeatedly combining adjacent values of the sequence by applying `operation` until only a single value is left. If `ordered` is present with the value true, the first value of the sequence is repeatedly combined with the next; otherwise any adjacent values could be combined (this is intended to permit efficient evaluation in parallel). If `mask` is present, the initial sequence consists only of those elements of `array` for which `mask` is true; elements not selected need not be defined. If the initial sequence is empty, the result is `identity` if it is present; otherwise, error termination is initiated.

---

[3]The operation need not be computationally associative – floating-point addition, for example.

If dim appears, the effect is as for the functions of Section 9.13.1. An alternative way to describe it is that the result is an array with rank one less than array, and with the shape of array with dimension dim removed, that is,

[size(array,1)...size(array,dim-1), size(array,dim+1) ...size(array,n)]

Each element $(i_1, ..., i_{dim-1}, i_{dim+1}, ..., i_n)$ of the result has the value of applying reduce to array $(i_1, ..., i_{dim-1}, :, i_{dim+1}, ..., i_n)$.

9.14 Array inquiry functions

9.14.1 Introduction

There are five functions for inquiries about the array properties (contiguity, bounds, shape, and size) of an object of any type. Because the result depends on only the array properties, the value of the array need not be defined.

9.14.2 Contiguity

Contiguity of the elements of an array can be tested.

is_contiguous (array) where array is an array of any type, returns a default logical scalar with the value true if array is contiguous, and false otherwise. If array is a pointer, it must be associated with a target.

9.14.3 Bounds, shape, and size

The following functions inquire about the array properties of an object. If allocatable, it must be allocated; and in the case of a pointer, it must be associated with a target. An array section or an array expression is taken to have lower bounds 1 and upper bounds equal to the extents (like an assumed-shape array with no specified lower bounds). If a dimension has size zero, the lower bound is taken as 1 and the upper bound is taken as 0. If it is of assumed rank and is associated with an assumed-size array, the final extent is considered to be -1.

lbound (array [, dim]) when dim is absent, returns a rank-one default integer array holding the lower bounds. When dim is present, it must be a scalar integer with value in the range $1 \leq \dim \leq n$ where n is the rank of array and the result is a scalar default integer holding the lower bound in dimension dim. As the rank of the result depends on whether dim is specified, the corresponding actual argument must not itself be an optional dummy argument, a disassociated pointer, or an unallocated allocatable object.

rank (a) returns a default integer scalar whose value is the rank of a, which may be a scalar or array of any type. It is intended for use with arguments of assumed rank (Section 7.18).

shape (source) returns a rank-one default integer array holding the shape of the array or scalar source, which is not permitted to be assumed-size. In the case of a scalar, the result has size zero.

size (array [,dim]) returns a scalar default integer that is the size of the array array or extent along dimension dim if the scalar integer dim is present. When dim is present, it must have a value in the range $1 \leq dim \leq n$ where n is the rank of array. If array is assumed-size, dim must be present and not have the value n.

ubound (array [,dim]) is similar to lbound except that it returns upper bounds and if array is assumed-size, dim must be present and not have the value equal to the rank of array.

Each of these functions has an optional kind argument at the end of its argument list. This argument specifies the kind of integer result the function returns. This is useful if a default integer is not big enough to contain the correct value (which may be the case on 64-bit machines). For example, in the code

```
real, allocatable :: a(:,:,:,:)
allocate (a(64,1024,1024,1024))
   ⋮
print *, size(a, kind=selected_int_kind(12))
```

the array a has a total of 2^{36} elements; on most machines this is bigger than huge(0), so the kind argument is needed to get the right answer from the reference to the intrinsic function size.

9.15   Array construction and manipulation functions

9.15.1   The merge elemental function

merge (tsource, fsource, mask) is an elemental function. The argument tsource may have any type and fsource must have the same type and type parameters. The argument mask must be of type logical. The result is tsource if mask is true and fsource otherwise. If tsource or fsource is not selected, it need not be defined.

The principal application of merge is when the three arguments are arrays having the same shape, in which case tsource and fsource are merged under the control of mask. Note, however, that tsource or fsource may be scalar while mask is an array, in which case the elemental rules effectively broadcast it to an array of the correct shape.

9.15.2   Packing and unpacking arrays

The transformational function pack packs into a rank-one array those elements of an array that are selected by a logical array of conforming shape, and the transformational function unpack performs the reverse operation. The elements are taken in array element order.

pack (array, mask [,vector]) returns a rank-one array of the type and type parameters of array. If vector is absent, the result contains the elements of array corresponding to true elements of mask in array element order; mask may be scalar with value true, in which case all elements are selected; the elements not selected need not be defined. If vector is present, it must be a rank-one array of the same type and type parameters as array and size at least equal to the number t of selected elements; the result has size equal to the size n of vector; if $t < n$, elements i of the result for $i > t$ are the corresponding elements of vector.

unpack (vector, mask, field) returns an array of the type and type parameters of vector and shape of mask. The argument mask must be a logical array and vector must be a rank-one array of size at least the number of true elements of mask; field must be of the same type and type parameters as vector and must either be scalar or be of the same shape as mask. The element of the result corresponding to the ith true element of mask, in array element order, is the ith element of vector; all others are equal to the corresponding elements of field if it is an array or to field if it is a scalar.

9.15.3 Reshaping an array

The transformational function reshape allows the shape of an array to be changed, with possible permutation of the subscripts.

reshape (source, shape [,pad] [,order]) returns an array with shape given by the rank-one integer array shape, and type and type parameters those of the array source. The size of shape must be constant, and its elements must not be negative. If pad is present, it must be an array of the same type and type parameters as source. If order is absent, the elements of the result, in array element order, are the elements of source in array element order followed by copies of pad in array element order. If order is present, it must be a rank-one integer array with a value that is a permutation of $(1,2,\ldots,n)$; the elements $r(s_1,\ldots,s_n)$ of the result, taken in subscript order for the array having elements $r(s_{order(1)},\ldots,s_{order(n)})$, are those of source in array element order followed by copies of pad in array element order. For example, if order has the value (3,1,2), the elements $r(1,1,1)$, $r(1,1,2)$, ..., $r(1,1,k)$, $r(2,1,1)$, $r(2,1,2)$, ... correspond to the elements of source and pad in array element order.

9.15.4 Transformational function for replication

spread (source, dim, ncopies) returns an array of type and type parameters those of source and of rank increased by one. The argument source may be scalar or array valued. The arguments dim and ncopies are integer scalars. The value of dim must be in the range $1 \leq dim \leq n+1$ where n is the rank of source. The result contains max (ncopies, 0) copies of source, and element $(r_1,\ldots,r_{n+1})$ of the result is source $(s_1,\ldots,s_n)$ where $(s_1,\ldots,s_n)$ is $(r_1,\ldots,r_{n+1})$ with subscript dim omitted (or source itself if it is scalar).

9.15.5 Array shifting functions

cshift (array, shift [,dim]) returns an array of the same type, type parameters, and shape as array. The argument shift is of type integer and must be scalar if array is of rank one. The argument dim is an integer scalar with value in the range $1 \leq \dim \leq n$ where n is the rank of array; if omitted, it is as if it were present with the value 1. If shift is scalar, the result is obtained by shifting every rank-one section that extends across dimension dim circularly shift times. The direction of the shift depends on the sign of shift, being to the left for a positive value and to the right for a negative value. Thus, for the case with shift=1 and array of rank one, lower bound one, and size m, the element i of the result is array(i+1), where $i = 1, 2, \ldots, m-1$, and element m is array(1). If shift is an array, it must have the same shape as that of array with dimension dim omitted, and it supplies a separate value for each shift. For example, if array is of rank three and shape (k, l, m) and dim has the value 2, shift must be of shape (k, m) and supplies a shift for each of the $k \times m$ rank-one sections in the second dimension of array.

eoshift (array, shift [,boundary] [,dim])   is identical to cshift except that an end-off shift is performed and boundary values are inserted into the gaps so created. The argument boundary may be omitted when array has intrinsic type, in which case the value zero is inserted for the integer, real, and complex cases; false in the logical case; and blanks in the character case. If boundary is present, it must have the same type and type parameters as array; it may be scalar and supply all needed values or it may be an array whose shape is that of array with dimension dim omitted and supply a separate value for each shift.

9.15.6   Matrix transpose

The transpose function performs a matrix transpose for any array of rank two.

transpose (matrix)   returns an array of the same type and type parameters as the rank-two array matrix. If matrix has shape $[m, n]$, the result has shape $[n, m]$. If matrix has lower bounds lb_1 and lb_2 (returned by lbound(matrix, 1) and lbound(matrix, 2)), element (i, j) of the result is matrix$(j + lb_1 - 1, i + lb_2 - 1)$.

9.16   Transformational functions for geometric location

There are three transformational functions that find the locations of the maximum and minimum values of an integer, real, or character array or of a given value in an array of any intrinsic type. They each have an optional logical argument mask that conforms with array and selects its elements; the elements not selected need not be defined. The functions each have an optional kind argument that must be a scalar integer constant expression and controls the kind of the result, which otherwise is default integer.

Also, they have a final, optional back argument to indicate whether the first or last occurrence is desired; it must be a scalar logical. For example, maxloc([1, 4, 4, 1]) is equal to 2, whereas maxloc([1, 4, 4, 1], back=.true.) is equal to 3.

maxloc (array [,mask] [,kind] [,back])   returns a rank-one integer array of size equal to the rank of array. Its value is the sequence of subscripts of an element of maximum value (among those corresponding to true values of the conforming logical variable `mask` if it is present), as though all the declared lower bounds of `array` were 1. If there is more than one such element, the first (or last if `back` is present with the value true) in array element order is taken. If there are no elements, the result has all elements zero.

maxloc (array, dim [,mask] [,kind] [,back]) returns an integer array of shape equal to that of `array` with dimension `dim` omitted, where `dim` is a scalar integer with value in the range $1 \leq \text{dim} \leq \text{rank(array)}$, or a scalar if the original rank is one. The value of each element of the result is the position of the first (or last if `back` is present with the value true) element of maximum value in the corresponding rank-one section spanning dimension `dim`, among those elements corresponding to true values of the conforming logical variable `mask` when it is present. If there are no such elements, the value is zero.

minloc (array [,mask] [,kind] [,back]) is identical to `maxloc (array [,mask] [,kind] [,back])` except that the position of an element of minimum value is obtained.

minloc (array, dim [,mask] [,kind] [,back]) is identical to `maxloc (array, dim [,mask] [,kind] [,back])` except that positions of elements of minimum value are obtained.

findloc (array, value [,mask] [,kind] [,back]) searches `array`, possibly masked by `mask`, for the first (or last if `back` is present with the value true) element with value `value`, and returns the vector of subscript positions identifying that element or a vector of zeros if there is no such element. The `array` argument must be of intrinsic type, and `value` must be a scalar such that testing with the operator `==` or `.eqv.` is valid (see Section 3.6). If present, `mask` must be of type logical and conformable with `array`, `kind` must be a scalar integer constant expression, and `back` must be a scalar logical. For example, `findloc([(i, i = 10, 1000, 10)], 470)` has the value 47.

findloc (array, value, dim [,mask] [,kind] [,back]) returns an integer array of shape equal to that of `array` with dimension `dim` omitted, where `dim` is a scalar integer with value in the range $1 \leq \text{dim} \leq \text{rank(array)}$, or a scalar if the original rank is one. The value of each element of the result is the position of the first (or last if `back` is present with the value true) element with value `value` in the corresponding rank-one section of `array` spanning dimension `dim`, masked by `mask` when it is present, or zero if there is no such element.

9.17 Transformational function for disassociated or unallocated entities

null ([mold]) returns a disassociated pointer or an unallocated allocatable entity. The argument `mold` is a pointer or allocatable. The type, type parameter, and rank of the result are those of `mold` if it is present and otherwise are those of the object with which it is associated. In an actual argument associated with a dummy argument of assumed character length, `mold` must be present.

9.18   Non-elemental intrinsic subroutines

9.18.1   Introduction

There are also non-elemental intrinsic subroutines in Fortran, which were chosen to be subroutines rather than functions because of the need to return information through the arguments.

9.18.2   Real-time clock

There are two subroutines that return information from the real-time clock, the first based on the ISO/IEC 8601 standard for the representation of dates and times. It is assumed that it is a basic system clock that is incremented by one for each clock count until a maximum count_max is reached and on the next count is set to zero. Default values are returned if there is no clock. All the arguments have intent out. Whether an image has no clock, has a single clock of its own, or shares a clock with another image, is processor dependent.

call date_and_time ([date] [,time] [,zone] [,values]) returns the following (with default values blank or -huge(0), as appropriate, when there is no clock):

> **date** is a default character scalar variable. It is assigned the value of the century, year, month, and day in the form *ccyymmdd*.

> **time** is a default character scalar variable. It is assigned the value of the time as hours, minutes, seconds, and milliseconds in the form *hhmmss.sss*.

> **zone** is a default character scalar variable. It is assigned the value of the difference between local time and UTC (also known as Greenwich Mean Time) in the form *Shhmm*, corresponding to sign, hours, and minutes. For example, a processor in New York in winter would return the value -0500.

> **values** is a rank-one integer array, of size at least eight and with a decimal exponent range of at least four, holding the following sequence of values: the year, the month of the year, the day of the month, the time difference in minutes with respect to UTC, the hour of the day, the minutes of the hour, the seconds of the minute, and the milliseconds of the second.

call system_clock ([count] [,count_rate] [,count_max]) returns the following:

> **count** is a scalar integer[4] holding a processor-dependent non-negative value based on the current value of the processor clock, or -huge(0) if there is no clock. On the first call, the processor may set an initial value that may be zero. It is incremented by one for each clock count until a maximum value is reached and is reset to zero at the next count.

---

[4]We recommend a kind that has a range of at least 18 decimal digits in order to accommodate the high clock rate that many systems have.

count_rate is a scalar integer[4] or real[5] holding an approximation to the number of clock counts per second, or zero if there is no clock.

count_max is a scalar integer[4] holding the maximum value that count may take, or zero if there is no clock.

9.18.3   CPU time

There is a non-elemental intrinsic subroutine that returns the processor time used.

call cpu_time (time) returns the following:

time is a real scalar with intent out that is assigned a processor-dependent approximation to the processor time in seconds, or a processor-dependent negative value if there is no clock. Whether the value assigned is an approximation to the amount of time used by an invoking image, or the amount of time used by the whole program, is processor dependent.

The exact definition of time is left imprecise because of the variability in what different processors are able to provide. The primary purpose is to compare different algorithms on the same computer or discover which parts of a calculation on a computer are the most expensive. Sections of code can be timed, as in the example

```
real :: t1, t2
⋮
call cpu_time(t1)
⋮                    ! Code to be timed.
call cpu_time(t2)
write (*,*) 'Time taken by code was ', t2-t1, ' seconds'
```

9.18.4   Random numbers

On each image, a sequence of pseudorandom numbers is generated from a seed that is held as a rank-one array of integers whose size is chosen by the processor. The subroutine random_init controls the initiation of this array, the subroutine random_number returns a pseudorandom number, and the subroutine random_seed allows an inquiry to be made about the size or value of the seed array, and the seed to be reset. The subroutines provide a portable interface to a processor-dependent sequence. They affect the value of the seed only on the invoking image.

call random_number (harvest) returns a pseudorandom number from the uniform distribution over the range $0 \leq x < 1$ or an array of such numbers; harvest has intent out, may be a scalar or an array, and must be of type real.

---

[5]This is to provide a more accurate result on systems whose clock does not tick an integral number of times each second.

`call random_seed ([size] [put] [get])` has the following arguments:

> **size** has intent out and is a scalar default integer that the processor sets to the size n of the seed array.
>
> **put** has intent in and is a default integer array of rank one and size at least n that is used by the processor to reset the seed. A processor may set the same seed value for more than one value of put.
>
> **get** has intent out and is a default integer array of rank one and size at least n that the processor sets to the current value of the seed. This value can be used later as put to replay the sequence from that point, or in a subsequent program execution to continue from that point.

No more than one argument may be specified; if no argument is specified, the seed is set to a processor-dependent value.

`call random_init (repeatable, image_distinct)`

> **repeatable** is a logical scalar of intent in.
>
> **image_distinct** is a logical scalar of intent in.

The call is equivalent to invoking random_seed with a processor-dependent value for put. There are four cases for the way the value of put is chosen:

> i) image_distinct true and repeatable true. The value of put is different on different invoking images but in each execution of the program with the same execution environment the value is the same if the invoking image index value in the initial team is the same.
>
> ii) image_distinct false and repeatable true. The value of put is the same on every invoking image and is the same in each execution of the program with the same execution environment.
>
> iii) image_distinct true and repeatable false. The value of put is different on each image and different on subsequent invocations.
>
> iv) image_distinct false and repeatable false. The value of put does not depend on the invoking image, and differs on subsequent invocations.

Which of these four cases applies if random_init has not been invoked is processor dependent.

9.18.5   Executing another program

The ability to execute another program from within a Fortran program is provided by the intrinsic subroutine execute_command_line; as its name suggests, this passes a 'command line' to the processor which will interpret it in a totally system-dependent manner. For example,

```
call execute_command_line ('ls -l')
```

is likely to produce a directory listing on Unix and an error message on Windows. The full syntax is as follows:

call execute_command_line (command `[,wait]` `[,exitstat]` `[,cmdstat]`
`[,cmdmsg]`) where the arguments are as follows:

command has intent `in` and is a scalar default character string containing the command line to be interpreted by the processor.

wait has intent `in` and is a scalar logical indicating whether the command should be executed asynchronously (`wait=.false.`), or whether the procedure should wait for it to terminate before returning to the Fortran program (the default).

exitstat has intent `inout` and is a scalar integer variable (with a decimal exponent range of at least nine) that, unless `wait` is false, will be assigned the 'process exit status' from the command (the meaning of this is also system dependent).

cmdstat has intent `out` and is a scalar integer variable, with a decimal exponent range of at least four, that is assigned zero if `execute_command_line` itself executed without error, −1 if the processor does not support command execution, −2 if `wait=.true.` was specified but the processor does not support asynchronous command execution, and a positive value if any other error occurred.

cmdmsg has intent `inout` and is a scalar default character string to which, if `cmdstat` is assigned a positive value, is assigned an explanatory message.

If any error occurs (such that a nonzero value would be assigned to `cmdstat`) and `cmdstat` is not present, the program is error-terminated.

Note that even if the processor supports asynchronous command execution, there is no mechanism provided for finding out later whether the command being executed asynchronously has terminated or what its exit status was.

9.19 Access to the computing environment

9.19.1 Environment variables

Most operating systems have some concept of an *environment variable*, associating names with values. Access to these is provided by an intrinsic subroutine.

call get_environment_variable (name `[,value]` `[,length]` `[,status]`
`[,trim_name]` `[,errmsg]`) where the arguments are defined as follows:

name has intent `in` and is a scalar default character string containing the name of the environment variable to be retrieved. Trailing blanks are not significant unless `trim_name` is present and false. Case may or may not be significant.

value has intent `out` and is a scalar default character variable; it receives the value of the environment variable (truncated or padded with blanks if the `value` argument is shorter or longer than the environment variable's value). If there is no such variable, there is such a variable but it has no value, or the processor does not support environment variables, this argument is set to blanks.

length has intent out and is a scalar integer variable with a decimal exponent range of at least four; if the specified environment variable exists and has a value, the length argument is set to the length of that value, otherwise it is set to zero.

status has intent out and is a scalar integer with a decimal exponent range of at least four; it receives the value 1 if the environment variable does not exist, 2 if the processor does not support environment variables, a number greater than 2 if an error occurs, −1 if the value argument is present but too short, and zero otherwise (indicating that no error or warning condition has occurred).

trim_name has intent in and is a scalar of type logical; if this is false, trailing blanks in name will be considered significant if the processor allows environment variable names to contain trailing blanks.

errmsg is a scalar intent inout argument of type default character and returns an error message if an error occurs. Otherwise, errmsg is left unchanged.

We note that it is processor dependent whether an environment variable that exists on an image also exists on another image and, if it does exist on both images, whether the values are the same.

9.19.2   Information about the program invocation

Two different methods of retrieving information about the command are provided, reflecting the two approaches in common use.

The Unix-like method is provided by two procedures: a function which returns the number of command arguments and a subroutine which returns an individual argument. These are:

command_argument_count () is transformational and returns, as a scalar default integer, the number of command arguments. If the result is zero, either there were no arguments or the processor does not support the facility. If the command name is available as an argument, it is not included in this count.

call get_command_argument (number [,value] [,length] [,status] [,errmsg]) where the arguments are defined as follows:

number has intent in and is a scalar integer indicating the number of the argument to return. If the command name is available as an argument, it is number zero.

value has intent out and is a scalar default character variable; it receives the value of the indicated argument (truncated or padded with blanks if the character variable is shorter or longer than the command argument) or blanks if there is no such argument.

length has intent out and is a scalar integer variable with a decimal exponent range of at least four; it receives the length of the indicated argument or zero if there is no such argument.

status has intent out and is a scalar integer variable with a decimal exponent range of at least four; it receives a positive value if that argument cannot be retrieved,

−1 to indicate that the `value` variable was shorter than the command argument, and zero otherwise.

errmsg is a scalar intent `inout` argument of type default character and returns an error message if an error occurs. In the case of a warning situation that would assign −1 to the argument `status`, `errmsg` is left unchanged.

The other paradigm for command processing provides a simple command line, not broken up into arguments. This is retrieved by the intrinsic subroutine

call get_command (*[command]* **[,length]** **[,status]** **[,errmsg]**) where the arguments are defined as follows:

command has intent `out` and is a scalar default character variable; it receives the value of the command line (truncated or padded with blanks if the variable is shorter or longer than the actual command line).

length has intent `out` and is a scalar integer variable with a decimal exponent range of at least four; it receives the length of the actual command line, or zero if the length cannot be determined.

status has intent `out` and is a scalar integer variable with a decimal exponent range of at least four; it receives a positive value if the command line cannot be retrieved, −1 if `command` was present but the variable was shorter than the length of the actual command line, and zero otherwise.

errmsg is a scalar intent `inout` argument of type default character and returns an error message if an error occurs. Otherwise, `errmsg` is left unchanged.

The result of calling one of the procedures in this section does not depend on the executing image.

9.20 Elemental functions for input/output status testing

Two elemental intrinsic functions are provided for testing the input/output status value returned through the `iostat=` specifier (Section 10.7). Both functions accept an argument of type integer, and return a default logical result.

is_iostat_end (i) returns the value true if `i` is an input/output status value that corresponds to an end-of-file condition, and false otherwise.

is_iostat_eor (i) returns the value true if `i` is an input/output status value that corresponds to an end-of-record condition, and false otherwise.

9.21 Size of an object in memory

storage_size (a *[,kind]*) returns the size, in bits, that would be taken in memory by an array element with the dynamic type and type parameters (Section 15.3.2) of a.

The argument a may be of any type or rank (including a scalar). It is permitted to be an undefined pointer unless it is polymorphic, and is permitted to be a disassociated pointer or unallocated allocatable unless it has a deferred type parameter or is unlimited polymorphic (Section 15.3.5).

The return type is integer with the specified kind, or default kind if kind is not present.

Note that the standard does not require the same size for named variables, array elements, and structure components of the same type; indeed, frequently these will have different padding to improve memory address alignment and thus performance.

Furthermore, if a is of a derived type with allocatable components or components whose size depends on the value of a length type parameter, the compiler is allowed to store those components separately from the rest of the variable, with a descriptor in the variable pointing to the additional storage. It is unclear whether storage_size will include the space taken up by such components, especially in the length type parameter case. Therefore, use of this function should be avoided for such problematic cases.

9.22   Miscellaneous procedures

The following procedures, introduced elsewhere, are fully described in the sections listed:

- the subroutine move_alloc, Section 6.8;
- the subroutine event_query, Section 17.15;
- the inquiry functions extends_type_of and same_type_as, Section 15.13;
- the atomic subroutines atomic_add, atomic_and, atomic_cas, atomic_or, atomic_fetch_add, atomic_fetch_and, atomic_fetch_or, atomic_fetch_xor, and atomic_xor, Section 17.23;
- the inquiry functions coshape, image_index, lcobound, and ucobound, Section 17.21.2;
- the transformational functions num_images and this_image, Section 17.21.3;
- the collective subroutines co_broadcast, co_max, co_min, co_reduce, and co_sum, Section 17.22;
- the transformational functions failed_images and stopped_images, and the elemental function image_status, Section 18.9;
- the functions get_team and team_number, Section 18.8.1;
- the double-precision functions dble and dprod, Section A.4; and
- the atomic subroutines atomic_define and atomic_ref, Section A.16.

9.23   Intrinsic modules

An intrinsic module is one that is provided by the Fortran processor instead of the user or a third party. A Fortran processor provides at least five intrinsic modules, iso_fortran_env, ieee_arithmetic, ieee_exceptions, ieee_features, and iso_c_binding, and may provide additional intrinsic modules.

The intrinsic module `iso_fortran_env` provides information about the Fortran environment, and is described in the next section. The IEEE modules provide access to facilities from the IEEE arithmetic standard and are described in Chapter 19. The intrinsic module `iso_c_binding` provides support for interoperability with C and is described in Chapter 20.

It is possible for a program to use an intrinsic module and a user-defined module of the same name, though they cannot both be referenced from the same scoping unit. To use an intrinsic module in preference to a user-defined one of the same name, the `intrinsic` keyword is specified on the `use` statement, for example

```
use, intrinsic :: ieee_arithmetic
```

Similarly, to ensure that a user-defined module is accessed in preference to an intrinsic module, the `non_intrinsic` keyword is used, for example:

```
use, non_intrinsic :: random_numbers
```

If both an intrinsic module and a user-defined module are available with the same name, a `use` statement without either of these keywords accesses the user-defined module. However, should the compiler not be able to find the user's module it would access the intrinsic one instead without warning; therefore we recommend that programmers avoid using the same name for a user-defined module as that of a known intrinsic module (or that the `non_intrinsic` keyword be used).

Intrinsic modules should always be used with an `only` clause (Section 8.13), as vendors or future standards could make additions to the module.

9.24   Fortran environment

9.24.1   Introduction

The intrinsic module `iso_fortran_env` provides information about the Fortran environment.

9.24.2   Named constants

The following named constants, which are scalars of type default integer, are useful in the contexts of data transfer (Chapter 10), external files (Chapter 12), and coarray processing (Chapter 17).

character_storage_size   The size in bits of a character storage unit, Section A.8.

current_team   The value that identifies the current team when it is used as the `level` argument to `get_team`, Section 18.8.1.

error_unit   The unit number for a preconnected output unit suitable for reporting errors.

file_storage_size   The size in bits of a file storage unit (the unit of measurement for the record length of an external file, as used in the `recl=` clause of an `open` or `inquire` statement).

initial_team The value that identifies the initial team when it is used as the `level` argument of `get_team`.

input_unit The unit number for the preconnected standard input unit (the same one that is used by `read` without a unit number, or with a unit specifier of `*`).

iostat_end The value returned by `iostat=` to indicate an end-of-file condition.

iostat_eor The value returned by `iostat=` to indicate an end-of-record condition.

iostat_inquire_internal_unit The value returned in an `iostat=` specifier by an `inquire` statement to indicate that a file unit number identifies an internal unit.

numeric_storage_size The size in bits of a numeric storage unit, Section A.8.

output_unit The unit number for the preconnected standard output unit (the same one that is used by `print`, or by `write` with a unit specifier of `*`).

parent_team The value that identifies the parent team when it is used as the `level` argument of `get_team`.

stat_failed_image This value is positive if failed image handling (Section 17.24) is supported and negative otherwise. It is the value returned in a `stat` specifier or a `stat` argument if a failed image is detected.

stat_locked The value returned by a `stat=` variable in a `lock` statement for a variable locked by the executing image, Section 17.13.

stat_locked_other_image The value returned by a `stat=` variable in a `lock` statement for a variable locked by another image, Section 17.13.

stat_stopped_image The value returned by a `stat=` variable or a `stat` argument to indicate an image that has initiated normal termination, Section 17.19. It is positive.

stat_unlocked The value returned by a `stat=` variable in a `unlock` statement for an unlocked variable, Section 17.13.

stat_unlocked_failed_image The value returned by a `stat=` variable in a `lock` statement if the lock variable is unlocked because of the failure of the image that locked it.

Unlike unit numbers chosen by the user, the special unit numbers might be negative, but they will not be -1 (this is because -1 is used by the `number=` clause of the `inquire` statement to mean that there is no unit number). The error reporting unit `error_unit` might be the same as the standard output unit `output_unit`.

9.24.3 Compilation information

Two transformational functions are available in the module `iso_fortran_env` to return information about the compiler (the so-called program translation phase).

`compiler_version ()` returns a string describing the name and version of the compiler used.

`compiler_options ()` returns a string describing the options used during compilation.

In each case the string is a default character scalar.

These functions may be used in constant expressions, for example

```
module my_module
   use iso_fortran_env, only: compiler_options, compiler_version
   private compiler_options, compiler_version
   character(*), parameter :: compiled_by = compiler_version()
   character(*), parameter :: compiled_with = compiler_options()
   :
end module
```

There are no actual requirements on the length of these strings or on their contents, but it is expected that they will contain something useful and informative. For example `compiler_version()` could return the string `'NAG Fortran 6.0(1273)'`, and `compiler_options()` could return the string `'-C=array -O3'`.

9.24.4 Names for common kinds

Named constants for some kind values for integer and real types are available in the module `iso_fortran_env`; these are the default integer scalars:[6]

`int8`	8-bit integer
`int16`	16-bit integer
`int32`	32-bit integer
`int64`	64-bit integer
`real32`	32-bit real
`real64`	64-bit real
`real128`	128-bit real

For example, in the code in Figure 9.1 the use of `int64` allows the subroutine to process very large arrays.

If the compiler supports more than one kind with a particular size, the standard does not specify which one will be chosen for the constant. If the compiler does not support a kind with a particular size, that constant will have a value of -2 if it supports a kind with a larger size, and -1 if it does not support any larger size.

This can be used together with `merge` to specify a desired size with a fallback to other predetermined sizes if that one is not available, as shown in Figure 9.2.

---

[6] In Fortran 2023, there are also four kind values for logical types and one for an additional real type.

Figure 9.1 Use of int64.

```
subroutine process(array)
  use iso_fortran_env
  real array(:, :)
  integer(int64) i, j
  do j=1, ubound(array, 2, int64)
    do i=1, ubound(array, 1, int64)
      : ! do something with array(i, j)
    end do
  end do
end subroutine
```

Figure 9.2 Kind selection with standard named constants.

```
subroutine process_bytes(bytes)
  use iso_fortran_env
  integer(merge(int8, merge(int16, int32, int16>=0), int8>=0)) bytes

  if (kind(bytes)==int8) then
    : ! process 8-bit bytes
  else if (kind(bytes)==int16) then
    : ! process 8-bit bytes in pairs
  else
    : ! process quadruples of 8-bit bytes
  end if
end subroutine
```

The named constants atomic_int_kind and atomic_logical_kind are default integer scalars available in the module iso_fortran_env for the kind values for integer and logical variables used as arguments for the intrinsic subroutines atomic_define and atomic_ref, Section A.16.

9.24.5 Kind arrays

Named array constants containing all the kind type parameter values for the intrinsic types that are supported by the processor are available in the intrinsic module iso_fortran_env. The named constants character_kinds, integer_kinds, logical_kinds, and real_kinds contain the supported kinds of type character, integer, logical, and real, respectively. These arrays are of type default integer, have a lower bound of one, and size equal to the number of kinds of the type. There is no array for type complex because there is a complex kind for each real kind. The order of values in each array is processor dependent.

9.24.6 Derived types for coarray programming

The following derived types are defined in the intrinsic module iso_fortran_env: lock_type, event_type, and team_type. Their properties are explained in Section 17.16.

9.25 Summary

In this chapter we introduced the six categories of intrinsic procedures, explained the intrinsic statement, and gave detailed descriptions of all the procedures. Intrinsic modules have also been introduced.

Exercises

1. Write a program to calculate the real roots or pairs of complex-conjugate roots of the quadratic equation $ax^2 + bx + c = 0$ for any real values of a, b, and c. The program should read these three values and print the results. Use should be made of the appropriate intrinsic functions.

2. Repeat Exercise 1 of Chapter 5, avoiding the use of do constructs.

3. Given the rules explained in Sections 3.13 and 9.2, what are the values printed by the following program?

```
program main
    real, target  :: a(3:10)
    real, pointer :: p1(:), p2(:)
!

    p1 => a(3:9:2)
    p2 => a(9:3:-2)
    print *, associated(p1, p2)
    print *, associated(p1, p2(4:1:-1))
end program main
```

4. In the following program, two pointer assignments, one to an array and the other to an array section, are followed by a subroutine call. Bearing in mind the rules given in Sections 3.13, 5.7.2, and 9.14.3, what values does the program print?

```
program main
    real, target  :: a(5:10)
    real, pointer :: p1(:), p2(:)
    p1 => a
    p2 => a(:)
    print *, lbound(a), lbound(a(:))
    print *, lbound(p1), lbound(p2)
    call what(a, a(:))
contains

    subroutine what(x, y)
        real, intent(in) :: x(:), y(:)
        print *, lbound(x), lbound(y)
    end subroutine what
end program main
```

5. Write a program that displays the sum of all the numbers on its command line.

6. Write a module that implements the standard `random_number` interface, for single and double precision real numbers, by the 'good, minimal standard' generator from 'Random Number Generators: Good Ones Are Hard to Find', S. K. Park and K. W. Miller, *Communications of the ACM* October 1988, 31 (10), 1192–1201. This is a parametric multiplicative linear congruential algorithm: $x_{\text{new}} = \text{mod}(16807 x_{\text{old}}, 2^{31} - 1)$.

10. Data transfer

10.1 Introduction

Fortran has, in comparison with many other high-level programming languages, a particularly rich set of facilities for input/output, but it is an area of Fortran into which not all programmers need to delve very deeply. For most small-scale programs it is sufficient to know how to read a few data records containing input variables, and how to transmit to a screen or printer the results of a calculation. In large-scale data processing, on the other hand, the programs often have to deal with huge streams of data to and from many files; in these cases it is essential that great attention be paid to the way in which the input/output is designed and coded, as otherwise both the execution time and the real time spent in the program can suffer dramatically. The term **external file** is used for a collection of data outside the main memory. A file is usually organized into a sequence of **records**, with access either sequential or direct. An alternative is provided by stream access, discussed in Section 10.16.

This chapter begins by discussing the various forms of **formatted input/output**, that is, input/output which deals with records that do not use the internal number representation of the computer but rather a character string that can be displayed. It is also the form usually needed for transmitting data between different kinds of computers. The so-called **edit descriptors**, which are used to control the translation between the internal number representation and the external format, are then explained. Finally, the topics of unformatted (or binary) input/output and direct-access files are covered.

10.2 Number conversion

The ways in which numbers are stored internally by a computer are the concern of neither the Fortran standard nor this book. However, if we wish to output values – to display them on a screen or to print them – then their internal representations must be converted into a character string that can be read in a normal way. For instance, the contents of a given computer word may be (in hexadecimal) 1d7dbf and correspond to the value -0.000450. For our particular purpose, we may wish to display this quantity as $-.000450$, or as $-4.5\text{E}-04$, or rounded to one significant digit as $-5\text{E}-04$. The conversion from the internal representation to the external form is carried out according to the information specified by an edit descriptor contained in a *format specification*. These will both be dealt with fully in the next chapter; for the moment, it is sufficient to give a few examples. For instance, to print an integer value in a field ten characters wide, we would use the edit descriptor i10, where i stands for integer

Modern Fortran Explained, 3rd Edition. M. Metcalf, J. Reid, M. Cohen, and R. Bader. Oxford University Press (2024). © M. Metcalf, J. Reid, M. Cohen, and R. Bader (2024). DOI 10.1093/oso/9780198876571.001.0010

conversion and 10 specifies the width of the output field. To print a real quantity in a field of ten characters, five of which are reserved for the fractional part of the number, we specify f10.5. The edit descriptor f stands for floating-point (real) conversion, 10 is the total width of the output field, and 5 is the width of the fractional part of the field. If the number given above were to be converted according to this edit descriptor, it would appear as *bb*-0.00045, where *b* represents a blank. To print a character variable in a field of ten characters we would specify a10, where a stands for alphanumeric conversion.

A format specification consists of a list of edit descriptors enclosed in parentheses, and can be coded either as a default character expression, for instance

```
'(i10, f10.3, a10)'
```

or as a separate format statement, referenced by a statement label, for example

```
10 format(i10, f10.3, a10)
```

To print the scalar variables j, b, and c, of types integer, real, and character, respectively, we may then write either

```
print '(i10, f10.3, a10)', j,b,c
```

or

```
print 10, j,b,c
10 format(i10, f10.3, a10)
```

The first form is normally used when there is only a single reference in an inclusive scope (Section 5.15) to a given format specification, and the second when there are several or when the format is complicated. The part of the statement designating the quantities to be printed is known as the *output list* and forms the subject of the following section.

10.3   Input/output lists

The quantities to be read or written by a program are specified in an input/output list. For output they may be expressions, but for input they must be variables. In both cases, list items may be implied-do lists of quantities. Examples are shown in Figure 10.1, where we note the use of a *repeat count* in front of those edit descriptors that are required repeatedly. A repeat count must be a positive integer literal constant and must not have a kind type parameter. Function references are permitted in an input/output list (but subject to conditions described in Section 11.7).

In all these examples, except the last one, the expressions consist of single variables and would be equally valid in input statements using the read statement, for example

```
read '(i10)', i
```

Such statements may be used to read values which are then assigned to the variables in the input list.

If an array appears as an item, it is treated as if the elements were specified in array element order. For example, the third of the print statements in Figure 10.1 could have been written

```
print '(3f10.3)', a(1:3)
```

Figure 10.1 Examples of formatted output.

```
integer                :: i
real, dimension(10)  :: a
character(len=20)    :: word
print '(i10)',      i
print '(10f10.3)', a
print '(3f10.3)',   a(1),a(2),a(3)
print '(a10)',      word(5:14)
print '(5f10.3)', (a(i), i=1,9,2)
print '(i10)',      i
print '(2f10.3)',   a(1)*a(2)+i, sqrt(a(3))
```

Figure 10.2 An invalid input item (array element appears twice).

```
integer :: j(10), k(3)
   :
k = (/ 1, 2, 1 /)
read '(3i10)', j(k)       ! Invalid because j(1) appears twice
```

However, no element of the array may appear more than once in an input item. Thus, the case in Figure 10.2 is not allowed.

If an allocatable object appears as an item, it must be allocated.

Any pointer in an input/output list must be associated with a target, and transfer takes place between the file and the target.

For formatted input/output,[1] unless defined input/output (Section 11.6) is in use, an item of derived type with no allocatable or pointer components at any level of component selection is treated as if the components were specified in the same order as in the type declaration. This rule is applied repeatedly for components of derived type, so that it is as if we specified the list of items of intrinsic type that constitute its ultimate components. For example, if p and t are of the types point and triangle of Figure 2.1, the statement

```
read '(8f10.5)', p, t
```

has the same effect as the statement

```
read '(8f10.5)', p%x, p%y, t%a%x, t%a%y, t%b%x,        &
                 t%b%y, t%c%x, t%c%y
```

During this expansion each component must be accessible (it may not, for example, be a private component of a public type).

An object in an input/output list is not permitted to be of a derived type that has an allocatable or pointer component at any level of component selection, unless that component is part of an item that would be processed by defined input/output. One reason for this restriction is that an input/output list item that is allocatable or a pointer is required to be

---

[1]Unformatted input/output handling of derived types is described in Section 10.12.

allocated or associated, so any unallocated or disassociated component would be problematic. Another is that these components usually vary in size dynamically, which could lead to ambiguous or confusing input/output. Finally, with a recursive data structure using pointers, either a disassociated pointer would be reached, or there would be a circular chain of pointers resulting in infinite input/output. Such types are better handled by doing the input/output individually for individual components with a procedure, or by using the defined input/output feature (see Section 11.6).

An input/output list may include an implied-do list, as illustrated by the fifth `print` statement in Figure 10.1. The general form is

$$(\textit{do-object-list}, \textit{do-var} = \textit{expr}, \textit{expr} [, \textit{expr}])$$

where each *do-object* is a variable (for input), an expression (for output), or is itself an implied-do list; *do-var* is a named scalar integer variable in the scope in which the statement lies;[2] and each *expr* is a scalar integer expression. The loop initialization and execution is the same as for a (possibly nested) set of do constructs (Section 4.4). For example, the value printed for i by the penultimate statement of Figure 10.1 is 11. In an input list, a variable that is an item in a *do-object-list* must not be a *do-var* of any implied-do list in which it is contained, nor be associated[3] with such a *do-var*. In an input or output list, no *do-var* may be a *do-var* of any implied-do list in which it is contained or be associated with such a *do-var*.

Note that a zero-sized array, or an implied-do list with a zero iteration count, may occur as an item in an input/output list. Such an item corresponds to no actual data transfer.

10.4   Format definition

In the `print` and `read` statements of the previous section, the format specification was given each time in the form of a character constant immediately following the keyword. In fact, there are three ways in which a format specification may be given. They are as follows.

A default character expression whose value commences with a format specification in parentheses:

> **A character constant** for example
>
> ```
> print '(f10.3)', q
> ```
>
> **A named character constant** for example
>
> ```
> character(len=*), parameter :: form='(f10.3)'
> ⋮
> print form, q
> ```
>
> **A character array** for example
>
> ```
> character :: carray(7) = (/ '(','f','1','0','.','3',')' /)
> ⋮
> print carray, q ! Elements of an array are concatenated.
> ```

---

[2]Unlike the implied-do in an array constructor (Section 7.15), no type declaration of *do-var* is permitted here.
[3]Such an association could be established by pointer association.

A character expression for example

```
character(4) :: carr1(10)
character(3) :: carr2(10)
integer      :: i, j
    :
    :
carr1(10) = '(f10'
carr2(3) = '.3)'
    :
    :
i = 10
j = 3
    :
    :
print carr1(i)//carr2(j), q
```

From these examples it may be seen that it is possible to program formats in a flexible way, and particularly that it is possible to use arrays, expressions, and also substrings in a way which allows a given format to be built up dynamically at execution time from various components. Any character data that might follow the trailing right parenthesis are ignored and may be undefined. In the case of an array, its elements are concatenated in array element order. However, on input *no* component of the format specification may appear also in the input list, or be associated with it. This is because the standard requires that the whole format specification be established *before* any input/output takes place. Further, no redefinition or undefinition of any characters of the format is permitted during the execution of the input/output statement.

An asterisk indicates a type of input/output known as **list-directed input/output**, in which the format is defined by the computer system at the moment the statement is executed, depending on both the type and magnitude of the entities involved. This facility is particularly useful for the input and output of small quantities of values, especially in temporary code which is used for test purposes, and which is removed from the final version of the program:

```
print *, 'Square-root of q = ', sqrt(q)
```

This example outputs a character constant describing the expression which is to be output, followed by the value of the expression under investigation. On the screen, this might appear as

```
Square-root of q = 4.392246
```

the exact format being dependent on the computer system used. Character strings in this form of output are normally undelimited, as if an a edit descriptor were in use, but an option in the `write` and `open` statements (Sections 10.8 and 12.4) may be used to require that they be delimited by apostrophes or quotation marks. Complex values are represented as two real values separated by a comma and enclosed in parentheses. Logical values are represented as T for true and F for false. Except for adjacent undelimited strings, values are separated by spaces or commas. The processor may represent a sequence of r identical values c by the form $r*c$. Further details of list-directed input/output are deferred until Section 10.9.

A statement label referring to a `format` statement containing the relevant specification between parentheses:

```
      print 100, q
      ⋮
  100  format(f10.3)
```

The `format` statement must appear in the same inclusive scope, before the `contains` statement if it has one. It is customary either to place each format statement immediately after the first statement which references it, or to group them all together just before the `contains` or end statement. A given `format` statement may be used by any number of formatted input/output statements, whether for input or for output.

Blank characters may precede the left parenthesis of a format specification, and may appear at any point within a format specification with no effect on the interpretation, except within a character string edit descriptor (Section 11.2).

10.5   Unit numbers

Input/output operations are used to transfer data between the variables of an executing program, as stored in the computer, and a file on an external medium. There are many types of external media: the screen, printer, hard disk, memory stick, and SSD are perhaps the most familiar. Whatever the device, a Fortran program regards each file from which it reads or to which it writes as a **unit**, and each unit, with two exceptions, has associated with it a **unit number**. Thus, we might associate the unit number 10 with a file on a memory stick from which we are reading, and the unit number 11 with a file on a hard disk to which we are writing. All program units of an executable program that refer to a particular unit number are referencing the same file. Many devices, such as a hard disk, may be referred to by more than one unit number, as they can hold many different files.

There are two input/output statements, `print` and a variant of `read`, that do not reference any unit number; these are the statements that we have used so far in examples, for the sake of simplicity. A `read` statement without a unit number will normally expect to read from the keyboard, unless the program is working in batch (non-interactive) mode, in which case there will be a disk file with a reserved name from which it reads. A `print` statement will normally expect to output to the screen, unless the program is in batch mode, in which case another disk file with a reserved name will be used. Such files are usually suitable for subsequent output on a physical output device. The system associates unit numbers with these default units (usually 5 for input and 6 for output) and for a unit suitable for reporting errors, and their actual values may be accessed from the intrinsic module `iso_fortran_env`, see Section 9.24.

Apart from these three special cases, all input/output statements must refer explicitly to a unit specified in an `open` statement (Section 12.4) in order to identify the file to which or from which data are to be transferred. The unit may be given in one of three forms. These are shown in the following examples, which use another form of read containing a unit specifier, *u*, and format specifier, *fmt*, in parentheses and separated by a comma, where *fmt* is a format specification as described in the previous section:

    read (u, fmt) list

The three forms of *u* are as follows:

A scalar integer expression that gives the unit number:

```
        read (4, '(f10.3)') q
        read (nunit, '(f10.3)') q
        read (4*i+j, 100) a
```

where the value may be any integer allowed by the system for this purpose.

An asterisk for example

```
        read (*, '(f10.3)') q
```

where the asterisk implies the standard input unit designated by the system (input_unit in iso_fortran_env), the same as that used for read without a unit number.

A character variable of default, ASCII, or ISO 10646 kind identifying an *internal file* (Section 10.6).

A long-standing inconvenience in Fortran programs has been the need to manually manage input/output unit numbers. This becomes a real problem when using older third-party libraries that perform input/output and for which the source code is unavailable; when opening a file, it is not difficult to find a unit number that is not currently in use, but it may be the same as one that is employed later by other code. This can be overcome by the newunit= specifier on the open statement (Section 12.4). This returns a unique negative unit number on a successful open. Being negative, it cannot clash with any user-specified unit number (these being required to be non-negative), and the processor will choose a value that does not clash with anything it is using internally. An example is:

```
    integer :: unit
    ⋮
    open (file='input.dat', newunit=unit)
    ⋮
    read (unit, '(f10.3)') q
```

10.6 Internal files

Data can be converted to/from character form by formatted input/output using a character variable as an **internal file**. Useful applications are to create a file name, create a character expression to use as a format, and prepare output lists containing mixed character and numerical data, all of which has to be prepared in character form, perhaps for output as a caption for a graph. The second application will now be described; the third will be dealt with in Section 10.8.

Imagine that we have to read a string of 30 digits, which might correspond to 30 one-digit integers, 15 two-digit integers, or 10 three-digit integers. The information as to which type of data is involved is given by the value of an additional digit, which has the value 1, 2, or 3, depending on the number of digits each integer contains. An internal file provides us with a mechanism whereby the 30 digits can be read into a character buffer area. The value of the final digit can be tested separately, and 30, 15, or 10 values read from the internal file, depending on this value. The basic code to achieve this might read as follows (no error recovery or data validation is included, for simplicity):

```
integer      :: ival(30), key, i
character(30):: buffer
character(6) :: form(3) = (/ '(30i1)', '(15i2)', '(10i3)' /)
read (*, '(a30,i1)')       buffer, key
read (buffer, form (key)) (ival(i), i=1,30/key)
```

Here, `ival` is an array which will receive the values, `buffer` a character variable of a length sufficient to contain the 30 input digits, and `form` a character array containing the three possible formats to which the input data might correspond. The first `read` statement reads 30 digits into `buffer` as character data, and a final digit into the integer variable `key`. The second `read` statement reads the data from `buffer` into `ival`, using the appropriate conversion as specified by the edit descriptor selected by `key`. The number of variables read from `buffer` to `ival` is defined by the implied-do loop, whose second specifier is an integer expression depending also on `key`. After execution of this code, `ival` will contain 30/`key` values, their number and exact format not having been known in advance.

If an internal file is a scalar, it has a single record whose length is that of the scalar. If it is an array, its elements, in array element order, are treated as successive records of the file and each has length equal to that of an array element. It must not be an array section with a vector subscript.

A record becomes defined when it is written. The number of characters sent must not exceed the length of the record. It may be less, in which case the rest of the record is padded with blanks. For list-directed output (Section 10.4), character constants are not delimited. A record may be read only if it is defined (which need not be by an output statement).

An internal file is always positioned at the beginning of its first record prior to data transfer (the array section notation may be used to start elsewhere in an array). Of course, if an internal file is allocatable or a pointer, it must be allocated or associated with a target. Also, no item in the input/output list may be in the file or associated with the file.

An internal file may be used for list-directed input/output (Section 10.9).

Numeric, logical, default character, ASCII character, and ISO 10646 (Section 3.7.3) character values may all be read from or written to such a variable. An example is shown in Figure 10.3.

Note that, although reading from an ISO 10646 internal file into a default character or ASCII character variable is possible, it is only allowed when the data being read is representable as default character or ASCII character.

Figure 10.3 Input/output of ISO 10646 characters.

```
subroutine japanese_date_stamp(string)
   integer, parameter :: ucs4 = selected_char_kind('ISO_10646')
   character(*, ucs4), intent(out) :: string
   integer                         :: val(8)
   call date_and_time(values=val)
   write (string, '(i0,a,i0,a,i0,a)')    &
                   val(1), '年', val(2), '月', val(3), '日'
end subroutine japanese_date_stamp
```

10.7 Formatted input

In the previous sections we have given complete descriptions of the ways that formats and units may be specified, using simplified forms of the read and print statements as examples. There are, in fact, two forms of the formatted read statement. Without a unit, it has the form

read *fmt* [, *input-list*]

and with a unit it has the form

read (*input-specifier-list*) [*input-list*]

where *input-list* is a list of variables and implied-do lists of variables. An *input-specifier* is one of:

[unit=]*u*	blank=*char-expr*	pad=*char-expr*
[fmt=]*fmt*	decimal=*char-expr*	pos=*int-expr*
[nml=]*ngn*	id=*int-variable*	rec=*int-expr*
advance=*char-expr*	iomsg=*char-variable*	round=*char-expr*
asynchronous=*char-expr*	iostat=*int-variable*	size=*int-variable*

where *u* and *fmt* are the unit and format specifiers described in Sections 10.4 and 10.5, *ngn* is a namelist group name, and each *char-expr* is a scalar expression of type default character and is interpreted without regard to case and with trailing blanks ignored. The unit specifier is required, and formatted input/output requires either a format specifier or a namelist group name; namelist is fully described in Section 10.10.

The *input-list* must include a unit specifier and must not include any specifier more than once. Specifier variables, as in iostat= and size=, must not be associated with each other (for instance, be identical), nor with any entity being transferred, nor with any *do-var* of an implied-do list of the same statement. If either of these variables is an array element, the subscript value must not be affected by the data transfer, implied-do processing, or the evaluation of any other specifier in the statement.

The keyword items may be specified in any order, although it is usual to keep the unit number and format specification as the first two. The unit number must be first if it does not have its keyword. If the format does not have its keyword, it must be second, following the unit number without its keyword.

The meanings of the specifiers other than *u*, *fmt*, and *ngn* are as follows:

- If `advance=` appears, *char-expr* must provide the value no for non-advancing input/output (Section 10.11) or yes for normal (advancing) input/output. If the specifier is absent, normal (advancing) input/output occurs.
- If `asynchronous=` appears, *char-expr* must be a constant expression that provides the value yes for asynchronous input/output (Section 10.15) or no for normal input/output. If the specifier is absent, normal input/output occurs.
- If `blank=` appears, *char-expr* must provide one of the values null and zero to specify the interpretation of embedded blanks in numeric fields (Section 11.5.2). If the specifier is absent, null is assumed.
- If `decimal=` appears, *char-expr* must provide one of the values point and comma to specify the decimal symbol in real numbers (Section 11.5.5). If the specifier is absent, point is assumed.
- If `id=` appears, `asynchronous=` must also appear and have the value yes; *int-variable* must be a scalar integer variable that is used to indicate whether the asynchronous input/output statement is complete (see Section 10.15).
- If `iomsg=` appears, *char-variable* identifies a scalar variable of type default character into which the processor places a message if an error, end-of-file, or end-of-record condition occurs during execution of the statement. If no such condition occurs, the value of the variable is not changed. Note that this is useful only for messages concerning error conditions, and an iostat= specifier is needed to prevent an error causing immediate termination.[4]
- If `iostat=` appears, *int-variable* must be a scalar integer variable with decimal exponent range at least four. After execution of the read statement, it has the negative value iostat_eor in iso_fortran_env if an end-of-record condition is encountered during non-advancing input (Section 10.11), the different negative value iostat_end in iso_fortran_env if an endfile condition was detected on the input device (Section 12.2.3), a positive value if an error was detected (for instance a formatting error), or the value zero otherwise. If an error or end-of-file condition occurs and iostat= is not specified, the program will terminate execution.[4]
- If `pad=` appears, *char-expr* must provide the value yes for padding of short input records (see Section 11.5.6) or no for no padding. The default is yes.
- If `pos=` appears the unit must be connected for stream access (Section 10.16) and *int-expr* must be an integer expression that provides a file position.
- If `rec=` appears the unit must be connected for direct access (Section 10.13) and *int-expr* must be an integer expression that provides a record number.
- If `round=` appears *char-expr* must provide one of the values up, down, zero, nearest, compatible, or processor_defined to specify the input/output rounding mode (see Section 11.5.3).
- If `size=` appears *int-variable* is a scalar integer variable that is defined with the number of characters read by the statement, not including any blanks inserted as padding.

---

[4]Program termination can also be avoided with err= or end= (Appendix A.12.3).

An example of a `read` statement with its associated error recovery is given in Figure 10.4, in which `last_file` and `error` are subroutines to deal with the exceptions. Subroutine `error` will normally be system dependent.

Figure 10.4 Testing for an error or the end of the file.

```
do
    read (nunit, '(3f10.3)', iostat=ios) a,b,c
    if (is_iostat_end(ios)) then
        call last_file(nunit, ios) ! End of file - if there is
            ! another file, open it on nunit and set ios to 0.
    else if (ios /= 0) then
        call error (ios) ! Error condition - take appropriate action.
        return
    end if
end do
! Successful read - continue execution.
```

If an error or end-of-file condition occurs on input, the statement terminates and all list items and any implied-do variables become undefined. If an end-of-file condition occurs for an external file, the file is positioned following the endfile record (Section 12.2.3); if there is otherwise an error condition, the file position is indeterminate. An end-of-file condition also occurs if an attempt is made to read beyond the end of an internal file.

It is a good practice to include some sort of error recovery in all `read` statements that are included permanently in a program. On the other hand, input for test purposes is normally sufficiently well handled by the simple form of `read` without a unit number, and without error recovery.

10.8 Formatted output

There are two types of formatted output statements, the `print` statement, which has appeared in many of the examples so far in this chapter,

   `print` *fmt* [*, output-list*]

and the `write` statement, whose syntax is similar to that of the `read` statement:

   `write` (*output-specifier-list*) [*output-list*]

where *output-specifier-list* is one of

[unit=]u	decimal=*char-expr*	pos=*int-expr*
[fmt=]fmt	delim=*char-expr*	rec=*int-expr*
[nml=]ngn	id=*int-variable*	round=*char-expr*
advance=*char-expr*	iomsg=*char-variable*	sign=*char-expr*
asynchronous=*char-expr*	iostat=*int-variable*	

and the specifiers that appear in the read statement (Section 10.7) have the same meaning here except that an asterisk for *u* specifies the standard output unit (output_unit in iso_fortran_env), as used by print.

The meanings of the additional specifiers are as follows:

- If delim= appears, either *u* must be an asterisk (list-directed formatting) or *ngn* present (namelist editing). *char-expr* must provide the value quote, apostrophe, or none. If apostrophe or quote is specified, the corresponding character will be used within the execution of the statement to delimit character constants and it will be doubled where it appears within a character constant; also, non-default character values will be preceded by kind values.

- If sign= appears, *char-expr* must provide one of the values suppress, plus, or processor_defined to specify how the signs of positive values are treated within the execution of the statement, see Section 11.5.4.

If an error condition occurs on output, execution of the statement terminates, any implied-do variables become undefined, and the file position becomes indeterminate.

An example of a write statement is

```
write (nout, '(10f10.3)', iostat=ios) a
```

An example using an internal file is given in Figure 10.5, which builds a character string from numeric and character components. The final character string might be passed to another subroutine for output, for instance as a caption for a graph. In this example we

---

Figure 10.5 Writing to an internal file.

```
integer         :: day
real            :: cash
character(len=50) :: line
 ⋮
!  write into line
write (line, '(a, i2, a, f8.2, a)')                        &
     'Takings for day ', day, ' are ', cash, ' dollars'
```

---

declare a character variable that is long enough to contain the text to be transferred to it. (The write statement contains a format specification with a edit descriptors without a field width. These assume a field width corresponding to the actual length of the character strings to be converted.) After execution of the write statement, line might contain the character string

```
Takings for day  3 are   4329.15 dollars
```

and this could be used as a string for further processing.

The number of characters written to line must not exceed its length.

10.9   List-directed input/output

In Section 10.4, the list-directed output facility using an asterisk as format specifier was introduced. We assumed that the list was short enough to fit into a single record, but for long lists the processor is free to output several records. Character constants may be split between records, and complex constants that are as long as, or longer than, a record may be split after the comma that separates the two parts. Apart from these cases, a value always lies within a single record. For historical reasons (carriage control, see Appendix B.2), the first character of each record is blank unless a delimited character constant is being continued. Note that when an undelimited character constant is continued, the first character of the continuation record is blank. The only blanks permitted in a numeric constant are within a split complex constant after the comma.

This facility is equally useful for input, especially of small quantities of test data. On the input record, the various constants may appear in most of their usual forms, just as if they were being read under the usual edit descriptors, as defined in Chapter 11. Exceptions are that complex values must consist of two numerical values separated by a comma and enclosed in parentheses, character constants may be delimited, a blank must not occur except in a delimited character constant or in a complex constant before or after a numeric field, blanks are never interpreted as zeros, and the optional characters that are allowed in a logical constant (those following t or f, see Section 11.3.7) must include neither a comma nor a slash. A complex constant spread over more than one record must have any end of record after the real part or before the imaginary part.

Character constants that are enclosed in apostrophes or quotation marks may be spread over as many records as necessary to contain them, except that a doubled quotation mark or apostrophe must not be split between records. Delimiters may be omitted for a default character constant if:

- it is of nonzero length;
- the constant does not contain a blank, comma, or slash;
- it is contained in one record;
- the first character is neither a quotation mark nor an apostrophe; and
- the leading characters are not numeric followed by an asterisk.

In this case, the constant is terminated when a blank, comma, slash, or end of record is encountered, and apostrophes or quotation marks appearing within the constant must not be doubled.

Whenever a character value has a different length from the corresponding list item, the value is truncated or padded on the right with blanks, as in the character assignment statement.

It is possible to use a repeat count for a given constant, for example 6*10 to specify six occurrences of the integer value 10. If it is possible to interpret the constant as either a literal constant or an undelimited character constant, the first corresponding list item determines which it is.

The (optionally repeated) constants are separated in the input by *separators*. A separator is one of the following, appearing other than in a character constant:

- a comma (or a semicolon if the decimal edit mode is comma, see Section 11.5.5), optionally preceded and optionally followed by one or more contiguous blanks;
- a slash (/), optionally preceded and optionally followed by one or more contiguous blanks; or

- one or more contiguous blanks between two non-blank values or following the last non-blank value.

An end of record not within a character constant is regarded as a blank and, therefore, forms part of a separator. A blank embedded in a complex constant or delimited character constant is not a separator. An input record may be terminated by a slash separator, in which case all the following values in the record are ignored, and the input statement terminates.

If there are no values between two successive separators, or between the beginning of the first record and the first separator, this is taken to represent a **null value** and the corresponding item in the input list is left unchanged, defined or undefined as the case may be. A null value must not be used for the real or imaginary part of a complex constant, but a single null value may be used for the whole complex value. A series of null values may be represented by a repeat count without a constant: , 6*,. When a slash separator is encountered, null values are given to any remaining list items.

An example of this form of the read statement is:

```
integer           :: i
real              :: a
complex           :: field(2)
logical           :: flag
character (len=12) :: title
character (len=4)  :: word
  ⋮
read *, i, a, field, flag, title, word
```

If this reads the input record

10*b*6.4*b*(1.,0.)*b*(2.,0.)*btbt*est/

(in which *b* stands for a blank, and blanks are used as separators), i, a, field, flag, and title will acquire the values 10, 6.4, (1.,0.) and (2.,0.), .true., and test, respectively, while word remains unchanged. For the input records

```
10,.64e1,2*,.true.
'histogramb10'/val1
```

(in which commas are used as separators), the variables i, a, flag, and title will acquire the values 10, 6.4, .true., and histogram*b*10, respectively. The variables field and word remain unchanged, and the input string val1 is ignored as it follows a slash. (Note the apostrophes, which are required as the string contains a blank. Without delimiters, this string would appear to be a string followed by the integer value 10.) Because of this slash, the read statement does not continue with the next record and the list is thus not fully satisfied.

10.10   Namelist input/output

It can be useful, especially for program testing, to input or output an annotated list of values (key–value pairs) to an external or internal file. The names of the objects involved are specified in a namelist group (Section 8.19), and the input/output is performed by a read or write statement that specifies a namelist group name *ngn* and does not have an input/output list. The sequence of records read or written by a namelist read or write statement begins with a record whose first non-blank character is an ampersand followed without an intervening blank by the group name, and ends with a record that terminates with a slash (/) that is not within a character constant. Each name or designator of an object is followed by an equals sign and a value or list of values, optionally preceded or followed by blanks. A value may be null. The objects with their values may appear in any order. The form of the list of values and null values is as that for list-directed input/output (Section 10.9), except that character constants must always be delimited in input records and logical constants must not contain an equals sign. A simple example is

```
integer    ::  no_of_eggs, litres_of_milk, kilos_of_butter
namelist/food/no_of_eggs, litres_of_milk, kilos_of_butter
read (5, nml=food)
```

to read the record

```
&food litres_of_milk=5, no_of_eggs=12 /
```

Note that the order of the two values given is not the same as their order in the namelist group. The value of kilos_of_butter remains unchanged.

On input, the designators of the objects are not case-sensitive and only those objects and subobjects that are specified in the input records and do not have a null value become defined. All other list items and parts of items remain in their existing state of definition or undefinition.

Where a subobject designator appears in an input record, all its substring expressions, subscripts, and strides must be scalar integer literal constants without specified kind parameters. All group names, object names, and component names are interpreted without regard to case. Blanks may precede or follow the name or designator, but must not appear within it.

If an object is scalar and of intrinsic type, the equals sign must be followed by one value. If it is an array, the equals sign must be followed by a list of values for its elements in array element order. If it is of derived type and defined derived-type input/output (Section 11.6) is not in use, the equals sign must be followed by a list of values of intrinsic type corresponding to its ultimate components in array element order. The list of values must not be too long, but it may be too short, in which case trailing null values are regarded as having been appended. If an object is of type character, the corresponding item must be of the same kind.

Zero-sized objects must not appear in a namelist input record. In any multiple occurrence of an object in a sequence of input records, the final value is taken.

Input records for namelist input may bear a comment following an object name/value separator other than a slash. As in the source form, it starts with an exclamation mark (!) and results in the rest of the record being ignored. This allows programmers to document the structure of a namelist input file line by line. For example, the input record of this section might be documented thus:

```
&food litres_of_milk=5,              ! For camping holiday
no_of_eggs=12 /
```

A comment line, with ! as the first non-blank character in an input record, is also permitted, but in a character context such a line will be taken as part of the character value.

On output, all the objects in the `namelist` group are represented in the output records with names in upper case and are ordered as in the `namelist` group. Thus, the statements

```
integer    :: number, list(10)
namelist/out/number, list
write (6, nml=out)
```

might produce the record

```
&OUT NUMBER=1, LIST=14, 9*0 /
```

Repetitions in the `namelist` group result in a value for each occurrence.

10.11   Non-advancing input/output

So far we have considered each `read` or `write` statement to perform the input or output of a complete record. There are, however, many applications, especially in screen management, where this would become an irksome restriction. What is required is the ability to read and write without always advancing the file position to ahead of the next record. This facility is provided by **non-advancing input/output**. To gain access to this facility, the optional `advance=` specifier must appear in the `read` or `write` statement and be associated with a scalar default character expression *advance* which evaluates, after suppression of any trailing blanks and conversion of any upper-case letters to lower case, to the value `no`. The only other allowed value is `yes`, which is the default value if the specifier is absent; in this case, normal (advancing) input/output occurs.

An advancing input/output statement always repositions the file after the last record accessed. A non-advancing input/output statement leaves the file positioned within the record, except that if it attempts to read data from beyond the end of the *current* record, an end-of-record condition occurs and the file is repositioned to follow the record. The `iostat` variable, if present, will acquire a negative value (`iostat_eor` in `iso_fortran_env`) that is different from the one indicating an end-of-file condition. In order to provide a means of controlling this process, the `size=` specifier, when present, sets the size variable to the number of characters actually read. A full example is thus

```
character(len=3) :: key
integer          :: unit, size, ios
read (unit, '(a3)', advance='no', size=size, iostat=ios) key
if (is_iostat_eor(ios)) then
   ! key is only partially defined
   key(size+1:) = ''
end if
```

As for error and end-of-file conditions, the program terminates when an end-of-record condition occurs unless `iostat=` is specified.[5]

If encountering an end-of-record on reading results in the input list not being satisfied, the `pad=` specifier will determine whether any padding with blank characters occurs. Blanks inserted as padding are not included in the `size=` count.

---

[5]Program termination can also be avoided with `eor=` (Appendix A.12.3).

It is possible to perform advancing and non-advancing input/output on the same record or file. For instance, a non-advancing read might read the first few characters of a record and an advancing read might read the remainder.

A particular application of this facility is to write a prompt to a screen and to read from the next character position on the screen without an intervening line-feed:

```
write (*, '(a)', advance='no') 'enter next prime number:'
read  (*, '(i10)') prime_number
```

Non-advancing input/output may be performed only on an external file, and may not be used for namelist or list-directed input/output. Note that, as for advancing input/output, several records may be processed by a single statement.

10.12   Unformatted input/output

This chapter has so far dealt with formatted input/output. The internal representation of a value may differ from the external form, which is always a character string contained in an input or output record. The use of formatted input/output involves an overhead for the conversion between the two forms, and often a round-off error too. There is also the disadvantage that the external representation usually occupies more space on a storage medium than the internal representation. These three drawbacks are all absent when unformatted input/output is used. In this form, the internal representation of a value is written exactly as it stands to the storage medium, and can be read back directly with neither round-off nor conversion overhead. Here, a value of derived type is treated as a whole and is not equivalent to a list of its ultimate components. This is another reason for the rule (Section 10.3) that it must not have an allocatable or pointer component at any level of component selection.

This type of input/output should be used in all cases where the records are generated by a program on one computer, to be read back on the same computer or another computer using the same internal number representations. Only when this is not the case, or when the data have to be visualized in one form or another, should formatted input/output be used. The records of a file must all be formatted or all be unformatted (apart from the endfile record).

Unformatted input/output has the incidental advantage of being simpler to program since no complicated format specifications are required. The forms of the read and write statements are the same as for formatted input/output, but without any format specifier *fmt* or namelist group name *ngn*:

```
read (4) q
write (nout, iostat=ios) a
```

The interpretation of the iostat= specifier is as for formatted input/output.[6]

Non-advancing input/output is not available (in fact, an advance= specifier is not allowed). Each read or write statement transfers exactly one record. The file must be an external file. On output to a file connected (Section 12.1) for sequential access, a record of sufficient length is created. On input, the type and type parameters of each entity in the list must agree with those of the value in the record, except that two reals may correspond to one complex when

---

[6]The interpretation of err= and end= specifiers (Appendix A.12.3) is also as for formatted input/output.

all three have the same kind parameter. The number of values specified by the input list of a `read` statement must not exceed the number of values available in the current record.

10.13   Direct-access files

The only type of file organization that we have so far dealt with is the sequential file which has a beginning and an end, and which contains a sequence of records, one after the other. Fortran permits another type of file organization known as **direct access** (or sometimes as random access or indexed). All the records have the same length, each record is identified by an index number, and it is possible to write, read, or rewrite any specified record without regard to position. (In a sequential file only the last record may be rewritten without losing other records; in general, records in sequential files cannot be replaced.) The records are either all formatted or all unformatted.

By default, any file used by a Fortran program is a sequential file. A direct-access file must be declared as such in its `open` statement (described in Section 12.4) with the `access= 'direct'` and `recl=rl` specifiers (*rl* is the length of a record in the file). By default, a direct-access file is unformatted; to change that, its `open` statement must specify `form='formatted'`. Once this declaration has been made, reading and writing, whether formatted or unformatted, proceeds as described for sequential files, except for the addition of a `rec=` specifier in the `read` or `write` statement to provide the index number of the record concerned. Usually, a data transfer statement for a direct-access file accesses a single record, but during formatted input/output any slash edit descriptor increases the record number by one and causes processing to continue at the beginning of this record. A sequence of statements to write, read, and replace a given record is given in Figure 10.6.

Figure 10.6 Write, read, and replace record 14. The `open` and `inquire` statements are explained in Sections 12.4 and 12.6.

```
integer, parameter :: nunit=2, len=100
integer            :: i, length
real               :: a(len), b(len+1:2*len)
   ⋮
inquire (iolength=length) a
open (nunit, access='direct', recl=length)
   ⋮
write (nunit, rec=14) b ! Write b to record 14 of direct-access file.
   ⋮
read (nunit, rec=14) a   ! Read the values back into array a.
   ⋮
do i = 1, len/2
   a(i) = i
end do
write (nunit, rec=14) a ! Replace record with new values.
```

The file must be an external file and namelist formatting, list-directed formatting, and non-advancing input/output are all unavailable.

Direct-access files are particularly useful for applications that involve lots of hopping around inside a file, or where records need to be replaced, for instance in database applications. A weakness is that the length of all the records must be the same.[7] On formatted output, the record is padded with blanks if necessary; for unformatted output, if the record is not filled, the remainder is undefined.

This simple and powerful facility allows much clearer control logic to be written than is the case for a sequential file which is repeatedly read, backspaced, or rewound. Only when sequential-access files become large may problems of long access times become evident on some computer systems, and this point should always be investigated before heavy investments are made in programming large direct-access file applications.

Some computer systems allow the same file to be regarded as sequential or direct access according to the specification in the open statement or its default. The standard, therefore, regards this as a property of the connection (Section 12.1) rather than of the file. In this case, the order of records, even for sequential input/output, is that determined by the direct-access record numbering.

10.14 UTF-8 files

The ISO 10646 standard specifies a standard encoding of UCS-4 characters into a stream of bytes, called UTF-8. Formatted files in UTF-8 format are supported by the encoding= specifier on the open statement (Section 12.4). For example,

```
open (20, file='output.file', action='write', encoding='utf-8')
```

The encoding= specifier on the inquire statement (Section 12.6.2) returns the encoding of a file, which will be UTF-8 if the file is connected for UTF-8 input/output or the processor can detect the format in some way, UNKNOWN if the processor cannot detect the format, or a processor-dependent value if the file is known to be in some other format (for example, UTF-16LE).

For the most part, UTF-8 files can be treated as ordinary formatted files. On output, all data is effectively converted to ISO 10646 characters for UTF-8 encoding.

On input, if data is being read into an ASCII character variable each input character must be in the range 0–127 (the ASCII subset of ISO 10646); if data is being read into a default character variable each input character must be representable in the default character set. These conditions will be satisfied if the data were written by numeric or logical formatting, or by character formatting from an ASCII or default character value; otherwise it would be safer to read the data into an ISO 10646 character variable for processing.

Figure 10.7 shows an example of a program that uses UCS-4 characters in a UTF-8 formatted internal file to write the date in Japanese style. It also illustrates changing a property of an existing file in an open statement.

---

[7]This deficiency is avoided with stream access, Section 10.16.

Figure 10.7 Program that uses UCS-4 characters in a UTF-8 formatted internal file.

```fortran
program iso10646
  use, intrinsic :: iso_fortran_env, only: output_unit
  implicit none
  integer, parameter :: ucs4 = selected_char_kind("iso_10646")
  character(len=11, kind=ucs4) :: string
  call  Japanese_date_string(string)
  open (output_unit, encoding='utf-8')
  write(output_unit,*) string
contains
  subroutine Japanese_date_string(string)
    character(len=*, kind=ucs4), intent(out) :: string
    intrinsic :: date_and_time
    character(1,ucs4),parameter::nen=char(int(z'5e74'),ucs4), & !year
          gatsu=char(int(z'6708'),ucs4), &                      !month
          nichi=char(int(Z'65e5'),ucs4)                         !day
    integer :: values(8)
    call date_and_time(values=values)
    write(string,"(3(i0,a))") &
          values(1),nen,values(2),gatsu,values(3),nichi
  end subroutine Japanese_date_string
end program iso10646
```

10.15   Asynchronous input/output

Asynchronous input/output allows other statements to execute while an input/output statement is in execution. It is permitted only for external files opened with `asynchronous=` specified as `'yes'` in its `open` statement and is indicated by `asynchronous` specified as `'yes'` in the `read` or `write` statement. By default, execution is synchronous even for a file opened as asynchronous. Execution of an asynchronous input/output statement initiates a 'pending' input/output operation and execution of other statements continues until it reaches a statement involving a wait operation for the file. This may be caused by an explicit `wait` statement such as

```fortran
    wait (10)
```

or by another statement that involves the file. The compiler is permitted to treat each asynchronous input/output statement as an ordinary input/output statement (this, after all, is just the limiting case of the input/output being fast).

Here is a simple example:

```
real :: a(100000), b(100000)
open (10, file='mydata', asynchronous='yes')
read (10, '(10f8.3)', asynchronous='yes') a
    ⋮   ! Computation involving the array b
wait (10)
    ⋮   ! Computation involving the array a
```

Further asynchronous input/output statements may be executed for the file before the wait operation is performed. The input/output statements for each file are performed in the same order as they would have been if they were synchronous.

An execution of an asynchronous input/output statement may be identified by a scalar integer variable in an id= specifier. Successful execution of the statement causes the variable to be given a processor-dependent value; the variable may be referenced as an id= specifier in a subsequent wait or inquire statement.

A wait statement has the form

wait (*wait-specifier-list*)

where *wait-specifier* is one of

[unit=]*u*	id=*int-variable*
iomsg=*char-variable*	iostat=*int-variable*

and the specifiers have the same meaning as in the read statement (Section 10.7).[8] If there is an id= specifier, only the identified pending operation is terminated; otherwise, all pending operations for the file are terminated in turn.

Execution of a wait statement specifying a unit that does not exist, has no file connected to it, or was not opened for asynchronous input/output is permitted, provided that the wait statement has no id= specifier; such a wait statement has no effect. A file positioning statement (backspace, endfile, rewind, Section 12.2) or a flush statement (Section 12.3) performs wait operations for all pending input/output operations for the unit. Execution of a read or write statement causes a wait operation to be performed for any pending read or write statements for the unit that have completed execution.

An inquire statement (Section 12.6.2) with a pending= specifier inquires about pending input/output statements. If an id= specifier is present, the pending= variable is given the value true if the particular operation is still pending and false otherwise. If no id= specifier is present, the variable is given the value true if any input/output operations for the unit are still pending and false otherwise. In the 'false' case, wait operations are performed for the file or files. Wait operations are not performed in the 'true' case, even if some of the input/output operations are complete.

Asynchronous input/output is not permitted in conjunction with defined derived-type input/output (Section 11.6) because the size of the data to be transferred through the unit is not known when the (parent) input/output data transfer statement is invoked.

A variable in a scoping unit is said to be an **affector** of a pending input/output operation if any part of it is associated with any part of an item in the input/output list, namelist, or

---

[8]The specifiers eor=, err=, and end= (Appendix A.12.3) may also occur here.

size= specifier. While an input/output operation is pending, an affector is not permitted to be redefined, become undefined, or have its pointer association status changed. While an input operation is pending, an affector is also not permitted to be referenced or associated with a dummy argument with the value attribute (Section 20.8).

The asynchronous attribute for a variable was introduced in Section 8.11.2. It warns the compiler that optimizations involving movement of code across statements that cause wait operations might lead to incorrect results. If a variable is a dummy argument or appears in an executable statement or a specification expression in a scoping unit and any statement of the scoping unit is executed while the variable is an affector, it must have the asynchronous attribute in the scoping unit.

A variable is automatically given this attribute if it or a subobject of it is an item in the input/output list, namelist, or size= specifier of an asynchronous input/output statement.

10.16   Stream access files

Stream access is a further method of organizing an external file. It allows great flexibility for both formatted and unformatted input/output. It is established by access being specified as 'stream' in the open statement. By default, a stream access file is unformatted; to change that, its open statement must specify form='formatted'.

The file is positioned by **file storage units**, usually bytes,[9] starting at position 1. The current position may be determined from a scalar integer variable in a pos= specifier of an inquire statement for the unit. A file may have the capability of positioning forwards or backwards, forwards only, or neither. If it has the capability, a required position may be indicated in a read or write statement by the pos= specifier, which accepts a scalar integer expression. In the absence of a pos= specifier, the file position is left unchanged.

It is the intention that unformatted stream input/output will read or write only the data to/from the file; that is, that there is no ancillary record length information (which is normally written for unformatted files). This allows easy interoperability with C binary streams, but the facility to skip or backspace over records is not available. If an output statement overwrites part of a file, the rest of the file is unchanged.

Here is a simple example of unformatted stream input/output:

```
real :: d
integer :: before_d
  :
open (unit, ..., access='stream', form='unformatted')
  :
inquire (unit, pos=before_d)
write (unit) d
  :
write (unit, pos=before_d) d + 1
```

---

[9]The exact size in bits is available as the constant file_storage_size in the module iso_fortran_env.

Assuming d occupies four bytes, the user could reasonably expect the first `write` to write exactly four bytes to the file. The use of the `pos=` specifier ensures that the second `write` will overwrite the previously written value of d.

Formatted **stream files** are very similar to ordinary (record-oriented) sequential files; the main difference is that there is no preset maximum record length (the `recl=` specifier in the `open` and `inquire` statements). If the file allows the relevant positioning, the value of a `pos=` specifier must be 1 or a value previously returned in an `inquire` statement for the file. As for a formatted sequential file, an output statement leaves the file ending with the data transferred.

Another difference from a formatted sequential file is that data-driven record termination in the style of C text streams is allowed. The intrinsic inquiry function `new_line(a)` (Section 9.8) returns the character that can be used to cause record termination. As an example, the following code will write two lines to the file `/dev/tty`:

```
open (28, file='/dev/tty', access='stream', form='formatted')
write (28, '(a)') 'Hello'//new_line('x')//'World'
```

10.17 Execution of a data transfer statement

So far, we have used simple illustrations of data transfer statements without dependencies. However, some forms of dependency are permitted and can be very useful. For example, the statement

```
read (*, *) n, a(1:n)                ! n is an integer
```

allows the length of an array section to be part of the data.

With dependencies in mind, the order in which operations are executed is important. It is as follows:

i) identify the unit;

ii) establish the format (if any);

iii) position the file ready for the transfer (if required);

iv) transfer data between the file and the input/output list or namelist;

v) position the file following the transfer (if required);

vi) cause the `iostat` and `size` variables (if present) to become defined.

The order of transfer of namelist input is that in the input records. Otherwise, the order is that of the input/output list or `namelist`. Each input item is processed in turn, and may affect later subobjects and implied-do indices. All expressions within an input/output list item are determined at the beginning of the processing of the item. If an entity is specified more than once during execution of a namelist input statement, the later value overwrites the earlier value. Any zero-sized array or zero-length implied-do list is ignored.

When an input item is an array, no element of the array is permitted to affect the value of an expression within the item. For example, the cases shown in Figure 10.8 are not permitted. This prevents dependencies occurring within the item itself.

Figure 10.8 Dependencies are not permitted within an input item.

```
integer :: j(10)
  :
read *, j(j)                        ! Not permitted
read *, j(j(1):j(10))               ! Not permitted
```

In the case of an internal file, an input/output item must not be in the file or associated with it. Nor may an input item contain or be associated with any portion of the established format.

Recursive input/output, that is, a function referenced in an input/output statement executing an input/output statement for the same unit, is discussed in Section 11.7.

10.18   Summary

This chapter has begun the description of Fortran's extensive input/output facilities. It has covered the formatted input/output statements, and the syntax of the read and write statements. It then turned to list-direct and namelist input/output, non-advancing input/output, unformatted input/output and direct-access, UTF-8, and asynchronous files, as well as stream input/output.

Exercises

1. Write statements to output the state of a game of tic-tac-toe (noughts and crosses) to a unit designated by the variable unit.

2. Write a program which reads an input record of up to 132 characters into an internal file and classifies it as a Fortran comment line with no statement, an initial line without a statement label, an initial line with a statement label, a continuation line, or a line containing multiple statements.

3. Write a subroutine get_char(unit,c,end_of_file) to read a single character c from a formatted, sequential file unit, ignoring any record structure; end_of_file is a logical variable that is given the value .true. if the end of the file is reached and the value .false. otherwise.

4. Write an input procedure that reads a variable number of characters from a file, stopping on encountering a character in a user-specified set or at end of record, returning the input in a deferred-length allocatable character string.

5. Write a program that reads a file (presumed to be a text file) as an unformatted stream, checking for Unix (LF) and DOS/Windows (CRLF) record terminators.

11. Edit descriptors

11.1 Introduction

In the description of number conversion in Section 10.2, a few examples of the edit descriptors were given. As mentioned there, edit descriptors give a precise specification of how values are to be converted into a character string on an output device or internal file, or converted from a character string on an input device or internal file to internal representations.

With certain exceptions noted in the following text, edit descriptors in a list are separated by commas.

Edit descriptors are interpreted without regard to case. This is also true for numerical and logical input fields; an example is 89AB as a hexadecimal input value. In output fields, any alphabetic characters are in upper case.

Edit descriptors fall into three classes: **character string**, **data**, and **control**.

11.2 Character string edit descriptor

A character literal constant without a specified kind parameter (thus of default kind) can be transferred to an output file by embedding it in the format specification itself, as in the example

```
print "('Square-root of q = ', f8.3)" , sqrt(q)
```

The string will appear each time the character string edit descriptor is encountered during format processing. In this descriptor, case is significant. Character string edit descriptors must not be used on input.

11.3 Data edit descriptors

11.3.1 Introduction

Data edit descriptors are edit descriptors that transfer data between the input/output list and an internal or external file. A format specification may contain no data edit descriptor if there is no data in the input/output list, for example, if the input/output list is empty or consists entirely of zero-sized arrays.

In a file, no value corresponding to a datum of intrinsic data type, on either input or output, may include a kind type parameter. For all the numeric edit descriptors, if an output field is too narrow to contain the number to be output, it is filled with asterisks.

Modern Fortran Explained, 3rd Edition. M. Metcalf, J. Reid, M. Cohen, and R. Bader. Oxford University Press (2024). © M. Metcalf, J. Reid, M. Cohen, and R. Bader (2024). DOI 10.1093/oso/9780198876571.001.0011

11.3.2   Repeat counts

A data edit descriptor may be preceded by a **repeat count** (a nonzero unsigned default integer literal constant), as in the example

```
10f12.3
```

A repeat count may also be applied directly to the slash edit descriptor (Section 11.4.4), but not to any other control edit descriptor or to a character string edit descriptor. However, a repeat count may be applied to a **subformat**, that is, a group of edit descriptors enclosed in parentheses:

```
print '(4(i5,f8.2))', (i(j), a(j), j=1,4)
```

(for integer i and real a). This is equivalent to writing

```
print '(i5,f8.2,i5,f8.2,i5,f8.2,i5,f8.2)', (i(j), a(j), j=1,4)
```

Repeat counts such as this may be nested:

```
print '(2(2i5,2f8.2))', i(1),i(2),a(1),a(2),i(3),i(4),a(3),a(4)
```

If a format specification without a subformat in parentheses is used with an input/output list that contains more elements than the number of edit descriptors, taking account of repeat counts, then a new record will begin, and the format specification will be repeated. Further records begin in the same way until the list is exhausted. To print an array of 100 integer elements, 10 elements to a line, the following statement might be used:

```
print '(10i8)', i(1:100)
```

Similarly, when reading from an input file, new records will be read until the list is satisfied, a new record being taken from the input file each time the specification is repeated *even if the individual records contain more input data than specified by the format specification*. These superfluous data will be ignored. For example, reading the two records (*b* again stands for a blank)

```
bbb10bbb15bbb20
bbb25bbb30bbb35
```

under control of the read statement

```
read '(2i5)', i,j,k,l
```

will result in the four integer variables i, j, k, and l acquiring the values 10, 15, 25, and 30, respectively.

If a format contains a subformat in parentheses, as in

```
'(2i5, 3(i2,2(i1,i3)), 2(2f8.2,i2))'
```

whenever the format is exhausted, a new record is taken and format control reverts to the repeat factor preceding the left parenthesis corresponding to the last-but-one right parenthesis, here 2(2f8.2,i2), or to the left parenthesis itself if it has no repeat factor. This we call **reversion**.

In order to make the writing of CSV (comma-separated values) files easier, **unlimited format repetition** can be used to repeat a format specification without any preset limit, as

long as there are still data left to transfer to or from the input/output list. This is specified by
* *(format-items)*, and is permitted only as the last item in a format specification. A colon edit
descriptor (Section 11.4.5) may be needed to avoid unwanted data from character string edit
descriptors after the final data item. Here is an example:

```
print '("List: ",*(g0,:,", "))', 10, 20, 30
```

will print

```
List: 10, 20, 30
```

(the g0 edit descriptor is described in Section 11.3.9).

11.3.3 Integer formatting

Integer values may be converted by means of the i edit descriptor. Its basic form is iw, where
w is a nonzero unsigned default integer literal constant that defines the width of the field. The
integer value will be read from or written to this field, adjusted to its right-hand side. If we
again designate a blank position by b then the value -99 printed under control of the edit
descriptor i5 will appear as bb-99, the sign counting as one position in the field.

For output, an alternative form of this edit descriptor allows the number of digits that are
to be printed to be specified exactly, even if some are leading zeros. The form i$w.m$ specifies
the width of the field, w, and that at least m digits are to be output, where m is an unsigned
default integer literal constant. The value 99 printed under control of the edit descriptor i5.3
would appear as bb099. The value of m is even permitted to be zero, and the field will be
then filled with blanks if the value printed is 0. On input, i$w.m$ is interpreted in exactly the
same way as iw.

In order to allow output records to contain as little unused space as possible, the i edit
descriptor may specify w to be zero, as in i0. This does not denote a zero-width field, but
a field that is of the minimum width necessary to contain the output value in question. The
programmer does not need to worry that a field with too narrow a width will cause an output
field to overflow and contain only asterisks.

Integer values may also be converted by the bw, b$w.m$, ow, o$w.m$, zw, and z$w.m$ edit
descriptors. These are similar to the i form, but are intended for integers represented in the
binary, octal, and hexadecimal number systems, respectively (Section 2.6.7). The external
form does not contain the leading letter (b, o, or z) or the delimiters. The $w.m$ form, with m
equal to w, is recommended on output, so that any leading zeros are visible.

11.3.4 Real formatting with e, en, es, and f edit descriptors

A real value may be converted by an e, en, es, ex, or f edit descriptor. We defer describing
the ex edit descriptor to the next subsection.

The f descriptor we have met in earlier examples. Its general form is f$w.d$, where w and
d are unsigned default integer literal constants which define, respectively, the field width and
the number of digits to appear after the decimal point in the output field. For input, w must
not be zero. The decimal point counts as one position in the field. On input, if the input string
has a decimal point, the value of d is ignored. Reading the input string b9.3729b with the

edit descriptor f8.3 would cause the value 9.3729 to be transferred. All the digits are used, but round-off may be inevitable because of the actual physical storage reserved for the value on the computer being used.

There are, in addition, two other forms of input string that are acceptable to the f edit descriptor. The first is an optionally signed string of digits without a decimal point. In this case, the *d* rightmost digits will be taken to be the fractional part of the value. Thus, *b*-14629 read under control of the edit descriptor f7.2 will transfer the value −146.29. The second form is the standard default real form of a literal constant, as defined in Section 2.6.3, and the variant in which the exponent is signed and e is omitted. In this case, the *d* part of the descriptor is again ignored. Thus, the value 14.629e-2 (or 14.629-2), under control of the edit descriptor f9.1, will transfer the value 0.146 29. The exponent letter may also be written in upper case.

Values are rounded on output following the normal rules of arithmetic. Thus, the value 10.9336, when output under control of the edit descriptor f8.3, will appear as *bb*10.934, and under the control of f4.0 as *b*11. For output, if *w* is zero, as in f0.3, this denotes a field that is of the minimum width necessary to contain the output value in question.

The e edit descriptor has two forms, e*w.d* and e*w.dee*, and is more appropriate for numbers with a magnitude below about 0.01, or above 1000. The value of *w* must not be zero. The rules for these two forms for input are identical to those for the f*w.d* edit descriptor. For output with the e*w.d* form of the descriptor, a different character string will be transferred, containing a significand with absolute value less than 1 and an exponent field of four characters that consists either of E followed by a sign and two digits or of a sign and three digits. Thus, for 1.234×10^{23} converted by the edit descriptor e10.4, the string *b*.1234E+24 or *b*.1234+024 will be transferred. The form containing the exponent letter E is not used if the magnitude of the exponent exceeds 99. For instance, e10.4 would cause the value 1.234×10^{-150} to be transferred as *b*.1234-149. Some processors print a zero before the decimal point.

In the second form of the e edit descriptor, e*w.dee*, *e* is an unsigned, nonzero default integer literal constant that determines the number of digits to appear in the exponent field. This form is obligatory for exponents whose magnitude is greater than 999. Thus, the value 1.234×10^{1234} with the edit descriptor e12.4e4 is transferred as the string *b*.1234E+1235. An increasing number of computers are able to deal with these very large exponent ranges. It can also be used if only one exponent digit is desired. For example, the value 1.211 with the edit descriptor e9.3e1 is transferred as the string *b*0.121E+1.

The en (**engineering**) edit descriptor is identical to the e edit descriptor except that on output the decimal exponent is divisible by three, a nonzero significand is greater than or equal to 1 and less than 1000, and the scale factor (Section 11.4.2) has no effect. Thus, the value 0.0217 transferred under an en9.2 edit descriptor would appear as 21.70E-03 or 21.70-003.

The es (**scientific**) edit descriptor is identical to the e edit descriptor, except that on output the absolute value of a nonzero significand is greater than or equal to 1 and less than 10, and the scale factor (Section 11.4.2) has no effect. Thus, the value 0.0217 transferred under an es9.2 edit descriptor would appear as 2.17E-02 or 2.17-002.

On a computer with IEEE arithmetic, all the edit descriptors for reals treat IEEE exceptional values in the same way and only the field width *w* is taken into account.

The output forms, each right justified in its field, are

i) `-Inf` or `-Infinity` for minus infinity;

ii) `Inf`, `+Inf`, `Infinity`, or `+Infinity` for plus infinity; and

iii) `NaN`, optionally followed by processor-dependent alphanumeric characters in parentheses (to hold additional information).

On input, upper- and lower-case letters are treated as equivalent. The forms are

i) `-Inf` or `-Infinity` for minus infinity;

ii) `Inf`, `+Inf`, `Infinity`, or `+Infinity` for plus infinity; and

iii) `NaN`, optionally followed by alphanumeric characters in parentheses for a NaN. With no such alphanumeric characters it is a quiet NaN.

If the value of the field width is zero, the processor selects the smallest positive actual field width that does not result in a field filled with asterisks. A zero value for the field width is not permitted for input.

The value 0 may be specified as the exponent width parameter e in the e$w.d$ee, en$w.d$ee, and es$w.d$ee edit descriptors. The effect of e0 is that the processor uses the minimum number of digits required to represent the exponent value, that is, without leading zeros. For example, writing the value -125.0 with `es0.3e0` produces the output `-1.250E+2`.

11.3.5 Hexadecimal significand input/output

The Fortran standard supports a hexadecimal significand character format and the `ex` edit descriptor for values in a file. The format begins with an optional sign followed by the hexadecimal indicator `0x` (or `0X`). This is followed by a hexadecimal significand, possibly containing a hexadecimal symbol (point or comma, the same as decimal), the exponent letter `p`, and the exponent, which is a power of two expressed as a decimal integer. For example, the decimal value 2.375 could be written as `0x2.6p+0` or `0x1.3p+1`. Note that the exponent is always present.

On input, hexadecimal significand character sequences are accepted as floating-point values by all edit descriptors for type real, as well as for list-directed and namelist input.

The `ex` edit descriptor has the forms e$xw.d$ and e$xw.d$ee. For input and for output of IEEE infinities and NaNs, this edit descriptor is treated identically to f$w.d$.

On output, the field width w may be zero to let the processor choose the width. The output is unaffected by the scale factor and always has one digit before the hexadecimal symbol; if d is nonzero, the fraction contains d digits (rounded if necessary), otherwise the fraction contains the minimum number of digits required to represent the internal value exactly. Note that d is not permitted to be zero if the radix of the internal value is not a power of two (for example, if the internal value is a decimal floating-point number).

If e appears, it specifies the number of digits in the exponent. If the value of e is zero, or if e does not appear, the exponent part contains the minimum number of digits to represent the exponent.

The exponent for the value zero is always zero, but for a non-zero value the choice of the binary exponent is processor dependent. For example, the value 11.5 could be written as any of 0X1.7P+3, 0X2.EP+2, 0X5.CP+1, or 0XB.8P+0.

Table 11.1 shows some additional examples of hexadecimal-significand output.

Table 11.1: Examples of hexadecimal-significand output.

Internal value	Edit descriptor	Possible output
1.375	ex0.1	0X1.6P+0
−15.625	ex14.4e3	−0X1.F400P+003
1 048 580.0	ex0.0	0X1.00004P+20

11.3.6   Complex formatting

Complex values may be edited under control of pairs of f, e, en, es, or ex edit descriptors. The two descriptors do not need to be identical. The complex value (0.1,100.) converted under control of f6.1,e8.1 would appear as *bbb*0.1*b*0.1E+03. The two descriptors may be separated by character string and control edit descriptors (Sections 11.2 and 11.4, respectively).

11.3.7   Logical formatting

Logical values may be edited using the l*w* edit descriptor. This defines a field of width *w* which on input consists of optional blanks, optionally followed by a decimal point, followed by t or f (or T or F), optionally followed by additional characters. Thus, a field defined by l7 permits the strings .true. and .false. to be input. The characters t or f will be transferred as the values true or false, respectively. On output, the character T or F will appear in the rightmost position in the output field.

11.3.8   Character formatting

Character values may be edited using the a edit descriptor in one of its two forms, either a or a*w*. In the first of the two forms, the width of the input or output field is determined by the actual width of the item in the input/output list, measured in number of characters of whatever kind. Thus, a character variable of length 10 containing the value STATEMENTS, when written under control of the a edit descriptor, would appear in a field ten characters wide, and the non-default character variable of length 4 containing the value 国際標準 would appear in a field four characters wide. If, however, the first variable were converted under an a11 edit descriptor, it would be printed with a leading blank: *b*STATEMENTS. Under control of a8, the eight leftmost characters only would be written: STATEMEN.

Conversely, with the same variable on input, an a11 edit descriptor would cause the ten rightmost characters in the 11-character-wide input field to be transferred, so that *b*STATEMENTS would be transferred as STATEMENTS. The a8 edit descriptor would cause the eight characters in the field to be transferred to the eight leftmost positions in the variable, and the remaining two would be filled with blanks: STATEMEN would be transferred as STATEMEN*bb*.

Whenever an input list and the associated format specify more data than appear in the record, any padding will be controlled by the pad= specifier currently in effect (Section 10.7).

All characters transferred under the control of an a or aw edit descriptor have the kind of the input/output list item, and we note that this edit descriptor is the *only* one which can be used to transmit non-default characters to or from a record. In the non-default case, the blank padding character is processor dependent.

11.3.9 General formatting

Any intrinsic data type values may be edited with the g*w.d* or g*w.dee* (**general**) edit descriptor. When used for real or complex types it is identical to the e*w.d* or e*w.dee* edit descriptor, except for output when the f format is suitable for representing exactly the decimal value N obtained by rounding the internal value to d significant decimal digits according to the input/output rounding mode in effect. For example, the edit descriptor g10.3 outputs the values 0.017 32, 1.732, 173.2, and 1732.0 as 0.173e-01, 1.73, 173., and 0.173e04, respectively. The formal rule for which form is employed is as follows. Let s be 1 if N is zero and otherwise be the integer such that

$$10^{s-1} \le N < 10^s.$$

If $0 \le s \le d$, then f$(w-n).(d-s)$ formatting followed by n blanks is used, where $n = 4$ for g*w.d* editing and $e + 2$ for g*w.dee* editing.

This form is useful for printing values whose magnitudes are not well known in advance, where f conversion is preferred when possible and e conversion otherwise.

When the g edit descriptor is used for integer, logical, or character types, it follows the rules of the i*w*, l*w*, and a*w* edit descriptors, respectively (any d or e is ignored).

In order to make the writing of CSV (comma-separated values) files easier, the g0 edit descriptor may be used; this transfers the user data as follows:

- integer data are transferred as if i0 had been specified;
- real and complex data are transferred as if es*w.dee* had been specified, where the compiler chooses the values of w, d, and e depending on the actual value to be output;
- logical data are transferred as if l1 had been specified;
- character data are transferred as if a had been specified.

For example,

```
print '(1x, 5(g0, "; "))', 17, 2.71828, .false., "Hello"
```

will print something like

```
17; 2.7183e+00; F; Hello;
```

(depending on the values for *w*, *d*, and *e* chosen by the compiler for the floating-point datum).

The g0.*d* edit descriptor is similar to the g0 edit descriptor but specifies the value to be used as *d* for floating-point data; as seen in the example above, this can be necessary if the value chosen by the compiler is unsuitable. The g0.*d* edit descriptor can also be used to specify the output of integer, logical, and character data. It follows the rules for the i0, l1, and a edit descriptors, respectively. Neither g0 nor g0.*d* is allowed in input.

The value 0 may be specified as the exponent width parameter *e* in the g*w.dee* edit descriptor. (But note that g*w.dee* does not permit *w* to be zero, unlike the others.) The effect of e0 is that the processor uses the minimum number of digits required to represent the exponent value, that is, without leading zeros.

11.3.10   Derived-type formatting

Derived-type values may be edited by the appropriate sequence of edit descriptors corresponding to the intrinsic types of the ultimate components of the derived type, provided none of them are allocatable or pointers. An example is:

```
type string
    integer           :: length
    character(len=20) :: word
end type string
type(string) :: text
read (*, '(i2, a)') text
```

They may also be edited by edit descriptors written by the programmer (see Section 11.6).

11.4   Control edit descriptors

11.4.1   Introduction

It is sometimes necessary to give instructions to an input/output device other than just the widths of fields and how the contents of these fields are to be interpreted. For instance, it may be that one wishes to position fields at certain columns or to start a new record without issuing a new `write` command. For this type of purpose, the control edit descriptors provide a means of telling the processor which action to take. Some of these edit descriptors contain information that is used as it is processed; others are like switches, which change the conditions under which input/output takes place from the point where they are encountered until the end of the execution of the input/output statement containing them. The latter descriptors are further divided into ones which alter the changeable connection modes (described in Section 11.5), and the irregular descriptor for changing the scale factor, which we will deal with first.

11.4.2   Scale factor

The scale factor applies to the input of real quantities under the e, f, en, es, ex, and g edit descriptors, and are a means of scaling the input values. Their form is *k*p, where *k* is a default integer literal constant. The scale factor is zero at the beginning of execution of the statement. The effect is that any quantity which does not have an exponent field will be reduced by a factor 10^k. Quantities with an exponent are not affected.

The scale factor kp also affects output with e, f, or g editing, but has no effect with en, es, or ex editing. Under control of an f edit descriptor, the quantity will be multiplied by a factor 10^k. Thus, the number 10.39 output by an f6.0 edit descriptor following the scale factor 2p will appear as $b1039..$ With the e edit descriptor, and with g where the e style editing is taken, the quantity is transferred as follows: if $0 < k < d + 2$, the output field contains exactly k significant digits to the left of the decimal point and $d - k + 1$ significant digits to its right; if $-d < k \leq 0$, the output field has a zero to the left of the decimal point, and to its right has $|k|$ zeros followed by $d - |k|$ significant digits. Other values of k are not permitted. Thus 310.0, written with 2p, e9.2 editing, will appear as 31.0E+01. This gives better control over the output style of real quantities which otherwise would have no significant digits before the decimal point.

The comma between a scale factor and an immediately following f, e, en, es, ex, or g edit descriptor (with or without a repeat count) may be omitted, but we do not recommend this since it suggests that the scale factor applies only to the next edit descriptor, whereas in fact it applies throughout the format until another scale factor is encountered.

11.4.3 Tabulation and spacing

Tabulation in an input or output field can be achieved using the edit descriptors tn, tr$n$, and tl$n$, where n is a positive default integer literal constant. These state, respectively, that the next part of the input/output should begin at position n in the current record (where the **left tab limit** is position 1), or at n positions to the right of the current position, or at n positions to the left of the current position (the left tab limit if the current position is less than or equal to n). Let us suppose that, following an advancing read, we read an input record $bb9876$ with the following statement:

```
read (*, '(t3, i4, tl4, i1, i2)') i, j, k
```

The format specification will move a notional pointer firstly to position 3, whence i will be read. The variable i will acquire the value 9876, and the notional pointer is then at position 7. The edit descriptor tl4 moves it left four positions, back to position 3. The quantities j and k are then read, and they acquire the values 9 and 87, respectively. These edit descriptors cause replacement on output, or multiple reading of the same items in a record on input. On output, any gaps ahead of the last character actually written are filled with spaces. If any character that is skipped by one of the descriptors is of other than default type, the positioning is processor dependent.

If the current record is the first one processed by the input/output statement and follows non-advancing input/output that left the file positioned within a record, the next character is the left tab limit; otherwise, the first character of the record is the left tab limit.

The nx edit descriptor is equivalent to the tr$n$ edit descriptor. It is often used to place spaces in an output record. For example, to start an output record with a blank by this method, one writes

```
fmt= '(1x, ....)'
```

Spaces such as this can precede a data edit descriptor, but `1x,i5` is not, for instance, exactly equivalent to `i6` on output, as any value requiring the full six positions in the field will not have them available in the former case.

The `t` and `x` edit descriptors never cause replacement of a character already in an output record, but merely cause a change in the position within the record such that such a replacement might be caused by a subsequent edit descriptor.

11.4.4   New records (slash editing)

New records may be started at any point in a format specification by means of the slash (`/`) edit descriptor. This edit descriptor, although described here, may in fact have repeat counts; to skip, say, three records one can write either `/,/,/` or `3/`. On input, a new record will be started each time a `/` is encountered, even if the contents of the current record have not all been transferred. Reading the two records

```
0009900010
0010000011
```

with the statement

```
read 'i5,i3,/,i5,i3,i2)', i, j, k, l, m
```

will cause the values 99, 0, 100, 0, and 11 to be transferred to the five integer variables, respectively. This edit descriptor does not need to be separated by a comma from a preceding edit descriptor, unless it has a repeat count; it does not ever need to be separated by a comma from a succeeding edit descriptor.

The result of writing with a format containing a sequence of, say, four slashes, as represented by

```
print '(i5,4/,i5)', i, j
```

is to separate the two values by three blank records (the last slash starts the record containing j); if i and j have the values 99 and 100, they would appear as

```
bbb99
b
b
b
bb100
```

A slash edit descriptor written to an internal file will cause the following values to be written to the next element of the character array specified for the file. Each such element corresponds to a record, and the number of characters written to a record must not exceed its length.

11.4.5   Colon editing

Colon editing is a means of terminating format control if there are no further items in an input/output list. In particular, it is useful for preventing further output of character strings used for annotation if the output list is exhausted. Consider the following output statement, for an integer array p (3):

```
print '(" p1 = ", i5, :, " p2 = ", i5, :," p3 = ", i5)', &
      (p(i) ,i=1,n)
```

If n has the value 3, then three values are printed. If n has the value 1, then, without the colons, the following output string would be printed:

```
p1 = 59 p2 =
```

The first colon, however, stops the processing of the format, so that the annotation for the absent second value is not printed. This edit descriptor need not be separated from a neighbour by a comma. It has no effect if there are further items in the input/output list.

11.5 Changeable file connection modes

11.5.1 Introduction

A file connected for formatted input/output has several modes that affect the processing of data edit descriptors. These modes can be specified (or changed) by an open statement (Section 12.4), overridden for the duration of a data transfer statement execution by a read or write statement (Sections 10.7 and 10.8), or changed within a format specification by an edit descriptor.

Note that, as usual, the specifiers on an open, read, or write statement take default character values, and are interpreted without regard to case.

For an internal file there is of course no open statement, so these have a predetermined value at the beginning of every read or write statement.

11.5.2 Embedded blank interpretation

Embedded blanks in numeric input fields are treated in one of two ways, either as zero, or as null characters that are squeezed out by moving the other characters in the input field to the right, and adding leading blanks to the field (unless the field is totally blank, in which case it is interpreted as zero). This is controlled by the **blank interpretation mode**.

The blank interpretation mode for a unit may be controlled by the blank= specifier on the open statement, which takes one of the values null or zero. If not specified, or for an internal file, the default is null. It may be overridden by a blank= specifier in a read statement.

There is a corresponding specifier in an inquire statement (Section 12.6.2) that is assigned the value NULL, ZERO, or UNDEFINED as appropriate.

The blank interpretation mode may be temporarily changed within a read statement by the bn (blanks null) and bz (blanks zero) edit descriptors.

Let us suppose that the mode is that blanks are treated as zeros. The input string *bb1b4* converted by the edit descriptor i5 would transfer the value 104. The same string converted by bn, i5 would give 14. A bn or bz edit descriptor switches the mode for the rest of that format specification, or until another bn or bz edit descriptor is met. The bn and bz edit descriptors have no effect on output.

11.5.3   Input/output rounding mode

Rounding during formatted input/output is controlled by the **input/output rounding mode**, which is one of up, down, zero, nearest, compatible, or processor_defined. The meanings are obvious except for the difference between nearest and compatible. Both refer to a closest representable value, but if two are equidistant, which is taken is processor dependent for nearest[1] and the value away from zero for compatible. An input/output rounding mode of processor_defined places no constraints on what kind of rounding will be performed by the processor.

The input/output rounding mode may be controlled by the round= specifier on the open statement, which takes one of the values up, down, zero, nearest, compatible, or processor_defined. If it is not specified, or for an internal file, the mode will be one of the above, but which one is processor dependent. It may be overridden by a round= specifier in a read or write statement with one of these values.

There is a corresponding specifier in the inquire statement that is assigned the value UP, DOWN, ZERO, NEAREST, COMPATIBLE, PROCESSOR_DEFINED, or UNDEFINED, as appropriate. The processor returns the value PROCESSOR_DEFINED only if the input/output rounding mode currently in effect behaves differently from the other rounding modes.

The input/output rounding mode may be temporarily changed within a data transfer statement to up, down, zero, nearest, compatible, or processor_defined by the ru, rd, rz, rn, rc, or rp edit descriptor, respectively.

11.5.4   Signs on positive values

Leading signs are always written for negative numerical values on output. For positive quantities other than exponents, the plus signs are optional, and whether the signs are written depends on the **sign mode**. If the mode is suppress, leading plus signs are suppressed, that is, the value 99 printed by i5 is *bbb*99 and 1.4 is printed by e10.2 as *bb*0.14E+01. If the mode is plus, leading plus signs are printed; the same numbers written in this mode become *bb*+99 and *b*+0.14E+01. If the mode is processor_defined (the default), it is processor dependent whether leading plus signs are printed or suppressed.

The sign mode may be specified by the sign= specifier on an open statement, which can take the value suppress, plus, or processor_defined. If no sign= specifier appears, or for output to an internal file, the sign mode is processor_defined.

The sign mode for a connection may be overridden by a sign= specifier in a write statement with one of these values.

There is a corresponding specifier in the inquire statement. If the connection is for formatted input/output, the specifier is assigned the value PLUS, SUPPRESS, or PROCESSOR_DEFINED, as appropriate; otherwise, it is assigned the value UNDEFINED.

The sign mode may be temporarily changed within an output statement to suppress, plus, or processor_defined by the ss, sp, or s edit descriptor, respectively. Thus, printing the values 99 and 1.4 with the edit descriptors ss,i5,sp,e10.2 will print *bbb*99

---

[1] If the processor supports IEEE arithmetic input/output conversions, nearest should be round to even, the same as IEEE arithmetic rounding.

and b+0.14E+01 regardless of the current sign mode. An ss, sp, or s will remain in force for the remainder of the format specification, unless another ss, sp, or s edit descriptor is met.

The sign= specifier and these edit descriptors provide complete control over sign printing, and are useful for producing coded outputs which have to be compared automatically on two different computers.

11.5.5 Decimal comma for input/output

Many countries use a decimal comma instead of a decimal point. Support for this is provided by the decimal= input/output specifier and by the dc and dp edit descriptors. These affect the **decimal edit mode** for the unit. While the decimal edit mode is point, decimal points are used in input/output. This is the default mode.

While the mode is comma, commas are used in place of decimal points both for input and for output. For example,

```
x = 22./7
print '(1x,f6.2)', x
```

would produce the output

```
3,14
```

in comma mode.

The decimal= specifier may appear on the open (Section 12.4), read, and write statements, and evaluates either to point or to comma. On the open statement it specifies the default decimal edit mode for the unit. With no decimal= clause on the open statement, or for an internal file, the default is point. On an individual read or write statement, the decimal= clause specifies the default mode for the duration of that input/output statement only.

The dc and dp edit descriptors change the decimal edit mode to comma and point, respectively. They take effect when they are encountered during format processing and continue in effect until another dc or dp edit descriptor is encountered or until the end of the current input/output statement. For example,

```
   write (*,10) x, x, x
10 format(1x,'Default ',f5.2,', English ',dp,f5.2,'Français',dc,f5.2)
```

would produce the value of x first with the default mode, then with a decimal point for English, and a decimal comma for French.

If the decimal edit mode is comma during list-directed or namelist input/output, a semicolon acts as a value separator instead of a comma.

11.5.6 Padding input records

By default, a formatted input record is regarded as padded out with blanks whenever an input list and the associated format specify more data than appear in the record. Control over this is provided by the pad= specifier in open and read statements. It evaluates either to yes, the default, or to no. If no is specified, the length of the input record must not be less than that specified by the input list and the associated format, except in the presence of an `advance='no'` specifier and an `iostat=` specifier.[2]

There is a corresponding specifier in the `inquire` statement. For a connection for formatted input/output, it is assigned the value `YES` or `NO` corresponding to the pad mode in effect. If there is no connection or if the connection is not for formatted input/output, the value returned is `UNDEFINED`.

11.5.7   Delimiting character values

By default, character values are written without delimiters in list-directed or `namelist` formatting. Control over this is provided by the `delim=` specifier in `open` and `write` statements. It must evaluate to `quote`, `apostrophe`, or `none`. If `apostrophe` or `quote` is specified, the corresponding character will be used to delimit character constants written with list-directed or `namelist` formatting, and it will be doubled where it appears within such a character constant; also, non-default character values will be preceded by kind values. No delimiting character is used if `none` is specified, nor does any doubling take place. The default value if the specifier is omitted is `none`. This specifier may appear only when opening a formatted file or writing with list-directed or `namelist` formatting.

11.6   Defined derived-type input/output

Defined derived-type input/output allows the programmer to provide formatting specially tailored to a type and to transfer structures with pointer, allocatable, or private components. Thus, it may be arranged that, when a derived-type object is encountered in an input/output list, the programmer's Fortran subroutine is called. Its first argument is a scalar of the type and it either reads some data from the file and constructs a value of the derived type or accepts a value of the derived type and writes some data to the file.

For formatted input/output, defined input/output for a type is always active for list-directed and namelist input/output, but for an explicit format specification, only when the `dt` data edit descriptor is used. For unformatted input/output, defined input/output for a type is always active.

Note that defined input/output fundamentally affects the way that list items of derived type are expanded into their components. For formatted input/output, an item that will be processed by defined input/output is not expanded into its components (note that for an explicit format specification, this depends on the edit descriptor being `dt`; if it is not, the item is expanded as usual).

For unformatted input/output, a derived-type list item for which a subcomponent will be processed by defined input/output is expanded into its components, unlike the usual case where the whole derived-type input/output list item is output as a single item.

The `dt` edit descriptor for formatted input/output provides an optional character string and an optional integer array that are passed to the Fortran subroutine to control its action. An example is

```
dt 'linked-list' (10, -4, 2)
```

---

[2]Or an `eor=` specifier (Appendix A.12.3).

If the character string is omitted, the effect is the same as if a string of length zero had been given. If the parenthetical list of integers is omitted, an array of size zero is passed. Note that if the parentheses appear, at least one integer must be specified.

Subroutines for defined derived-type input/output may be bound to the type as generic bindings (see Section 15.8.3) of the forms

```
generic :: read(formatted) => r1, r2
generic :: read(unformatted) => r3, r4, r5
generic :: write(formatted) => w1
generic :: write(unformatted) => w2, w3
```

which makes the subroutines r1–w3 accessible wherever an object of the type of their first argument is accessible. An alternative is an interface block such as

```
interface read(formatted)
    module procedure r1, r2
end interface
```

The form of such a subroutine depends on whether it is for formatted or unformatted input/output:

subroutine formatted_io(dtv,unit,iotype,v_list,iostat,iomsg)
subroutine unformatted_io(dtv,unit,   iostat,iomsg)

dtv is a scalar of the derived type. It must be polymorphic if the type is extensible; otherwise, it must be non-polymorphic. All length type parameters must be assumed. For output, it is of intent in and holds the value to be written. For input, it is of intent inout and is altered in accord with the values read.

unit is a scalar of intent in and type default integer. Its value is the unit on which input/output is taking place or a negative value generated by the processor if on an internal file.

iotype is a scalar of intent in and type character(*). Its value is 'LISTDIRECTED', 'NAMELIST', or 'DT'//*string*, where *string* is the character string from the dt edit descriptor.

v_list is a rank-one assumed-shape array of intent in and type default integer. Its value comes from the parenthetical list of the edit descriptor.

iostat is a scalar of intent out and type default integer. If an error condition occurs, it is given a positive value. Otherwise, if an end-of-file or end-of-record condition occurs it is given, respectively, the value iostat_end or iostat_eor of the intrinsic module iso_fortran_env. Otherwise, it must be given the value zero.

iomsg is a scalar of intent inout and type character(*). If iostat is given a nonzero value, iomsg must be set to an explanatory message. Otherwise, it must not be altered.

The names of the subroutine and its arguments are not significant when they are invoked as part of input/output processing.

Within the subroutine, input/output to external files is limited to the specified unit and in the specified direction. Such a data transfer statement is called a **child** data transfer statement and the original statement is called the **parent**. No file positioning takes place at the beginning or end of the execution of a child data transfer statement. Any advance= specifier is ignored. Input/output to an internal file is permitted. An input/output list may include a dt edit descriptor for a component of the dtv argument, with the obvious meaning. Execution of any of the statements open, close, backspace, endfile, and rewind is not permitted. Also, the procedure must not alter any aspect of the parent input/output statement, except through the dtv argument.

The file position on entry is treated as a left tab limit and there is no record termination on return. Therefore, positioning with rec= (for a direct-access file, Section 10.13) or pos= (for stream access, Section 10.16) is not permitted in a child data transfer statement.

This feature is not available in combination with asynchronous input/output (Section 10.15).

A simple example of derived-type formatted output follows. The derived-type variable chairman has two components. The type and an associated write-formatted procedure are defined in a module called person_module and might be invoked as shown in Figure 11.1.

Figure 11.1 A program with a dt edit descriptor.

```
program main
  use person_module
  integer id, members
  type (person) :: chairman
    ⋮
  write (6, fmt="(i2, dt(15,4), i5)" ) id, chairman, members
! This writes a record with four fields, with lengths 2, 15, 6, 5,
! respectively
end program
```

The module that implements this is shown in Figure 11.2. From the edit descriptor dt(15,6), it constructs the format (a15,i 6) in the local character variable pfmt and applies it. It would also be possible to check that iotype indeed has the value 'DT' and to set iostat and iomsg accordingly.

In the Figure 11.3 example we illustrate the output of a structure with a pointer component and show a child data transfer statement itself invoking derived-type input/output. Here, we show the case where the same (recursive) subroutine is invoked in both cases. The variables of the derived type node form a chain, with a single value at each node and terminating with a null pointer. The subroutine pwf is used to write the values in the list, one per line.

Figure 11.2 A module containing a write(formatted) subroutine.

```
module person_module
   type :: person
      character (len=20) :: name
      integer :: age
   contains
      procedure :: pwf
      generic   :: write(formatted) => pwf
   end type person
contains
   subroutine pwf (dtv, unit, iotype, vlist, iostat, iomsg)
   ! Arguments
      class(person), intent(in)       :: dtv
      integer, intent(in)             :: unit
      character (len=*), intent(in)   :: iotype
      integer, intent(in)             :: vlist(:)
      ! vlist(1) and (2) are to be used as the field widths
      ! of the two components of the derived type variable.
      integer, intent(out)            :: iostat
      character (len=*), intent(inout) :: iomsg
      ! Local variable
      character (len=9) :: pfmt
      ! Set up the format to be used for output
      write (pfmt, '(a,i2,a,i2,a)' ) &
         '(a', vlist(1), ',i', vlist(2), ')'
      ! Now the child output statement
      write (unit, fmt=pfmt, iostat=iostat) dtv%name, dtv%age
   end subroutine pwf
end module person_module
```

11.7 Recursive input/output

A recursive input/output statement is one that is executed while another input/output statement is in execution. We met this in connection with derived-type input/output (Section 11.6); a child data transfer statement is recursive since it always executes while its parent is in execution. Recursive input/output is allowed for input/output to/from an internal file where the statement does not modify any internal file other than its own.

Recursive input/output is also permitted for external files, provided only that the same unit is not involved in both input/output actions. This is particularly useful while debugging and also for logging and error reporting. For example,

```
  print *, invert_matrix(x)
  ⋮
contains
  function invert_matrix(a)
    ⋮
    if (singular) then
      write (error_unit,*) &
        'Cannot invert singular matrix - continuing!'
      return
    end if
    ⋮
```

Figure 11.3 A module containing a recursive `write(formatted)` subroutine.

```
module list_module
  type node
    integer                :: value = 0
    type (node), pointer :: next_node => null ( )
  contains
    procedure :: pwf
    generic   :: write(formatted) => pwf
  end type node
contains
  recursive subroutine pwf (dtv, unit, iotype, vlist, iostat, iomsg)
  ! Write the chain of values, each on a separate line in I9 format.
    class(node), intent(in)          :: dtv
    integer, intent(in)              :: unit
    character (len=*), intent(in)    :: iotype
    integer, intent(in)              :: vlist(:)
    integer, intent(out)             :: iostat
    character (len=*), intent(inout) :: iomsg
    write (unit, '(i9,/)', iostat = iostat) dtv%value
    if (iostat/=0) return
    if (associated(dtv%next_node)) &
        write (unit, '(dt)', iostat=iostat) dtv%next_node
  end subroutine pwf
end module list_module
```

11.8   Summary

This chapter has described in full detail the edit descriptors used for formatting in input/output statements. Furthermore, recursive and user-defined derived-type input/output were covered.

Exercises

1. Write suitable `print` statements to print the name and contents of each of the following arrays:

    i) `real :: grid(10,10)`, ten elements to a line (assuming the values are between 1.0 and 100.0);

    ii) `integer :: list(50)`, the odd elements only;

    iii) `character(len=10) :: titles(20)`, two elements to a line;

    iv) `real :: power(10)`, five elements to a line in engineering notation;

    v) `logical :: flags(10)`, on one line;

    vi) `complex :: plane(5)`, on one line.

2. Write separate list-directed input statements to fill each of the arrays of Exercise 1. For each statement write a sample first input record.

3. Write a function that formats a real input value, of a kind that has a decimal precision of 15 or more, in a suitable form for display as a monetary value in Euros. If the magnitude of the value is such that the 'cent' field is beyond the decimal precision, a string consisting of all asterisks should be returned.

4. Write a program that displays the effects of the `sign=` specifier and the `ss`, `sp`, and `s` edit descriptors. What output would you expect if the file is opened with `sign=' suppress'`?

12. Operations on external files

12.1 Introduction

So far we have discussed the topic of external files in a rather superficial way. In the examples of the various input/output statements in the previous chapter, an implicit assumption has always been made that the specified file was actually available, and that records could be written to it and read from it. For sequential files, the file control statements described in the next section further assume that it can be positioned. In fact, these assumptions are not necessarily valid. In order to define explicitly and to test the status of external files, three file status statements are provided: open, close, and inquire. Before beginning their description, however, two new definitions are required.

A computer system contains, among other components, a CPU and a storage system. Modern storage systems are usually based on some form of disk, which is used to store files for long or short periods of time. The execution of a computer program is, by comparison, a transient event. A file may exist for years, whereas programs run for only seconds or minutes. In Fortran terminology, a file is said to *exist* not in the sense we have just used, but in the restricted sense that it exists as a file *to which the program might have access*. In other words, if the program is prohibited from using the file because of a password protection system, or because some other necessary action has not been taken, the file 'does not exist'.

A file which exists for a running program may be empty and may or may not be *connected* to that program. The file is connected if it is associated with a unit number known to the program. Such connection is usually made by executing an open statement for the file, but many computer systems will *preconnect* certain files which any program may be expected to use, such as terminal input and output. Thus, we see that a file may exist but not be connected. It may also be connected but not exist. This can happen for a preconnected new file. The file will only come into existence (be *created*) if some other action is taken on the file: executing an open, write, print, or endfile statement. A unit must not be connected to more than one file at once.

There are a number of other points to note with respect to files.

- The set of allowed names for a file is processor dependent.

- Both sequential and direct access may be available for some files, but normally a file is limited to sequential, direct, or stream access.

- A file never contains both formatted and unformatted records.

Modern Fortran Explained, 3rd Edition. M. Metcalf, J. Reid, M. Cohen, and R. Bader. Oxford University Press (2024). © M. Metcalf, J. Reid, M. Cohen, and R. Bader (2024). DOI 10.1093/oso/9780198876571.001.0012

- It can be convenient to connect a file to more than one unit at a time. For example, the two units might be reading different parts of the same sequential file.[1]

Finally, we note that no statement described in this chapter applies to internal files.

12.2 Positioning statements for sequential files

When reading or writing an external file that is connected for sequential access, whether formatted or unformatted, it is sometimes necessary to perform other control functions on the file in addition to input and output. In particular, one may wish to alter the current position, which may be within a record, between records, ahead of the first record (at the *initial point*), or after the last record (at its *terminal point*). The following three statements are provided for these purposes.

12.2.1 The `backspace` statement

It can happen in a program that a series of records is being written and that, for some reason, the last record written should be overwritten by a new one. Similarly, when reading records, it may be necessary to reread the last record read, or to check-read a record which has just been written. For this purpose, Fortran provides the `backspace` statement, which has the syntax

```
backspace u
```
or
```
backspace ([unit=]u [,iostat=ios] [,iomsg=char-variable])
```

where *u* is a scalar integer expression whose value is the unit number, and the other optional specifiers have the same meaning as for a `read` statement.[2] Again, keyword specifiers may be in any order, but the unit specifier must come first as a positional specifier.

The action of this statement is to position the file before the current record if it is positioned within a record, or before the preceding record if it is positioned between records. An attempt to backspace when already positioned at the beginning of a file is not regarded as an error and results in no change in the file's position. If the file is positioned after an endfile record (Section 12.2.3), it becomes positioned before that record. It is not possible to backspace a file that does not exist, nor to backspace over a record written by a list-directed or namelist output statement (Sections 10.9 and 10.10). A series of `backspace` statements will backspace over the corresponding number of records. This statement is often very costly in computer resources and should be used as little as possible.

12.2.2 The `rewind` statement

In an analogous fashion to rereading, rewriting, or check-reading a record, a similar operation may be carried out on a complete file. For this purpose the `rewind` statement,

---

[1]However, there is no requirement on the processor to provide this functionality. For a given file and action choice, whether or not it is available is processor dependent; however, most operating systems will permit a file to be open with `action='read'` on more than one unit at the same time.

[2]The obsolescent `err=` (Appendix A.12.3) may also be specified here.

```
rewind u
```
or
```
rewind ([unit=]u [,iostat=ios] [,iomsg=char-variable])
```

may be used to reposition a file, whose unit number is specified by the scalar integer expression u.[2] Again, keyword specifiers may be in any order, but the unit specifier must come first as a positional specifier. Their meanings are as for backspace. If the file is already at its beginning, this is not regarded as an error and there is no change in its position. The statement is permitted for a file that does not exist, and has no effect.

12.2.3   The endfile statement

The end of a file connected for sequential access is normally marked by a special record which is identified as such by the computer hardware, and computer systems ensure that all files written by a program are correctly terminated by such an **endfile record**. In doubtful situations, or when a subsequent program step will reread the file, it is possible to write an endfile record explicitly using the endfile statement:

```
endfile u
```
or
```
endfile ([unit=]u [,iostat=ios] [,iomsg=char-variable])
```

where u, once again, is a scalar integer expression specifying the unit number.[2] Again, keyword specifiers may be in any order, but the unit specifier must come first as a positional specifier. Their meanings are as for backspace. The file is then positioned after the endfile record. A subsequent read on the file will result in an end-of-file condition. Prior to data transfer, a file must not be positioned after an endfile record, but it is possible to backspace or rewind across an endfile record, which allows further data transfer to occur. An endfile record is written automatically whenever either a backspace or rewind operation follows a write operation as the next operation on the unit, or the file is closed by execution of a close statement (Section 12.5), by an open statement for the same unit (Section 12.4), or by normal program termination.

If the file may also be connected for direct access, only the records ahead of the endfile record are considered to have been written and only these may be read during a subsequent direct-access connection.

Note that if a file is connected to a unit but does not exist for the program, it will be made to exist by executing an endfile statement on the unit.

12.2.4   Data transfer statements

Execution of a data transfer statement (read, write, or print) for a sequential file also affects the file position. If it is between records, it is moved to the start of the next record unless non-advancing access is in operation. Data transfer then takes place, which usually moves the position. No further movement occurs for non-advancing access. For advancing access, the position finally moves to follow the last record transferred.

12.3   The flush statement

Execution of a flush statement for an external file causes data written to it to be available to other processes, or causes data placed in it by means other than Fortran to be available to a read statement. The syntax is just like that of the file positioning statements (Section 12.2), for instance

```
flush(6)
```

In combination with advance='no' or stream access (Section 10.16), it permits the program to ensure that data written to one unit are sent to the file before requesting input on another unit; that is, that 'prompts' appear promptly.

12.4   The open statement

The open statement is used to connect an external file to a unit, create a file that is preconnected, create a file and connect it to a unit, or change certain properties of a connection. The syntax is

> open (*olist*)

where *olist* is a list of optional specifiers.[2] A specifier must not appear more than once. In the specifiers, all entities are scalar and all characters are of default kind. In character expressions, any trailing blanks are ignored and, except for file=, any upper-case letters are converted to lower case. The specifiers are as follows:

[unit=]*u* where *u* is a scalar integer expression specifying the external file unit number. If the unit is specified without unit=, it must appear first in *olist*. One, but not both, of [unit=]*u* and newunit= must appear.

access=*acc* where *acc* is a character expression that provides one of the values sequential, direct, or stream. For a file which already exists, this value must be an allowed value. If the file does not already exist, it will be brought into existence with the appropriate access method. If this specifier is omitted, the value sequential will be assumed.

action=*act* where *act* is a character expression that provides the value read, write, or readwrite. If read is specified, the write, print, and endfile statements must not be used for this connection; if write is specified, the read statement must not be used (and backspace and position='append' may fail on some systems); if readwrite is specified, there is no restriction. If the specifier is omitted, the default value is processor dependent.

asynchronous=*asy* where *asy* is a character expression that provides the value yes or no. If yes is specified, asynchronous input/output on the unit is allowed. If no is specified, asynchronous input/output on the unit is not allowed. If the specifier is omitted, the default value is no.

blank=*bl* where *bl* is a character expression that provides the value null or zero. This connection must be for formatted input/output. This specifier sets the default for the interpretation of blanks in numeric input fields, as discussed in the description of the bn and bz edit descriptors (Section 11.5.2). If the value is null, such blanks will be ignored (except that a completely blank field is interpreted as zero). If the value is zero, such blanks will be interpreted as zeros. If the specifier is omitted, the default is null.

decimal=*decimal* where *decimal* is a character expression that specifies the decimal edit mode (see Section 11.5.5).

delim=*del* where *del* is a character expression that provides the value quote, apostrophe, or none. It specifies how character values are delimited in files, see Section 11.5.7. The default value if the specifier is omitted is none. This specifier may appear only for formatted files.

encoding=*enc* where *enc* is a character expression that provides the value utf-8 or default. If utf-8 is specified, UTF-8 coding is used for the file (see Section 10.14). If default is specified, the encoding is processor dependent.

file=*fln* where *fln* is a character expression that provides the name of the file. If this specifier is omitted and the unit is not connected to a file, the status= specifier must be specified with the value scratch and the file connected to the unit will then depend on the computer system. Whether the interpretation is case sensitive varies from system to system.

form=*fm* where *fm* is a character expression that provides the value formatted or unformatted, and determines whether the file is to be connected for formatted or unformatted input/output. For a file which already exists, the value must be an allowed value. If the file does not already exist, it will be brought into existence with an allowed set of forms that includes the specified form. If this specifier is omitted, the default is formatted for sequential access and unformatted for direct or stream access.

iomsg=*message* where *message* is a scalar variable of type default character in which the processor places a message if an error condition occurs, and is unchanged otherwise.

iostat=*ios* where *ios* is a default integer variable which is set to zero if the statement is correctly executed, and to a positive value otherwise.

newunit=*nu* where *nu* is an integer variable which is set on a successful open statement to a negative unit value chosen by the processor for the file, avoiding a clash with anything it is using internally. To avoid any confusion in the result of the number= specifier of the inquire statement, where -1 indicates a file that is not connected, newunit= will never return -1. One, but not both, of [unit=]*u* and newunit= must appear.

pad=*pad* where *pad* is a character expression that provides the value yes or no to control whether a formatted input record will be regarded as padded out with blanks whenever an input list and the associated format specify more data than appear in the record, see Section 11.5.6.

position=*pos* where *pos* is a character expression that provides the value asis, rewind, or append. The access method must be sequential, and if the specifier is omitted the default value asis will be assumed. A new file is positioned at its initial point. If asis is specified and the file exists and is already connected, the file is opened without changing its position; if rewind is specified, the file is positioned at its initial point; if append is specified and the file exists, it is positioned ahead of the endfile record if it has one (and otherwise at its terminal point). For a file which exists but is not connected, the effect of the asis specifier on the file's position is unspecified.

recl=*rl* where *rl* is an integer expression whose value must be positive. For a direct-access file, it specifies the length of the records, and is obligatory. For a sequential file, it specifies the maximum length of a record and is optional, with a default value that is processor dependent. It must not appear for a stream-access file. For formatted files, the length is the number of characters for records that contain only default characters; for unformatted files it is system dependent but the inquire statement (see Section 12.6.3) may be used to find the length of an input/output list. In either case, for a file which already exists, the value specified must be allowed for that file. If the file does not already exist, the file will be brought into existence with an allowed set of record lengths that includes the specified value.

round=*rnd* where *rnd* is a character expression that specifies the rounding mode (see Section 11.5.3).

sign=*sign* where *sign* is a character expression that specifies the sign mode (see Section 11.5.4).

status=*st* where *st* is a character expression that provides the value old, new, replace, scratch, or unknown. The file= specifier must be present if new or replace is specified or if old is specified and the unit is not connected; the file= specifier must not be present if scratch is specified. If old is specified, the file must already exist; if new is specified, the file must not already exist, but will be brought into existence by the action of the open statement. The status of the file then becomes old. If replace is specified and the file does not already exist, the file is created; if the file does exist, the file is deleted and a new file is created with the same name. In each case the status is changed to old. If the value scratch is specified, the file is created and becomes connected, but it cannot be kept after completion of the program or execution of a close statement (Section 12.5). If unknown is specified, the status of the file is system dependent. This is the default value of the specifier, if it is omitted.

An example of an open statement is

```
open (2, iostat=ios, file='cities', status='new',  &
     access='direct', recl=100)
```

which brings into existence a new, direct-access, unformatted file named cities, whose records have length 100. The file is connected to unit number 2. The value of ios should be checked to ensure that the statement has been executed correctly.

The open statements in a program are best collected together in one place, so that any changes which might have to be made to them when transporting the program from one system to another can be carried out without having to search for them. Regardless of where they appear, the connection may be referenced in any program unit of the program.

The purpose of the open statement is to connect a file to a unit. If the unit is, however, already connected to a file then the action may be different. If the file= specifier is omitted, the default is the name of the connected file. If the file in question does not exist, but is preconnected to the unit, then all the properties specified by the open statement become part of the connection. If the file is already connected to the unit, then of the existing attributes only the blank=, decimal=, delim=, pad=, round=, and sign= specifiers may have values different from those already in effect. If the unit is already connected to another file, the effect of the open statement includes the action of a prior close statement on the unit (without a status= specifier, see next section).

In general, by repeated execution of the open statement on the same unit, it is possible to process in sequence an arbitrarily high number of files, whether they exist or not, as long as the restrictions just noted are observed.

12.5   The close statement

The purpose of the close statement is to disconnect a file from a unit. Its form is

```
close ([unit=]u [,iostat=ios] [,iomsg=message] [,status=st])
```

where *u*, *ios*, and *message* have the same meanings as described in the previous section for the open statement. Again, keyword specifiers may be in any order, but the unit specifier must come first as a positional specifier.[3]

The function of the status= specifier is to determine what will happen to the file once it is disconnected. The value of *st*, which is a scalar default character expression, may be either keep or delete, ignoring any trailing blanks and converting any upper-case letters to lower case. If the value is keep, a file that exists continues to exist after execution of the close statement, and may later be connected again to a unit. If the value is delete, the file no longer exists after execution of the statement. In either case, the unit is free to be connected again to a file. The close statement may appear anywhere in the program, and if executed for a non-existing or unconnected unit, acts as a 'do nothing' statement. The value keep must not be specified for files with the status scratch.

If the status= specifier is omitted, its default value is keep unless the file has status scratch, in which case the default value is delete. On normal termination of execution, all connected units are closed, as if close statements with omitted status= specifiers were executed.

An example of a close statement is

```
close (2, iostat=ios, status='delete')
```

---

[3]The obsolescent err= (Appendix A.12.3) may also be specified here.

12.6 The `inquire` statement

12.6.1 Introduction

The status of a file can be defined by the operating system prior to execution of the program, or by the program itself during execution, either by an `open` statement or by some action on a preconnected file which brings it into existence. At any time during the execution of a program it is possible to inquire about the status and attributes of a file using the `inquire` statement. Using a variant of this statement, it is similarly possible to determine the status of a unit, for instance whether the unit number exists for that system (that is, whether it is an allowed unit number), whether the unit number has a file connected to it, and, if so, which attributes that file has. Another variant permits an inquiry about the length of an output list when used to write an unformatted record. The three variants are known as `inquire` by file, `inquire` by unit, and `inquire` by output list.

12.6.2 Inquire by file or unit

Some of the attributes that may be determined by use of the `inquire` statement for a file or unit are dependent on others. For instance, if a file is not connected to a unit, it is not meaningful to inquire about the form being used for that file. If this is nevertheless attempted, the relevant specifier is undefined.

The form of this `inquire` statement is

> `inquire ([unit=]u,` *ilist*`)`

for `inquire` by unit, where *u* is a scalar integer expression specifying an external unit, and

> `inquire ( file=`*fln*`,` *ilist*`)`

for `inquire` by file, where *fln* is a scalar character expression whose value, ignoring any trailing blanks, provides the name of the file concerned. Whether the interpretation is case sensitive is system dependent. If the unit or file is specified by keyword, it may appear in *ilist*. A specifier must not occur more than once in the list of optional specifiers, *ilist*. All assignments occur following the usual rules, and all values of type character, apart from that for the name= specifier, are in upper case. The specifiers, in which all variables are scalar and those of integer or logical type may be of of any kind, are as follows:

access=*acc* where *acc* is a character variable that is assigned one of the values SEQUENTIAL, DIRECT, or STREAM depending on the access method for a file that is connected, and UNDEFINED if there is no connection.

action=*act* where *act* is a character variable that is assigned the value READ, WRITE, or READWRITE, according to the connection. If there is no connection, the value assigned is UNDEFINED.

asynchronous=*asynch* where *asynch* is a character variable that is assigned the value YES if the file is connected and asynchronous input/output on the unit is allowed; it is assigned the value NO if the file is connected and asynchronous input/output on the unit is not allowed. If there is no connection, it is assigned the value UNDEFINED.

blank=*bl* where *bl* is a character variable that is assigned the value NULL or ZERO, depending on whether the blanks in numeric fields are by default to be interpreted as null fields or zeros, respectively, and UNDEFINED if there is either no connection, or if the connection is not for formatted input/output.

decimal=*dec* where *dec* is a character variable that is assigned the value COMMA or POINT, corresponding to the decimal edit mode in effect for a connection for formatted input/output. If there is no connection, or if the connection is not for formatted input/output, it is assigned the value UNDEFINED.

delim=*del* where *del* is a character variable that is assigned the value QUOTE, APOSTROPHE, or NONE, as specified by the corresponding open statement (or by default). If there is no connection, or if the file is not connected for formatted input/output, the value assigned is UNDEFINED.

direct=*dir* where *dir* is a character variable that is assigned the value YES, NO, or UNKNOWN, depending on whether or not the file *may* be opened for direct access, or whether this cannot be determined.

encoding=*enc* where *enc* is a character variable that is assigned the value utf-8 or default corresponding to the encoding in effect.

exist=*ex* where *ex* is a logical variable. The value true is assigned to *ex* if the file (or unit) exists, and false otherwise.

form=*frm* where *frm* is a character variable that is assigned one of the values FORMATTED or UNFORMATTED, depending on the form for which the file is actually connected, and UNDEFINED if there is no connection.

formatted=*fmt* where *fmt* is a character variable that is assigned the value YES, NO, or UNKNOWN, depending on whether or not the file *may* be opened for formatted or access, or whether this cannot be determined.

id=*id* where *id* is an integer variable that identifies an asynchronous input/output operation (see pending=).

iomsg=*message* where *message* is a scalar variable of type default character into which the processor places a message if an error, end-of-file, or end-of-record condition occurs, and is unchanged otherwise. The specifier iomsg= is useful only if iostat= is also specified, as otherwise the program would be immediately terminated.

iostat=*ios* where *ios* is a default integer variable which is set to zero if the statement is correctly executed, and to a positive value otherwise. The iostat= variable is the only one which is defined if an error condition occurs during the execution of the statement.

name=*nam* where *nam* is a character variable. If the file has a name, *nam* will be assigned that name; otherwise it becomes undefined. The file name is not necessarily the same as that given in the file specifier, if used, but may be qualified in some way. However, in all cases it is a name which is valid for use in a subsequent open statement, and so inquire can be used to determine the actual name of a file before connecting it. Whether the file name is case sensitive is system dependent.

named=*nmd* where *nmd* is a logical variable that is assigned the value true if the file has a name, and false otherwise.

nextrec=*nr* where *nr* is an integer variable that is assigned the value of the record number of the last record read or written, plus one. If no record has yet been read or written, it is assigned the value 1. If the file is not connected for direct access or if the position is indeterminate because of a previous error, *nr* becomes undefined.

number=*num* where *num* is an integer variable that is assigned the value of the unit number connected to the file, or -1 if no unit is connected to the file.

opened=*open* where *open* is a logical variable. The value true is assigned to *open* if the file (or unit) is connected to a unit (or file), and false otherwise.

pad=*pad* where *pad* is a character variable that is assigned the value YES or NO, as specified by the corresponding open statement (or by default). If there is no connection, or if the file is not connected for formatted input/output, the value assigned is UNDEFINED.

pending=*pending* where, if an id= specifier is present, the scalar default logical variable *pending* is given the value true if the asynchronous input/output operation identified by id= is pending and false otherwise.

pos=*pos* where *pos* is a scalar integer variable that is assigned the current position in a stream-access file. If there are pending data transfer operations for the unit, the value assigned is computed as if all the pending data transfers had already been completed. If the file is not connected for stream access or if the position of the file is indeterminate because of previous error conditions, the variable becomes undefined.

position=*pos* where *pos* is a character variable that is assigned the value REWIND, APPEND, or ASIS, as specified in the corresponding open statement, if the file has not been repositioned since it was opened. If there is no connection, or if the file is connected for direct access, the value is UNDEFINED. If the file has been repositioned since the connection was established, the value is processor dependent (but is not REWIND or APPEND unless that corresponds to the true position).

read=*rd* where *rd* is a character variable that is assigned the value YES, NO, or UNKNOWN, according to whether read is allowed, not allowed, or is undetermined for the file.

readwrite=*rw* where *rw* is a character variable that is assigned the value YES, NO, or UNKNOWN, according to whether read/write is allowed, not allowed, or is undetermined for the file.

recl=*rec* where *rec* is an integer variable that is assigned the value of the record length of a file connected for direct access, or the maximum record length allowed for a file connected for sequential access. The length is the number of characters for formatted records containing only characters of default type, and system dependent otherwise.

If there is no connection or the connection is for stream access, the variable becomes undefined. It is assigned the value -1 if there is no connection and the value -2 if the connection is for stream access.

round=*rnd* where *rnd* is a character variable that is assigned the value UP, DOWN, ZERO, NEAREST, COMPATIBLE, or PROCESSOR_DEFINED according to the rounding mode in effect (see Section 11.5.3). If there is no connection, or if the file is not connected for formatted input/output, the value assigned is UNDEFINED.

sequential=*seq* where *seq* is a character variable that is assigned the value YES, NO, or UNKNOWN, depending on whether or not the file *may* be opened for sequential access or whether this cannot be determined.

sign=*sign* where *sign* is a character variable that is assigned the value PLUS, SUPPRESS, or PROCESSOR_DEFINED, according to the sign mode in effect (see Section 11.5.4). If there is no connection, or if the file is not connected for formatted input/output, the value assigned is UNDEFINED.

size=*variable* where *variable* is an integer variable that is assigned the size of the file in file storage units. If the file size cannot be determined, the variable is assigned the value -1. For a file that may be connected for stream access, the file size is the number of the highest-numbered file storage unit in the file. For a file that may be connected for sequential or direct access, the file size may be different from the number of storage units implied by the data in the records; the exact relationship is processor dependent. If there are pending data transfer operations for the specified unit, the value assigned is computed as if all the pending data transfers had already completed.

stream=*stm* where *stm* is a character variable that is assigned the value YES, NO, or UNKNOWN, depending on whether or not the file *may* be opened for stream access, or whether this cannot be determined.

unformatted=*unf* where *fmt* is a character variable that is assigned the value YES, NO, or UNKNOWN, depending on whether or not the file *may* be opened for unformatted access, or whether this cannot be determined.

write=*wr* where *wr* is a character variable that is assigned the value YES, NO, or UNKNOWN, according to whether write is allowed, not allowed, or is undetermined for the file.

A variable that is a specifier on an inquire statement or is associated with one must not appear in another specifier in the same statement.

12.6.3   Inquire by input/output list

The statement for inquire by input/output list has the form

        inquire (iolength=*length*) *olist*

where *length* is a scalar integer variable of default kind and is used to determine the length of an unformatted output list in file storage units. It might be used to establish whether, for instance, an output list is too long for the record length given in the recl= specifier of an open statement, or be used as the value of the length to be supplied to a recl= specifier (see Figure 10.6 in Section 10.13).

An example of the inquire statement, for the file opened as an example of the open statement in Section 12.4, is

```
logical                :: ex, op
character (len=11) :: nam, acc, seq, frm
integer                :: ios, irec, nr
inquire (2, iostat=ios, exist=ex, opened=op, name=nam, access=acc, &
     sequential=seq, form=frm, recl=irec, nextrec=nr)
```

After successful execution of this statement, the variables provided will have been assigned the following values:

ios	0
ex	.true.
op	.true.
nam	cities*bbbbb*
acc	DIRECT*bbbbb*
seq	NO*bbbbbbbbb*
frm	UNFORMATTED
irec	100
nr	1

(assuming no intervening read or write operations).

12.7   Summary

This chapter has completed the description of the input/output features begun in the previous two chapters, and together they provide a complete reference to all the facilities available.

Exercises

1. A direct-access file is to contain a list of names and initials, to each of which there corresponds a telephone number. Write a program which opens a sequential file and a direct-access file, and copies the list from the sequential file to the direct-access file, closing it for use in another program.

    Write a second program which reads an input record containing either a name or a telephone number (from a terminal if possible), and prints out the corresponding entry (or entries) in the direct-access file if present, and an error message otherwise. Remember that names are as diverse as Wu, O'Hara, and Trevington-Smythe, and that it is insulting for a computer program to corrupt or abbreviate people's names. The format of the telephone numbers should correspond to your local numbers, but the actual format used should be readily modifiable to another.

13. Further type parameter features

In this chapter we consider features for type parameter inquiries and for parameterizing derived types.

13.1   Type parameter inquiry

The (current) value of a type parameter of a variable can be discovered by a **type parameter inquiry**. This uses the same syntax as for component access, but the value is always scalar, even if the object is an array; for example, in

```
real(selected_real_kind(10,20)) :: z(100)
   ⋮
print *,z%kind
```

a single value is printed, that being the result of executing the reference to the intrinsic function `selected_real_kind`. This particular case is equivalent to `kind(z)`. However, the type parameter inquiry may be used even when the intrinsic function is not available; for example, in

```
subroutine write_centered(ch, len)
   character(*), intent(inout) :: ch
   integer, intent(in)         :: len
   integer                     :: i
   do i=1, (len-ch%len)/2
```

it would not be possible to replace the type parameter inquiry `ch%len` with a reference to the intrinsic function `len(ch)` because `len` is the name of a dummy argument.

Note that this syntax must not be used to alter the value of a type parameter, say by appearing on the left-hand side of an assignment statement.

13.2   Parameterized derived types

A derived type can have type parameters, in exact analogy with type parameters of intrinsic types. Like intrinsic type parameters, derived type parameters come in two flavours: those that must be known at compile time (like the `kind` parameter for type `real`), and those whose evaluation may be deferred until run time (like the `len` parameter for type `character`). The former are known as **kind** type parameters (because, for the intrinsic types, these are all named `kind`), and the latter as **length** type parameters (by analogy with character length).

Modern Fortran Explained, 3rd Edition. M. Metcalf, J. Reid, M. Cohen, and R. Bader. Oxford University Press (2024). © M. Metcalf, J. Reid, M. Cohen, and R. Bader (2024). DOI 10.1093/oso/9780198876571.001.0013

13.2.1   Defining a parameterized derived type

To define a derived type that has type parameters, the type parameters are listed on the type definition statement and must also be explicitly declared at the beginning of the derived-type definition. For example,

```
type matrix(real_kind, n, m)
   integer, kind   :: real_kind
   integer, len    :: n, m
   real(real_kind) :: value(n, m)
end type matrix
```

defines a derived type `matrix` with one kind type parameter named `real_kind` and two length type parameters named n and m. All type parameters must be explicitly declared to be of type `integer` with the attribute `kind` or `len` to indicate a kind or length parameter, respectively. Within the derived-type definition a kind type parameter may be used in both constant and specification expressions, but a length type parameter may only be used in a specification expression (that is, for array bounds and for other length type parameters such as character length). There is, however, no requirement that a type parameter be used at all.

If a component is default-initialized, its type parameters and array bounds must be constant expressions. For example, if a component is declared as

```
character(n) :: ch(m) = 'xyz'
```

both n and m must be named constants or kind type parameters.

Examples of valid and invalid parameterized derived types are shown in Figure 13.1.

Figure 13.1 A valid and an invalid parameterized derived type.

```
type goodtype(p1, p2, p3, p4)
   integer, kind     :: p1, p3
   integer, len      :: p2, p4
   real(kind=p1)     :: c1      ! ok, p1 is a kind type parameter
   character(len=p2) :: c2      ! ok, this is a specification expr
   complex           :: c3(p3)  ! ok, p3 can be used anywhere
   integer           :: c4 = p1 ! ok, p1 can be used anywhere
   ! p4 has not been used, but that is ok.
end type goodtype

type badtype(p5)
   integer, len :: p5
   real(kind=p5) :: x      ! Invalid, p5 is not a kind type parameter
   integer       :: y = p5 ! Invalid, p5 is not a kind type parameter
end type badtype
```

When declaring an entity of a parameterized derived type, its name is qualified by the type parameters in a type declaration statement of the form

```
type( derived-type-spec )
```

where *derived-type-spec* is
    derived-type-name (*type-param-spec-list*)
in which *derived-type-name* is the name of the derived type and *type-param-spec* is
    [keyword =] type-param-value
The keyword must be the name of one of the type parameters of the type. Like keyword
arguments in procedure calls, after a *type-param-spec* that includes a *keyword* = clause, any
further type parameter specifications must include a keyword. Note that this is consistent
with the syntax for specifying type parameters for intrinsic types. Here are some examples
for variables of our type matrix:

```
type(matrix(kind(0.0), 10, 20)) :: x
type(matrix(real_kind=kind(0d0), n=n1, m=n2)) :: y
```

13.2.2   Assumed and deferred type parameters

As for a dummy argument of the intrinsic type character, a length type parameter for a
derived type dummy argument may be **assumed**. In this case, its value is indicated by a
type-param-value that is an asterisk and is taken from that of the actual argument, as in the
example:

```
subroutine print_matrix(z)
   type(matrix(selected_real_kind(30,999), n=*, m=*)) :: z
     ⋮
```

An asterisk may also be used for an assumed type parameter in the allocate statement (see
Section 15.4) and the select type statement (see Section 15.7).
   As for the intrinsic type character, a length *type-param-value* for a derived type may be
deferred. For example, in

```
type(matrix(selected_real_kind(30,999), n=:, m=:)), pointer :: mp
type(matrix(selected_real_kind(30,999), n=100, m=200)), target :: x
mp => x
```

the values for mp of both n and m are deferred until association or allocation. After execution
of the pointer assignment, the n and m type parameter values of mp are equal to those of x
(100 and 200, respectively).

13.2.3   Default type parameter values

All type parameters for intrinsic types have default values. Similarly, a type parameter for a
derived type may have a default value; this is declared using the same syntax as for default
initialization of components, for example

```
type char_with_max_length(maxlen, kind)
   integer, len        :: maxlen = 255
   integer, kind       :: kind = selected_char_kind('default')
   integer             :: len
   character(maxlen, kind) :: value
end type char_with_max_length
```

When declaring objects of type `char_with_max_length`, it is not necessary to specify the `kind` or `maxlen` parameters if the default values are acceptable.

This also illustrates that, in many simple cases that have only one kind type parameter, the natural name for the type parameter may be `kind` (just as it is for the intrinsic types). That name was chosen in this particular example because `char_with_max_length` was meant to be as similar to the intrinsic type `character` as possible. Note that this choice does not conflict with the attribute keyword `kind`.

13.2.4 Derived type parameter inquiry

The value of a type parameter of a variable can be discovered by a type parameter inquiry, as with intrinsic types (see Section 13.1). For example, using the type of Figure 13.2, in

```
type(character_with_max_length(...,...)) :: x, y(100)
   ⋮
print *,x%kind
print *,y%maxlen
```

the values of the `kind` type parameter of x and the `maxlen` type parameter of y will be printed.

Figure 13.2 Using default parameter values.

```
type character_with_max_length(maxlen, kind)
    integer, len    :: maxlen
    integer, kind   :: kind = selected_char_kind('default')
    integer         :: length = 0
    character(maxlen, kind) :: value
end type character_with_max_length
   ⋮
type(character_with_max_length(100)) :: name
   ⋮
name = character_with_max_length(100)(value='John Hancock')
```

Because component syntax is used to access the value of a type parameter, a type is not allowed to have a component whose name is the same as one of the parameters of the type.

13.2.5 Structure constructor

In a structure constructor for a derived type that has type parameters, the type parameters are specified in parentheses immediately after the type name. Further, if the type parameters have default values, they may be omitted, as in the example in Figure 13.2.

Exercises

1. Write a replacement for the intrinsic type complex, which is opaque (has private components), uses polar representation internally, and has a single `kind` parameter that has the same default as the intrinsic type.

2. Write replacements for the character concatenation operator (`//`) and the intrinsic function `index` which work on type `char_with_max_length` (defined in Section 13.2.3).

14. Abstract interfaces and procedure pointers

14.1 Abstract interfaces

We have seen that to declare a dummy or an external procedure one needs to use an interface block. This is fine for a single procedure, but is somewhat verbose for declaring several procedures that have the same interface (apart from the procedure names). Also, there are several situations where this becomes impossible (procedure pointer components or abstract type-bound procedures). For these situations the **abstract interface** is available. An abstract interface gives a name to a set of characteristics and argument keyword names that would constitute an explicit interface to a procedure, without declaring any actual procedure to have those characteristics. This abstract interface name may be used in the `procedure` statement to declare procedures which might be external procedures, dummy procedures, procedure pointers, or deferred type-bound procedures.

An abstract interface block contains the `abstract` keyword, and each procedure body declared therein defines a new abstract interface. For example, given the abstract interface block

```
abstract interface
    subroutine boring_sub_with_no_args
    end subroutine boring_sub_with_no_args
    real function r2_to_r(a, b)
        real, intent(in) :: a, b
    end function r2_to_r
end interface
```

the declaration statements

```
procedure(boring_sub_with_no_args) :: sub1, sub2
procedure(r2_to_r) :: modulus, xyz
```

declare `sub1` and `sub2` to be subroutines with no actual arguments, and `modulus` and `xyz` to be `real` functions of two `real` arguments. The names `boring_sub_with_no_args` and `r2_to_r` are local to the scoping unit in which the abstract interface block is declared, and do not represent procedures or other global entities in their own right.

As well as with abstract interfaces, the `procedure` statement may be used with any specific procedure. For example,

Modern Fortran Explained, 3rd Edition. M. Metcalf, J. Reid, M. Cohen, and R. Bader. Oxford University Press (2024). © M. Metcalf, J. Reid, M. Cohen, and R. Bader (2024). DOI 10.1093/oso/9780198876571.001.0014

```
procedure(fun) :: fun2
```

declares `fun2` to be a procedure with an identical interface to that of `fun`.

The `procedure` statement is not available for a set of generic procedures, but can be used for a specific procedure that is a member of a generic set. All the intrinsic procedures are generic so cannot be named in a `procedure` statement.[1]

14.2   Procedure pointers

14.2.1   Introduction

The pointers that were discussed in previous chapters were **data pointers** that can be associated with a data object. A **procedure pointer** is a pointer that can be associated with a procedure instead of a data object. Its association with a target is as for the association of an actual and dummy procedure, so its interface is not permitted to be generic or elemental.

14.2.2   Named procedure pointers

A procedure pointer is declared by specifying that it is both a procedure and has the `pointer` attribute. For example,

```
pointer :: sp
interface
   subroutine sp(a, b)
      real, intent(inout) :: a
      real, intent(in)    :: b
   end subroutine sp
end interface
```

declares `sp` to be a pointer to a subroutine with the specified explicit interface.

If a procedure pointer is currently associated (is neither disassociated nor undefined), its target may be invoked by referencing the pointer. For example,

```
sp => sub
call sp(a, b)    ! calls sub
```

A procedure pointer is usually declared with the `procedure` statement specifying the `pointer` attribute:

```
procedure(sp), pointer :: p1 ! Pointer with the interface of sp
```

The statement may also be used to declare an external or dummy procedure. The full syntax of the `procedure` declaration statement is

procedure (*proc-interface*) [[, *proc-attr-spec*] ... ::] *proc-decl-list*

---

[1] A few intrinsics also have obsolescent specific versions that may be passed as actual arguments and are listed in Table B.2. An intrinsic may be named in a `procedure` statement only if the name appears in this table.

where *proc-interface* is an interface name and *proc-attr-spec* is one of

```
bind(c[,name=character-string])   intent(in)   intent(out)   intent(inout)
optional                          pointer      private       protected
public                            save
```
and a *proc-decl* is

  procedure-name [=> procedure-pointer-init]

where *procedure-pointer-init* is a reference to the intrinsic function `null` with no arguments, or the name of a non-elemental external or module procedure.[2]; it initializes pointer procedures. The `bind` attribute for procedures is described in Section 20.9. Each *proc-attr-spec* gives all the procedures declared in that statement the corresponding attribute. The `intent`, `protected`, and `save` attributes, and the initialization (to being a null pointer or associated with another procedure), may only appear if the procedures are pointers.

14.2.3   Procedure pointer components

A component of a derived type is permitted to be a procedure pointer. It must be declared using the `procedure` statement. For example, to define a type for representing a list of procedures (each with the same interface) to be called at some time, a procedure pointer component can be used, see Figure 14.1.

Figure 14.1 A type with a procedure pointer component.

```
type process_list
   procedure(process_interface), pointer :: process
   type(process_list), pointer           :: next => null()
end type process_list
abstract interface
   subroutine process_interface( ... )
      ⋮
   end subroutine process_interface
end interface
```

A procedure pointer component may be pointer-assigned to a procedure pointer, passed as an actual argument, or invoked directly. For example,

```
type(process_list) :: x, y(10)
procedure(process_interface), pointer :: p
   ⋮
p => x%process
call another_subroutine(x%process)
call y(i)%process(...)
```

---
[2]Or a specific intrinsic name, see Section B.1.15

Note that, just as with a data pointer component, in a reference to a procedure pointer component, the object of which the pointer is a component must be scalar (because there are no arrays of pointers in Fortran). Note also that if the object is allocatable it must be allocated and if it is a pointer it must have a target.

When a procedure is called through a pointer component of an object there is often a need to access the object itself; this is the topic of the next section.

14.2.4 The pass attribute

When a procedure pointer component (or a type-bound procedure, Section 15.8) is invoked, the object through which it is invoked is normally passed to the procedure as an additional argument. By default, this is the first argument, but another argument may be specified with the pass attribute as illustrated in Figure 14.2. The pass attribute naming the first dummy argument or without an argument name in parentheses may be used to confirm the default.

Figure 14.2 Associating the invoking object with the dummy argument x.

```
type t
   procedure(obp), pointer, pass(x) :: p
end type
abstract interface
   subroutine obp(w, x)
      import   :: t
      integer  :: w
      class(t) :: x
   end subroutine
end interface
   :
type(t) a
a%p => my_obp_sub
   :
call a%p(32)   ! equivalent to 'call my_obp_sub(32, a)'
```

The dummy argument to which the object is to be passed is known as the *passed-object dummy argument*. Obviously, it must be scalar and of the same declared type as the object. The dummy argument is required not to be a pointer or allocatable. This allows the object to be allocatable, a pointer, or neither. It also disallows the object becoming deallocated during the execution of the procedure. Also, the dummy argument is not permitted to have the value attribute and if it has any length type parameters they are required to be assumed.

If the type is extensible, the actual argument might be polymorphic. To ensure that the complete argument is passed to the procedure, a passed-object dummy argument of extensible type is required to be polymorphic, see Figure 14.2.

Note that the pass attribute applies to the procedure pointer component and not to the procedure with which it is associated. For example, a procedure might be associated with two different procedure pointers; the object might be passed as the first argument in the first case and as the second argument in the second case. However, if the associated procedure is invoked through some other means, there is no passed-object dummy argument, so an explicit actual argument must be provided in the reference (as in 'call my_obp_sub(32, a)' in Figure 14.2).

If it is not desired to pass the invoking object to the procedure at all, the nopass attribute must be used to override the default.

14.2.5   Internal procedures as targets of a procedure pointer

An internal procedure can be used as the target of a procedure pointer. When it is invoked via the corresponding procedure pointer, it has access to the variables of its host procedure as if it had been invoked there. When the host procedure returns, the procedure pointer will become undefined because the environment necessary for the evaluation of the internal procedure will have disappeared. For example, in Figure 14.3, on return from sub the variable n no longer exists for f to refer to.

Figure 14.3 Unsafe pointer to internal procedure.

```
module unsafe
   procedure(real), pointer :: funptr
contains
   subroutine sub(n)
      funptr => f    ! Associates funptr with internal function f.
      call process   ! funptr will remain associated with f during the
                     ! execution of subroutine "process".
      return         ! Returning from sub makes funptr become undefined.
   contains
      real function f(x)
         real, intent(in) :: x
         f = x**n
      end function
   end subroutine
end module
```

Exercises

1. Design a derived type to specify user-defined error handling; it should have a place for returning an error code and error message, and a procedure pointer component to allow for an 'error callback'.

2. Write an action queue (data structure) and dispatcher (procedure) using procedure pointer components. Each action should have a time and a procedure to be invoked; the action procedures should take the time as an argument. There should be a schedule procedure which, given a time and a procedure, queues an action for that time. If the time has already passed, the procedure should still be enqueued for immediate activation. The dispatcher procedure itself should, on invocation, process each event in the queue in time order (including extra events scheduled during this process) until the queue is empty.

15. Object-oriented programming

15.1   Introduction

The object-oriented approach to programming and design is characterized by its focus on the data structures of a program rather than the procedures. Often, invoking a procedure with a data object as its principal argument is thought of as 'sending a message' to the object. Typically, special language support is available for collecting these procedures (sometimes known as 'methods') together with the definition of the type of the object.

This approach is supported in Fortran by type extension, polymorphic variables, and type-bound procedures.

15.2   Type extension

15.2.1   Introduction

Type extension was introduced in Section 2.14. It creates a new derived type by extending an existing derived type. For example, the type

```
type person
   character(len=10)  :: name
   real               :: age
   integer            :: id
end type person
```

can be extended to form a new type thus

```
type, extends(person) :: employee
   integer :: national_insurance_number
   real    :: salary
end type employee
```

The old type must be extensible and is known as the parent type. The new type inherits all the components of the parent type by a process known as inheritance association and may have additional components. Where the order matters, that is, in a structure constructor that does not use keywords[1] and in default derived-type input/output (Section 10.3), the components inherited from the parent come first in their order, followed by the new components in their order.

---

[1]The use of keywords in structure constructors was described in Section 8.18.

Modern Fortran Explained, 3rd Edition. M. Metcalf, J. Reid, M. Cohen, and R. Bader. Oxford University Press (2024). © M. Metcalf, J. Reid, M. Cohen, and R. Bader (2024). DOI 10.1093/oso/9780198876571.001.0015

An extended type is itself extensible. We will use the term **ancestor type** to refer to the type itself, its parent type, or the parent of the parent type, etc., and the term **descendant type** for any type for which the given type is an ancestor. The components of a type are ordered by applying recursively the rule that parent type components are followed by type components.

15.2.2   Parent component

Additionally, an extended type has a **parent component**; this is a component that has the type and type parameters of the parent type and its name is that of the parent type. It allows the inherited portion to be referenced as a whole. Thus, an `employee` variable has a component called `person` of type `person`, associated with the inherited components. For example, given

```
type(employee) :: director
```

the component `director%name` is the same as `director%person%name`, and so on. The parent component is particularly useful when invoking procedures that operate on the parent type but which were not written with type extension in mind. For example, the procedure

```
subroutine display_older_people(parray, min_age)
    type(person), intent(in) :: parray(:)
    integer, intent(in)      :: min_age
    intrinsic                :: size
    do i=1, size(parray)
        if (parray(i)%age >= min_age) print *, parray(i)%name
    end do
end subroutine display_older_people
```

may be used with an array of `type(employee)` by passing it the parent component of the array, for example

```
type(employee) :: staff_list(:)
    :
! Show the employees eligible for early retirement
call display_older_people(staff_list%person, 55)
```

The parent component is not ordered with respect to the other components, so no value for it can appear in a structure constructor unless keywords are used (see Section 8.18).

The parent component is itself inherited if the type is further extended (becoming a 'grandparent component'); for example, with

```
type, extends(employee) :: salesman
    real :: commission_rate
end type salesman
type(salesman) :: traveller
```

the `traveller` has both the `employee` and `person` components, and `traveller%person` is exactly the same as `traveller%employee%person`.

15.2.3   Extension without adding components

A type can be extended without adding components, for example

```
type, extends(employee) :: clerical_staff_member
end type clerical_staff_member
```

Although a `clerical_staff_member` has the same ultimate components as an `employee`, it is nonetheless considered to be a different type.

Extending a type without adding components can be useful in several situations, in particular:

- to create a type with additional operations (as type-bound procedures, see Section 15.8);
- to create a type with different effects for existing operations, by overriding (Section 15.8.4) specific type-bound procedures; and
- for classification, that is, when the only extra information about the new type is the fact that it is of that type (for example, as in the `clerical_staff_member` type above).

15.2.4   Type extension and type parameters

When a type is extended, the new type inherits all of the type parameters. New type parameters may also be added, as illustrated in Figure 15.1, where the variable x has four type parameters: `real_kind`, n, m, and `max_label_length`.

Figure 15.1 Extending with added type parameters.

```
type matrix(real_kind, n, m)
   integer, kind   :: real_kind
   integer, len    :: n, m
   real(real_kind) :: value(n, m)
end type matrix
   ⋮
type, extends(matrix) :: labelled_matrix(max_label_length)
   integer, len                 :: max_label_length
   character(max_label_length) :: label
end type labelled_matrix
   ⋮
type(labelled_matrix(kind(0.0), 10, 20, 200)) :: x
```

15.3   Polymorphic entities

15.3.1   Introduction

Polymorphic pointer and allocatable variables were introduced in Section 2.15 and polymorphic dummy arguments were introduced in Section 5.7.4. A polymorphic variable must be a pointer, allocatable, or a dummy argument, and is declared using the `class` keyword in place of the `type` keyword. A type component may be polymorphic; it must be allocatable or a pointer.

The type named in the `class` attribute must be an extensible derived type. This type is called the **declared type** of the polymorphic entity, and the type of the object to which it refers is called the **dynamic type**. The dynamic type is always a descendant of the declared type (of the same declared type or any of the type's extensions).

We say that the polymorphic object is **type compatible** with any object of a descendant type.[2] A polymorphic pointer may only be pointer associated with a type-compatible target, a polymorphic allocatable variable may only be allocated to have a type-compatible allocation (see Section 15.4.3), and a polymorphic dummy argument may only be argument-associated with a type-compatible actual argument. Furthermore, if a polymorphic dummy argument is allocatable or a pointer, the actual argument must be of the same declared type; this is to ensure that the type-compatibility relationship (the dynamic type is a descendant of the declared type) is enforced.

However, even when a polymorphic entity is referring to an object of an extended type, it provides access via component notation only to components, type parameters, and bindings (see Section 15.8) of the declared type. This is because the compiler only knows about the declared type of the object, it cannot know about the dynamic type (which may vary at run time). While it is very useful to be able to write code (often in subprograms) that will execute correctly for any extension of a given declared type, there are situations where it is desirable to gain access to components, etc. that are in the dynamic type but not the declared type, or to execute alternative code for different extensions. This facility is provided by dynamic dispatch, which allows a polymorphic variable of a given dynamic type to be associated with a polymorphic variable whose declared type is that type (see Section 15.9). Alternatively, access to components, etc. that are in the dynamic type but not the declared type is provided by the `select type` construct (see Section 15.7).

15.3.2 Establishing the dynamic type

A polymorphic dummy argument that is neither allocatable nor a pointer has its dynamic type and the values of its type parameters established by argument association and they do not vary during a single execution of the procedure, though they may differ on different invocations. However, the dynamic type and type parameter values of a polymorphic allocatable or pointer variable can be altered at any time, as follows:

- it can be allocated to be of the type and type parameters specified on the `allocate` statement, see Section 15.4;
- using the `source=` or `mold=` specifier on the `allocate` statement, it can be allocated to have the type and type parameters of another variable;

the dynamic type of a polymorphic allocatable variable can be altered:

- when an allocation is transferred from one allocatable variable to another using the intrinsic subroutine `move_alloc` (see Section 6.8), the receiving variable takes on the dynamic type and type parameters of the sender;

---

[2] A non-polymorphic object is type compatible only with objects of the same declared type.

- in an intrinsic assignment statement, see Section 15.5;

and the dynamic type of a polymorphic pointer variable can be altered:

- via pointer association because a polymorphic pointer has the dynamic type and type parameters of its target.

Note that an `allocate` statement without a type specification, a `source=` specifier, or a `mold=` specifier will allocate the variable to be of its declared type.

The dynamic type and type parameters of a disassociated pointer or unallocated allocatable variable are those of its declared type. A pointer with undefined association status has no defined dynamic type: it is not permitted to be used in any context where its dynamic type would be relevant.

15.3.3   Limitations on the use of a polymorphic variable

A polymorphic variable may appear in an input/output list only if it is processed by defined derived-type input/output (Section 11.6).

The variable in an intrinsic assignment statement is not permitted to be polymorphic unless it is allocatable. However, if it is associated with a non-polymorphic variable, assigning to the non-polymorphic variable will have the desired effect, for example

```
type(person) :: Jones = person("John",50.,4567)
type(person), target :: Smith
class(person), pointer :: pupil
   :
pupil => Smith
Smith = Jones   ! Valid assignment to target of pupil
```

In a pure procedure, a polymorphic variable is not permitted to be an intent out dummy argument and no statement that might result in the deallocation of a polymorphic entity is permitted. A polymorphic variable is not permitted to be an actual argument corresponding to an intent out assumed-size dummy argument (see Section 20.5).

15.3.4   Polymorphic arrays and scalars

A polymorphic variable can be either an array or a scalar (including also an allocatable scalar).

A polymorphic array is always homogeneous; that is, each array element has the same dynamic type and type parameters. This is by language design: every method for establishing the dynamic type of a polymorphic variable provides a single type for the entire array. The reason for this is both to make reasoning about programs simpler and to ensure that accessing an element of a polymorphic array is reasonably efficient.

If a heterogeneous polymorphic array is required, an array of a derived type with a scalar polymorphic component may be used, for example

```
type poly_array
    class(person), allocatable :: comp
end type
type (poly_array) :: p_array(10)
type(person) :: Jones = person("John",50.,4567)
p_array(5)%comp = Jones
```

15.3.5   Unlimited polymorphic entities

Sometimes one wishes to have a polymorphic variable that has no declared type but may be given by the usual mechanisms of association or allocation any dynamic type, including a non-extensible or intrinsic type. For example, one might wish to have a 'universal' list of variables (pointer targets), each of which might be of any type. This can be done with an **unlimited polymorphic** object. These are declared using * as the class specifier, for example

```
class(*), pointer :: up
class(*), allocatable :: down
```

declares up to be an unlimited polymorphic pointer. This could be associated with a real target, for instance:

```
real, target :: x
    :
up => x
```

The value of an unlimited polymorphic object cannot be accessed directly, but the object as a whole can be used; for example, it can be copied with source=, passed as an argument, used in pointer assignment, and most importantly it can be the selector in a select type statement (Section 15.7).

Type information is maintained for an unlimited polymorphic pointer while it is associated with an intrinsic type or an extensible derived type, but not when it is associated with a non-extensible derived type. (This is because different non-extensible types are considered to be the same if they have the same structure and names.) To prevent a pointer of intrinsic or extensible type from becoming associated with an incompatible target, such a pointer is not permitted to be the left-hand side of a pointer assignment if the target is unlimited polymorphic. This is shown in Figure 15.2.

Instead of the invalid pointer assignment, a select type construct must be used to associate a pointer of intrinsic or extensible type with an unlimited polymorphic target. For example, if the unlimited polymorphic pointer up is associated with the real target x, the execution of

```
select type(up)
type is (real)
    up = 3.5
    rp => up
end select
```

Figure 15.2 Compatible target and incompatible target.

```
use iso_c_binding
type, bind(c) :: triplet
   real(c_double) :: values(3)
end type triplet
class(*), pointer      :: univp
type(triplet), pointer :: tripp
real, pointer          :: realp
:

univp => tripp           ! Valid
univp => realp           ! Valid
:

tripp => univp           ! Valid when the dynamic type matches
realp => univp           ! Always invalid
```

---

assigns the value of 3.5 to x and associates the `real` pointer rp with x. Note that the syntax of the `select type` construct does not allow a `type is` statement to specify a `sequence` or `bind` derived type.

When an unlimited polymorphic pointer is allocated, the required type and type parameter values must be specified in the `allocate` statement (Section 15.4).

A longer example, showing the use of unlimited polymorphic in constructing a generic vector list package, is shown in Figure 15.3.

15.3.6   Polymorphic entities and generic resolution

Because a polymorphic dummy argument may be associated with an actual argument of an extended type, a polymorphic dummy argument is not distinguishable from a dummy argument of an extended type in the rules for distinguishing procedures in a generic set (Section 5.18). For example, the procedure

```
real function data_distance(a, b)
   class(data_point) :: a, b
   data_distance = ...
end function data_distance
```

is not permitted in the same generic set as the function `distance` defined in Section 5.7.4. Where such an effect is required, type-bound procedures (Section 15.8.4) may be employed.

In the case of an unlimited polymorphic dummy argument, because it is type-compatible with any type (any type is a descendant type), it is indistinguishable from any argument of the same rank (except as noted in Section 5.18).

Figure 15.3 Generic vector list and type selection.

```
type generic_vector_list_elt
   class(*), allocatable :: element_vector(:)
   type(generic_vector_list_elt),pointer :: next => null()
end type generic_vector_list_elt

type(generic_vector_list_elt), pointer :: p
: ! Code to construct list associated with p
do
   if (.not.associated(p)) exit
   select type(q => p%element_vector)
   type is (integer(selected_int_kind(9)))
      call special_process_i9(q)
   type is (real)
      call special_process_default_real(q)
   type is (character(*))
      call special_process_character(q)
   class default
      stop "Type not supported"
   end select
   p => p%next
end do
```

15.4 Typed and sourced allocation

15.4.1 Introduction

As well as determining array size, the allocate statement can determine type parameter values, dynamic type (for a polymorphic variable), and value. The general form of the allocate statement (Section 6.5) is

allocate *[type-spec ::] allocation-list [, alloc-spec] ...*)

If *type-spec* is present, it takes the form of the type name followed by the type parameter values in parentheses, if any, for both intrinsic and derived types. Alternatively, an *alloc-spec* may be source=*expr* or mold=*expr*, where *expr* is an expression that is type-compatible with the declared type of each of the objects in *allocation-list* (a descendant type of each type). If each *allocation* is for an array of the same rank, *expr* may be an array of that rank, otherwise *expr* must be scalar.

An allocate statement with a *type-spec* is a **typed allocation**, and an allocate statement with a source= or mold= clause is a **sourced allocation**. Only one of these clauses is allowed, so an allocate statement cannot both be a typed allocation and a sourced allocation. We now explain these features.

15.4.2   Typed allocation and deferred type parameters

A length type parameter that is deferred (indicated by a colon in the *type-spec*) has no
defined value until it is given one by an `allocate` statement or by pointer assignment (a
type parameter that is not deferred cannot be altered by `allocate` or pointer assignment).
For example, in

```
character(:), allocatable :: x(:)
   :
allocate (character(n) :: x(m))
```

the array x will have m elements and each element will have character length n after execution
of the `allocate` statement.

    If a length parameter of an item being allocated is assumed, it must be specified as an
asterisk in the *type-spec*. For example, the type parameter `string_dim` in Figure 15.4 must
be specified as * because it is assumed.

Figure 15.4 Allocating a dummy argument with an assumed type parameter.

```
type string_vector(string_dim, space_dim)
   integer, len             :: string_dim, space_dim
   type(string(string_dim)) :: value(space_dim)
end type string_vector
   :
subroutine allocate_string_vectors(vp, n, m)
   type(string_vector(*,:)), pointer :: vp(:)
   integer, intent(in)               :: n, m
   allocate (string_vector(string_dim=*, space_dim=n) :: vp(m))
end subroutine allocate_string_vectors
```

Note that there is only one *type-spec* in an `allocate` statement, so it must be suitable for
all the items being allocated. In particular, if any one of them is a dummy argument with an
assumed type parameter, they must all be dummy arguments that assume this type parameter.

    If any type parameter is neither assumed nor deferred, the value specified for it by the
type-spec must be the same as its current value. For example, in

```
subroutine allocate_string3_vectors(vp, n, m)
   type(string_vector(3,:)), pointer :: vp(:)
   integer, intent(in)               :: n, m
   allocate (string_vector(string_dim=3, space_dim=n) :: vp(m))
end subroutine allocate_string3_vectors
```

the expression provided for the `string_dim` type parameter must be equal to 3.

15.4.3   Polymorphic variables and typed allocation

For polymorphic variables, the *type-spec* specifies not only the values of any deferred type
parameters, but also the dynamic type. If an item is unlimited polymorphic, it can be allocated

to be any type (including intrinsic types); otherwise the type specified in the `allocate` statement must be a descendant of the declared type of the item.

For example,

```
class(*), pointer :: ux, uy(:)
class(t), pointer :: x, y(:)
   ⋮
allocate (t2 :: ux, x, y(10))
allocate (real :: uy(100))
```

allocates ux, x, and y to be of type `t2` (an extension of `t`), and uy to be of type default `real`.

15.4.4 Sourced allocation

Instead of allocating a variable with an explicitly specified type (and type parameters), it is possible to take the type, type parameters, and value from another variable or expression. This effectively produces a 'clone' of the source expression, and is done by using the `source=` clause in the `allocate` statement. For example, in

```
subroutine s(b)
   class(t), allocatable :: a
   class(t)              :: b
   allocate (a, source=b)
```

the variable a is allocated with the same dynamic type and type parameters as b, and will have the same value.

This is useful for copying heterogeneous data structures such as lists and trees, as in the example in Figure 15.5. Note that the procedure in which the allocation is performed does not need access to the definition of the source object's dynamic type.

Figure 15.5 Allocating an object with the type and type parameters of another object.

```
type singly_linked_list
   class(singly_linked_list), pointer :: next => null()
   ! No data - the user of the type should extend it to include
   ! desired data.
end type singly_linked_list
   ⋮
recursive function sll_copy(source) result(copy)
   class(singly_linked_list), pointer     :: copy
   class(singly_linked_list), intent(in) :: source
   allocate (copy, source=source)
   if (associated(source%next)) copy%next => sll_copy(source%next)
end function sll_copy
```

If the allocated item is an array, its bounds and shape are specified in the usual way and are not taken from the source. This allows the source to be a scalar whose value is given to every element of the array. Alternatively, it may be an array of the same shape.

As we have seen, making a clone of an array can be done as follows:

```
class(t), allocatable :: a(:), b(:)
   ⋮
allocate (a(lbound(b,1):ubound(b,1)), source=b)
```

However, the bounds may be omitted, in which case they will be taken from the source= specifier, allowing the much simpler

```
allocate (a, source=b)
```

Further, there is a facility to allocate a variable to the shape, type, and type parameters of an expression without copying its value. This is done with the mold= specifier, for example

```
allocate (a, mold=b)
```

The shape of the allocated array is that of the source= or mold= expression and its lower bounds are those returned by the intrinsic lbound for the expression. After the allocation any relevant default initialization will be applied to the allocated array.

Finally, either mold= or source= may be used when allocating multiple objects, for example,

```
allocate (a(10), b(20), source=173)
```

15.5   Assignment for allocatable polymorphic variables

Intrinsic assignment to an allocatable polymorphic variable is allowed, and this extends the automatic reallocation feature that permits array shape and deferred type parameters to be changed.

If the variable is allocated and its dynamic type and type parameters differ from that of the expression, the variable is deallocated (just as if it were an array with different shape or had different deferred type parameter values). If the variable was unallocated, or is deallocated by the previous step, it is allocated to have the dynamic type of the expression (and array bounds or type parameter values, if applicable). Finally, the value is copied just as in normal assignment (with shallow copying for any pointer components and deep copying for any allocatable components).

An example is

```
class(*), allocatable :: x
   ⋮
x = 3
```

The effect of automatic reallocation is similar to that of

```
if (allocated(variable)) deallocate (variable)
allocate (variable, source=expression)
```

except that, in the intrinsic assignment case,

- the variable may appear in the expression, and any reallocation occurs after evaluation of the expression and before the copying of the value; and

- if the variable is already allocated with the correct type (and shape and deferred type parameter values, if applicable), no reallocation is done; apart from performance, this only matters when the variable also has the `target` attribute and there is a pointer associated with it: instead of the pointer becoming undefined, it will remain associated and will see the new value.

15.6 The `associate` construct

The `associate` construct allows one to associate a name either with a variable or with the value of an expression, for the duration of a block. Any entity with this name outside the construct is separate and inaccessible inside it. The association is known as **construct association**. During execution of the block, the *associate-name* remains associated with the variable (or retains the value) specified, and takes its type, type parameters, and rank from its association. This construct is useful for simplifying multiple accesses to a variable which has a lengthy description (subscripts and component names). For example, given a nested set of derived-type definitions, the innermost of which is

```
type one
   real, allocatable, dimension(:) :: xvec, levels
   logical                         :: tracing
end type one
```

then the association as specified in

```
associate(point_qfstate => master_list%item(n)%qfield%posn(i,j)%state)
   point_qfstate%xvec = matmul(transpose_matrix, point_qfstate%xvec)
   point_qfstate%levels = timestep(point_qfstate%levels, input_field)
   if (point_qfstate%tracing) call show_qfstate(point_qfstate, stepno)
end associate
```

would be even less comprehensible if `point_qfstate` were written out in full in each occurrence.

Formally, the syntax is

> *[name:]* `associate` (*association-list*)
>   *block*
> `end associate` *[name]*

where each *association* is

> *associate-name* => *selector*

and *selector* is either a variable or an expression. As with other constructs, the `associate` construct can be named; if *name:* appears on the `associate` statement, the same name must appear on the `end associate` statement.

If the association is with a variable that does not have a vector subscript, the *associate-name* may be used as a variable within the block. The association is as for argument association of a dummy argument that does not have the `pointer` or `allocatable` attribute, but the *associate-name* has the `target` attribute if the variable does. If the association is with an expression, the *associate-name* may be used only for its value. If the association is with an array, the bounds of *associate-name* are given by the intrinsic functions lbound and ubound applied to the array.

If the *selector* is polymorphic, *associate-name* is also polymorphic. If *selector* is a pointer or has the target attribute, *associate-name* has the target attribute. The only other attributes that *associate-name* receives from the *selector* are the asynchronous and volatile attributes; in particular, if *selector* has the optional attribute, *associate-name* does not and so *selector* must be present when the construct is executed.

Multiple associations may be established within a single associate construct. For example, in

```
associate ( x => arg(i)%ground%coordinates(1), &
             y => arg(i)%ground%coordinates(2) )
   distance = sqrt((myloc%x-x)**2+(myloc%y-y)**2)
   bearing = atan2(myloc%y-y, myloc%x-x)
end associate
```

the simplifying names x and y improve the readability of the code.

Without this construct, to make this kind of code readable either a procedure would need to be used, or pointers (requiring, in addition, the target attribute on the affected variables). This could adversely affect the performance of the program (and indeed would probably still not attain the readability shown here).

The construct may be nested with other constructs in the usual way.

15.7   The select type construct

To execute alternative code depending on the dynamic type and type parameters of a polymorphic entity and to gain access to the dynamic parts, the select type construct is provided, for example,

```
class(person) :: a
select type (out_a => a)
type is (employee)
   write(*,*) "Employee ",out_a%name, "has salary ", out_a%salary
type is (person)
   write(*,*) out_a%name, "is not an employee"
end select
```

The construct takes the general form

```
[ name: ] select type ( [ associate-name =>] selector)
[ type-guard-stmt [ name ]
     block ] ...
end select [ name ]
```

where each *type-guard-stmt* is one of

```
type is (derived-type-spec)
type is (intrinsic-type [ (type-parameter-value-list) ] )
class is (derived-type-spec)
class default
```

where *derived-type-spec* is defined in Section 13.2.1.

The *selector* is a variable or an expression and the *associate-name* is associated with it within each block by construct association in exactly the same way as for an associate construct (Section 15.6). At most one of the blocks is executed and it is chosen according to the dynamic type of *selector* as follows:

i) The block following a type is guard is executed if the dynamic type of the *selector* is exactly the derived type specified, and the kind type parameter values match. The associated variable is not polymorphic and has that type and those type parameter values.

ii) Failing this, the block following a class is guard is executed if it is the only one for which the dynamic type is the derived type specified, or an extension thereof, and the kind type parameter values match. If there is more than one such guard, one of them must be of a type that is an extension of the types of all the others, and its block is executed. The associated variable is polymorphic with the dynamic type of the *selector* and the declared type and type parameters given in *derived-type-spec*.

iii) Failing this, the block following a class default guard is executed. The associated variable has the declared and dynamic type and type parameters of *selector*.

Each *derived-type-spec* is required to be an extensible type that is type-compatible with the *selector* (a descendant of the type of the *selector*). A type guard that specifies an intrinsic type is permitted only if the *selector* is unlimited polymorphic (Section 15.3.5).

In the (frequently occurring) case where the *selector* is a simple name and the same name is suitable for the *associate-name*, the '*associate-name*=>' may be omitted.

As with other constructs, the select type construct can be named; if *name*: appears on the select type statement, the same name must appear on the end select statement and may appear on a type guard statement.

If the *derived-type-spec* contains a *type-param-spec-list*, values corresponding to kind type parameters must be constant expressions and those for length type parameters must be asterisks. This is so that length type parameters do not participate in type parameter matching, but are always assumed from the *selector*. Here is an example, based on the types in Figure 15.1.

```
subroutine out(a)
    class(matrix(kind(0.0),*,*)) :: a
    select type (a)
    type is (labelled_matrix(kind(0.0),*,*,*))
      write(*,*) "Matrix a has label ", trim(a%label)
    end select
end subroutine
```

The example in Figure 15.6 shows a typical use of select type. Each type guard statement that specifies an extended type provides access via component notation to the extended components.

Figure 15.6 Using the `select type` construct for polymorphic objects of class `particle`.

```
subroutine describe_particle(p)
   class(particle) :: p

! These attributes are common to all particles.
   call describe_vector('Position:',p%position)
   call describe_vector('Velocity:',p%velocity)
   print *,'Mass:',p%mass
! Check for other attributes.
   select type (p)
   type is (charged_particle)
      print *,'Charge:',p%charge
   class is (charged_particle)
      print *,'Charge:',p%charge
      print *,'... may have other (unknown) attributes.'
   type is (particle)
      ! Just the basic particle type, there is nothing extra.
   class default
      print *,'... may have other (unknown) attributes.'
   end select
end subroutine describe_particle
```

15.8   Type-bound procedures

15.8.1   Introduction

In object-oriented programming one often wishes to invoke a procedure to perform a task whose nature varies according to the dynamic type and type parameters of a polymorphic object.

This is the purpose of **type-bound procedures**. These are procedures which are invoked through an object, and the actual procedure executed depends on the dynamic type and type parameters of the object. This is sometimes known as dynamic dispatch.

They are called type-bound because the selection of the procedure depends on the type and type parameters of the object, in contrast to procedure pointer components which depend on the value of the object (one might call the latter object-bound).

In some other languages type-bound procedures are known as methods, and invocation of a method is thought of as 'sending a message' to the object.

However, type-bound procedures can be used even when there is no intention to extend the type. We will first describe how to define and use type-bound procedures in the simple case, and later explain how they are affected by type extension.

15.8.2 Specific type-bound procedures

The type-bound procedure section of a derived-type definition is separated from the component section by the `contains` statement, analogous to the way that module variables are separated from the module procedures. The default accessibility of type-bound procedures is separate from the default accessibility for components; that is, even with `private` components, each type-bound procedure is `public` unless a `private` statement appears in the type-bound procedure section or unless it is explicitly declared to be `private`.

Figure 15.7 A type with two type-bound procedures.

```
module mytype_module
    type mytype
        private
        real :: myvalue(4) = 0.0
    contains
        procedure :: write => write_mytype
        procedure :: reset
    end type mytype
    private :: write_mytype, reset
contains
    subroutine write_mytype(this, unit)
        class(mytype)    :: this
        integer, optional :: unit
        if (present(unit)) then
            write (unit, *) this%myvalue
        else
            print *,this%myvalue
        end if
    end subroutine write_mytype
    subroutine reset(variable)
        class(mytype) :: variable
        variable%myvalue = 0.0
    end subroutine reset
end module mytype_module
```

Each type-bound procedure declaration specifies the name of the **binding**, and the name of the actual procedure to which it is bound. (The latter may be omitted if it is the same as the type-bound procedure name.) For example, in Figure 15.7 objects of type `mytype` have two type-bound procedures, `write` and `reset`. These are invoked as if they were component procedure pointers of the object, and the invoking object is normally passed to the procedure as its first argument. For example, the procedure references

```
call x%write(6)
call x%reset
```

are equivalent to

```
call write_mytype(x,6)
call reset(x)
```

However, because they are public, the type-bound procedures (`write` and `reset`) can be referenced anywhere in the program that has a `type(mytype)` variable, whereas, because the module procedures (`write_mytype` and `reset`) are private, they can only be directly referenced from within `mytype_module`.

The full syntax of the statement declaring specific type-bound procedures is

procedure (*interface-name*), *binding-attr-list* :: *binding-name-list*

or

procedure *[[, binding-attr-list] ::] type-bound-proc-decl-list*

where each *binding-attr* is one of

```
public or private
deferred
non_overridable
nopass or pass [ (arg-name) ]
```

each *type-bound-proc-decl* is *tbp-name [=> proc-name]* and each *interface-name* or *proc-name* is the name of a procedure with an explicit interface. The `public` and `private` attributes are permitted only in the specification part of a module. The `pass` and `nopass` attributes are described in Section 14.2.4. The form with *interface-name* is for declaring deferred bindings, so it must contain the `deferred` attribute; these are described in Section 15.10. The form without *interface-name* declares ordinary type-bound procedures, so it must not contain the `deferred` attribute. An example of the case where it is not desired to pass the invoking object is shown in Figure 15.8.

A type-bound procedure declaration statement may take a list of procedure bindings, so that multiple type-bound procedures can be declared in a single statement, as in

```
type mycomplex
   ⋮
contains
   procedure :: i_plus_myc, myc_plus_i, myc_plus_myc=>myc_plus, &
                myc_plus_r, r_plus_myc
   ⋮
end type
```

This can be a significant improvement when a type has many type-bound procedures.

If the `non_overridable` attribute appears, that type-bound procedure cannot be overridden during type extension (see Section 15.8.4). Note that `non_overridable` is incompatible with `deferred`, since that requires the type-bound procedure to be overridden.

15.8.3   Generic type-bound procedures

Type-bound procedures may be generic. A generic type-bound procedure is defined with the `generic` statement within the type-bound procedure part. This statement takes the form

generic *[[, access-spec] ::] generic-spec => tbp-name-list*

Figure 15.8 Two type-bound procedures with the `nopass` attribute.

```
module utility_module
    private
    type, public :: utility_access_type
    contains
        procedure, nopass :: startup
        procedure, nopass :: shutdown
    end type
contains
    subroutine startup
        print *,'Process started'
    end subroutine
    subroutine shutdown
        stop 'Process stopped'
    end subroutine
end module
    :
use utility_module
type(utility_access_type) :: process_control
call process_control%startup
```

---

and can be used for named generics as well as for operators, assignment, and user-defined derived-type input/output specifications. Each *tbp-name* specifies an individual (specific) type-bound procedure to be included in the generic set. As always, procedures with the same *generic-spec* must all be subroutines or all be functions.

For example, in Figure 15.9 the type-bound procedure `extract` is generic, being resolved to one of the specific type-bound procedures `xi` or `xc`, depending on the data type of the argument. Thus, in

```
use container_module
type(container) v
integer ix
complex cx
    :
call v%extract(ix)
call v%extract(cx)
```

one of the 'extract_*something*_from_container' procedures will be invoked.

A generic type-bound procedure need not be named; it may be an operator, assignment, or a user-defined derived-type input/output specification. In this case, the object through which the type-bound procedure is invoked is whichever of the operands corresponds to the passed-object dummy argument. For this reason, the specific type-bound procedures for an unnamed generic must not have the `nopass` attribute. Like other type-bound procedures, unnamed generics that are public are accessible wherever the type or an object of the type is accessible.

Figure 15.9 A named generic type-bound procedure.

```
module container_module
   private
   type, public :: container
      integer, private :: i = 0
      complex, private :: c = (0.,0.)
   contains
      private
      procedure :: xi => extract_integer_from_container
      procedure :: xc => extract_complex_from_container
      generic, public :: extract => xi, xc
   end type
contains
   subroutine extract_integer_from_container(this, val)
      class(container), intent(in) :: this
      integer, intent(out)         :: val
      val = this%i
   end subroutine extract_integer_from_container
   subroutine extract_complex_from_container(this, val)
      class(container), intent(in) :: this
      complex, intent(out)         :: val
      val = this%c
   end subroutine extract_complex_from_container
end module container_module
```

Inside a derived-type definition, the generic statement can be used to declare type-bound generic interfaces. This statement may also be used to declare generic interfaces outside a derived-type definition. The syntax is similar to that of the type generic statement in a derived-type definition:

  generic [[, *access-spec*] ::] *generic-spec* => *specific-procedure-list*

where *access-spec* is public or private. This is more concise than using an interface block, reducing three statements (four if an *access-spec* is needed) to a single statement. An interface block still needs to be used when a specific procedure needs to be declared with an interface body.

All this is useful for packaging-up a type and its operations, because the only clause of a use statement does not affect the accessibility of type-bound operators, unlike operators defined by an interface block. This prevents the accidental omission of required operators by making a mistake in the use statement. This is particularly germane when using defined assignment between objects of the same type, since omitting the defined assignment would cause an unwanted intrinsic assignment to be used without warning.

For example, Figure 15.10 shows the overloading of the operator (+) for operations on type(mycomplex); these operations are available even if the user has done

```
use mycomplex_module, only: mycomplex
```

Figure 15.10 A generic type-bound operator.

```
module mycomplex_module
   type mycomplex
      private
      :   ! data components not shown
   contains
      procedure          :: mycomplex_plus_mycomplex
      procedure          :: mycomplex_plus_real
      procedure, pass(b) :: real_plus_mycomplex
      generic, public :: operator(+) => mycomplex_plus_mycomplex, &
                         mycomplex_plus_real, real_plus_mycomplex
      procedure          :: conjg => mycomplex_conjg
      :   ! many other operations and functions...
   end type
contains
   :   ! procedures which implement the operations
end module
```

15.8.4   Type extension and type-bound procedures

When a type is extended, the new type usually inherits all the type-bound procedures of the old type, as illustrated in Figure 15.11, where the new type charged_particle inherits not only the components of particle, but also its type-bound procedures momentum and energy.

Specific type-bound procedures defined by the new type are either additional bindings (with a new name), or may **override** type-bound procedures that would otherwise have been inherited from the old type. (However, overriding a type-bound procedure is not permitted if the inherited one has the non_overridable attribute.) An overriding type-bound procedure binding must have exactly the same interface as the overridden procedure except for the

Figure 15.11 Extending a type with type-bound procedures.

```
type particle
   type(vector) :: position, velocity
   real         :: mass
contains
   procedure :: momentum => particle_momentum
   procedure :: energy => particle_energy
end type particle

type, extends(particle) :: charged_particle
   real :: charge
end type charged_particle
```

type of the passed-object dummy argument; if there is a passed-object dummy argument, the overriding procedure must specify its type to be class (*new-type*). Furthermore, if the overridden procedure is pure, the overriding one must also be pure.

Generic type-bound procedures defined by the new type always extend the generic set; the complete set of generic bindings for any particular generic identifier (including both the inherited and newly defined generic bindings) must satisfy the usual rules for generic disambiguation (Sections 5.18.2 and 15.3.6). A procedure that would be part of an inherited generic set may be overridden using its specific name. For example, in Figure 15.12 the three specific type-bound procedures have been overridden; when the generic operation of (+) is applied to entities of type instrumented_mycomplex, one of the overriding procedures will be invoked.

Figure 15.12 Extending the type of Figure 15.10 with overriding of type-bound procedures.

```
type, extends(mycomplex) :: instrumented_mycomplex
   integer, public :: plus_op_count = 0
contains
   procedure :: mycomplex_plus_mycomplex => instrumented_myc_plus_myc
   procedure :: mycomplex_plus_real => instrumented_myc_plus_r
   procedure :: real_plus_mycomplex => instr_r_p_myc
end type instrumented_mycomplex
```

15.9   Design for overriding

When designing a type that can be overridden, there are several important points that should be considered.

Firstly, the specific procedures in a generic set need to be public, otherwise they cannot be overridden. This is why mycomplex_plus_mycomplex in Figure 15.10 is public; not because the user would want to invoke it directly, but to name the operation so it can be overridden.

Secondly, because the declared type of a function result cannot be changed when overriding, if a function has a result of the type, and when extended would be expected to return the extended type, the function result should be polymorphic and allocatable. For example, the result of mycomplex_conjg should be class(mycomplex), allocatable; in instrumented_mycomplex, the overriding function can then allocate its result to have a dynamic type of instrumented_mycomplex.

Thirdly, if a function implements a binary operation, because dynamic dispatch only occurs via one of the arguments, it needs to check whether the other argument is further extended, and if so, dispatch via that argument. For a commutative operation like addition, this is simply a matter of swapping the operands and invoking the type-bound addition again.

An example of mycomplex_plus_mycomplex that illustrates these points is in Figure 15.13, and an example of instrumented_myc_plus_myc with this design is in Figure 15.14.

Figure 15.13 Providing for multiple dispatch.

```
recursive function mycomplex_plus_mycomplex(a, b) result(r)
  class(mycomplex), intent(in) :: a, b
  class(mycomplex), allocatable :: r
  if (.not.same_type_as(a, b) .and. extends_type_of(b, a)) then
    ! Dispatch via b in case it has overridden this procedure.
    r = b + a
  else
    allocate(r)
    :  ! Do the addition and assign to the components of r.
  end if
end function
```

Figure 15.14 Overriding and multiple dispatch.

```
function instrumented_myc_plus_myc (a,b) result(r)
  class(instrumented_mycomplex), intent(in) :: a
  class(mycomplex), intent(in) :: b
  class(mycomplex), allocatable :: r
  if (.not.same_type_as(a,b) .and. extends_type_of(b,a)) then
    ! Dispatch via b in case it has overridden this procedure.
    r = b + a
  else
    r = a
    select type(r)
    class is (instrumented_mycomplex) ! Always true.
      r%plus_op_count = a%plus_op_count + 1
      select type(b)
      class is (instrumented_mycomplex)
        r%plus_op_count = r%plus_op_count + b%plus_op_count
        ! Use the mycomplex plus operation to do the actual addition.
        r%mycomplex = r%mycomplex + b%mycomplex
      class default
        ! Use the mycomplex plus operation to do the actual addition.
        r%mycomplex = r%mycomplex + b
      end select
    end select
  end if
end function
```

A fourth consideration is that if an operation is not commutative, providing for multiple dispatch will require the 'reverse' function also to be provided. For example, in the case of subtraction (a − b), because b − a would give the wrong result, we would need a function that computes 'b.rsub.a'. Finally, although the implied extra memory allocation can easily be eliminated by the compiler for r = b + a, this is not the case for the additions to r%mycomplex. These could be avoided if mycomplex also provided an 'add to' function, that adds a mycomplex value to a mycomplex variable; in this case the actual addition invocations in Figure 15.14 would be replaced by call r%mycomplex%addto(...), where '...' is b%mycomplex for the first actual addition call, and b for the second. (Although making the components public so that instrumented_myc_plus_myc could do the addition itself would also avoid the extra memory allocation overhead, that would break the encapsulation of mycomplex, making it difficult if not impossible to later change its internal representation.)

15.10   Deferred bindings and abstract types

Sometimes a type is defined not for the purpose of creating objects of that type, but only to serve as a base type for extension. In this situation, a type-bound procedure in the base type might have no default or natural implementation, but rather only a well-defined purpose and interface. This is supported by the abstract keyword on the type definition and the deferred keyword in the procedure statement.

A simple example is shown in Figure 15.15. Here, the intention is that extensions of the type would have components that hold data about the file, and that my_open and my_close would be overridden by a procedures that uses these data to open and close it.

Figure 15.15 An abstract type.

```
type, abstract :: file_handle
contains
    procedure (manage_file), deferred, pass :: my_open, my_close
    ⋮
end type file_handle
abstract interface
    subroutine manage_file(handle)
        import                              :: file_handle
        class(file_handle), intent(inout) :: handle
    end subroutine manage_file
end interface
```

Such a procedure is known as a **deferred** type-bound procedure. An interface is required, which may be an abstract interface or that of a procedure with an explicit interface.

No ordinary variable is permitted to be of an abstract type, but a polymorphic variable may have it as its declared type. When an abstract type is extended, the new type may be a normal extended type or may itself be abstract. Deferred bindings are allowed only in abstract types. (But an abstract type is not required to have any deferred binding.)

Figure 15.16 shows the definition of an abstract type `my_numeric_type`, and the creation of the normal type `my_integer_type` as an extension of it. Variables that are declared to be `my_numeric_type` must be polymorphic, and if they are pointer or allocatable the `allocate` statement must specify a normal type (see Section 15.4).

Figure 15.16 Abstract numeric type.

```
type, abstract :: my_numeric_type
contains
   private
   procedure(op2), deferred :: add
   procedure(op2), deferred :: subtract
   :  ! procedures for other operations not shown
   generic, public :: operator(+) => add, ...
   generic, public :: operator(-) => subtract, ...
   :  ! generic specs for other operations not shown
end type my_numeric_type
abstract interface
   function op2(a, b) result(r)
      import :: my_numeric_type
      class(my_numeric_type), intent(in)  :: a, b
      class(my_numeric_type), allocatable :: r
   end function op2
end interface
type, extends(my_numeric_type) :: my_integer
   integer, private :: value
contains
   procedure :: add => add_my_integer
   procedure :: subtract => subtract_my_integer
   :
end type my_integer
```

The use of the `abstract` and `deferred` attributes ensures that objects of insufficient type cannot be created, and that when extending the abstract type to create a normal type, the programmer can expect a diagnostic from the compiler if he or she has forgotten to override any inherited deferred type-bound procedures.

15.11 Finalization

15.11.1 Introduction

When variables are deallocated or otherwise cease to exist, it is sometimes desirable to execute some procedure which 'cleans up' after the variable, perhaps releasing some resource (such as closing a file or deallocating a pointer component). This process is known as

finalization and is provided by **final subroutines**. Finalization is only available for extensible types. Array constructors and structure constructors are not finalized.

The set of final subroutines for a derived type is specified by statements of the form

`final [ :: ]` *subroutine-name-list*

in the type-bound procedure section; however, they are not type-bound procedures, and have no name which can be accessed through an object of the type. Instead, they execute automatically when an object of that type ceases to exist.

A final subroutine for a type must be a module procedure with a single dummy argument of that type. All the final subroutines for that type form a generic set and must satisfy the rules for unambiguous generic references; since they each have exactly one dummy argument of the same type, this simply means that the dummy arguments must have different kind type parameter values or rank. Each such dummy argument must be a variable without the `allocatable`, `intent(out)`, `optional`, `pointer`, or `value` attribute, and any length type parameter must be assumed (the value must be '`*`').

A non-pointer object is **finalizable** if its type has a final subroutine whose dummy argument matches the object. When a finalizable object is about to cease to exist (for example, by being deallocated or the execution of a `return` statement), the final subroutine is invoked with the object as its actual argument. This also occurs (in the called procedure) when the object is passed to an intent `out` dummy argument, or is the variable on the left-hand side of an intrinsic assignment statement. In the latter case, the final subroutine is invoked after the expression on the right-hand side has been evaluated, but before it is assigned to the variable.

An example is shown in Figure 15.17. When subroutine s returns, the subroutine `close_scalar_file_handle` will be invoked with x as its actual argument, and `close_rank1_file_handle` will be invoked with y as its actual argument. The order in which these will be invoked is processor dependent.

Termination of a program by an error condition, or by execution of a `stop` statement or the `end` statement in the main program, does not invoke any final subroutines.

If an object contains any (non-pointer) finalizable components, the object as a whole will be finalized before the individual components. That is, in Figure 15.18, when `ovalue` is finalized, `destroy_outer_ftype` will be invoked with `ovalue` as its argument before `destroy_inner_ftype` is invoked with `ovalue%ivalue` as its argument.

15.11.2   Type extension and final subroutines

When a type is extended, the new type does not inherit any of the final subroutines of the old type. The new type is, however, still finalizable, and when it is finalized any applicable final subroutines of the old type are invoked on the parent component.

If the new type defines any final subroutine, it will be invoked before any final subroutines of the old type are invoked. (Which is to say, the object as a whole is finalized, then its parent component is finalized, etc.) This operates recursively, so that when x is deallocated in the code of Figure 15.19, `destroy_bottom_type` will be invoked with x as its argument, then `destroy_top_type` will be invoked with `x%top_type` as its argument.

Figure 15.17 An example of finalization.

```
module file_handle_module
   type file_handle
      private
         ⋮
   contains
      final :: close_scalar_file_handle, close_rank1_file_handle
   end type file_handle
contains
   subroutine close_scalar_file_handle(h)
      type(file_handle) :: h
         ⋮
   end subroutine close_scalar_file_handle
   ⋮
end module file_handle_module
⋮
subroutine s(n)
   type(file_handle) :: x, y(n)
      ⋮
end subroutine s
```

Figure 15.18 A finalizable type with a finalizable component.

```
type inner_ftype
   ⋮
contains
   final :: destroy_inner_ftype
end type inner_ftype
type outer_ftype
   type(inner_ftype) :: ivalue
contains
   final :: destroy_outer_ftype
end type outer_ftype
⋮
type(outer_ftype) :: ovalue
```

Figure 15.19 Nested extensions of finalizable types.

```
type top_type
    ⋮
contains
    final :: destroy_top_type
end type
type, extends(top_type) :: middle_type
    ⋮
end type
type, extends(middle_type) :: bottom_type
    ⋮
contains
    final :: destroy_bottom_type
end type

type(bottom_type), pointer :: x
allocate (x)
    ⋮
deallocate (x)
```

15.12   Procedure encapsulation example

A procedure may require its user to define the problem to be solved by providing a function as well as data. The example that we will consider here is that of multi-dimensional quadrature, where the function to be integrated must be specified. This function may depend on other data in some complicated way that was not anticipated by the writer of the quadrature procedure.

Previously available solutions for problems of this kind have been:

i) for the quadrature procedure to accept an extra argument, typically a real vector, and pass that to the user-defined function when it is called;

ii) for the program to pass the information to the function via module variables;[3] or

iii) the use of 'reverse communication' techniques, where the program repeatedly calls the quadrature procedure providing the extra information it requested after the previous call, until the quadrature procedure is satisfied.

These all have disadvantages; the first is not very flexible (a real vector might be a poor way of representing the data), the second requires global data (recognized as being poor practice) and is not thread-safe, while the third is flexible and thread-safe but very complicated to use, particularly for the writer of the quadrature procedure.

---

[3]Or common blocks (Section B.1.7).

With type extension, the quadrature procedure is written to access the user's function as an abstract type-bound function of an abstract type. The user extends the type and overrides the function, and can include any kind of required data in components of the type. The quadrature procedure will pass the data back for the function to use. Figure 15.20 shows the definition of the types concerned and an outline of the quadrature procedure. Details not relevant to the function evaluation (such as the definition of the types for passing options to the procedure, and for receiving the status of the integration) have been omitted.

Figure 15.21 shows how the user could extend the abstract type to include the necessary data components for an arbitrary polynomial function of the form $\sum_{i=1}^{m} \sum_{j=0}^{n} c_{ij} x_i^j$ and to bind the function that evaluates this to the type.

To actually perform an integration, the user merely loads a local variable of this type with the required data, and calls the quadrature procedure as shown in Figure 15.22.

15.13 Type inquiry functions

There are two intrinsic functions for comparing dynamic types. These are intended for use on polymorphic variables, but may also be used on non-polymorphic variables.

extends_type_of(a, mold) returns, as a scalar default logical, whether the dynamic type of a is an extension of the dynamic type of mold. Both a and mold must either be unlimited polymorphic or of extensible type.

This returns true if mold is unlimited polymorphic and is either a disassociated pointer or an unallocated allocatable variable; otherwise, if a is unlimited polymorphic and is either a disassociated pointer or an unallocated allocatable variable, it will return false.

Otherwise, if both a and mold are unlimited polymorphic and neither has extensible dynamic type, the result is processor dependent.

same_type_as(a, b) returns, as a scalar default logical, whether the dynamic type of a is the same as the dynamic type of b. Both a and b must either be unlimited polymorphic or of extensible type.

If both a and b are unlimited polymorphic and neither has extensible dynamic type, the result is processor dependent.

For both functions, neither argument is permitted to be a pointer with undefined association status, unless it is non-polymorphic. Because the dynamic type of a non-polymorphic pointer is always well defined, such arguments to these two intrinsic functions need not have defined pointer association status.

These two functions are not terribly useful, because knowing the dynamic type of a (or how it relates to the dynamic type of b or mold) does not in itself allow access to the extended components. Therefore, we recommend that select type be used for testing the dynamic types of polymorphic entities.

Figure 15.20 Outline of a quadrature module.

```
module quadrature_module
    integer, parameter :: wp = selected_real_kind(15)
    type, abstract :: bound_user_function
        ! No data components
    contains
        procedure(user_function_interface), deferred :: eval
    end type bound_user_function
    type quadrature_options
        ! Details omitted
    end type
    type quadrature_status
        ! Details omitted
    end type
    abstract interface
        real(wp) function user_function_interface(data, coords)
            import                    :: wp, bound_user_function
            class(bound_user_function) :: data
            real(wp), intent(in)      :: coords(:)
        end function user_function_interface
    end interface
    ⋮
contains
    real(wp) function ndim_integral(hyper_rect, userfun, options, &
                                    status)
        real(wp), intent(in)                   :: hyper_rect(:)
        class(bound_user_function)             :: userfun
        type(quadrature_options), intent(in) :: options
        type(quadrature_status), intent(out) :: status
        ⋮
        ! This is how the user function is invoked
        single_value = userfun%eval(coordinates)
        ⋮
    end function ndim_integral
    ⋮
end module
```

Figure 15.21 Extending the type of Figure 15.20 for polynomial integration.

```
module polynomial_integration
   use quadrature_module

   type, extends(bound_user_function) :: my_bound_polynomial
      integer                :: degree, dimensionality
      real(wp),allocatable :: coeffs(:,:)
   contains
      procedure :: eval => polynomial_evaluation
   end type
contains
   real(wp) function polynomial_evaluation(data, coords) result(r)
      class(my_bound_polynomial) :: data
      real(wp), intent(in)      :: coords(:)
      integer                    :: i, j
      r = 0
      do i=1, data%dimensionality
         r = r + sum([ (data%coeffs(i, j)*coords(i)**j, &
                                       j=1, data%degree) ])
      end do
   end function polynomial_evaluation
end module polynomial_integration
```

Figure 15.22 Performing polynomial integration.

```
use polynomial_integration
type(my_bound_polynomial) :: poly
real(wp)                   :: integral
real(wp), allocatable     :: hyper_rectangle(:)
type(quadrature_options)  :: options
type(quadrature_status)   :: status

! Read the data into the local variable
read (...) poly%degree, poly%dimensionality
allocate (poly%coeffs(poly%dimensionality, poly%degree))
read (...) poly%coeffs

! Read the hyper-rectangle information
allocate (hyper_rectangle(poly%dimensionality))
read (...) hyper_rectangle
: ! Option-setting omitted
! Evaluate the integral
integral = ndim_integral(hyper_rectangle, poly, options, status)
```

Exercises

1. Define a polygon type where each point is defined by a component of class `point` (defined in Section 15.3). A function to test whether a position is within the polygon would be useful. A typical extension of such a type could have a label and some associated data; define such an extension.

2. Define a data logging type. This should contain type-bound procedures to initialize logging to a particular file, and to write a log entry. The file should automatically be closed if the object ceases to exist.

3. Use pointer functions to implement a vector that counts how many times it is accessed as a whole vector and how many times a single element from it is accessed.

4. Elaborate the type definition from Figure 15.15 as follows:

   i) Place the type definition in an appropriately named module and add deferred type-bound procedures for sending array data to the file, getting array data from the file, and closing the file.

   ii) Add to the module an extended type for opening and closing a file, with procedures for the formatted reading and writing of a real-valued array, and a private component for storing the unit value. Include a constructor for an object of this type.

   iii) Create a program that uses the above facilities. For simplicity use a fixed name for the file.

   iv) What changes are needed for the program to work in case another type extension is added?

16. Submodules

16.1 Introduction

The module facilities that we have described so far are adequate for programs of modest size, but they have some shortcomings for very large programs. These shortcomings all arise from the fact that, although modules are an aid to modularization of the program, they are themselves difficult to modularize. As a module grows larger, perhaps because the concept it is encapsulating is large, we can break it into several modules, but this would expose the internal structure, raising the potential for unnecessary name clashes and giving the user of the module access to what ought to be private. Worse, if the subfeatures of the module are interconnected, they must remain together in a single module, however large.

Another significant shortcoming of this approach is that if a change is made to the code inside a module procedure, even a private one, typical use of make or a similar tool results in the recompilation of every file which used that module, directly or indirectly.

Submodules address these problems. They allow module procedures to be defined in separate program units, which can be in separate files, while their interfaces remain defined in the module. A change in a submodule cannot alter an interface, and so does not cause the recompilation of program units that use the module. A submodule has access via host association to entities in the module, and may have entities of its own in addition to providing implementations of module procedures.

Submodules give other benefits, which we can explain more easily once we have described the feature.

16.2 Separate module procedures

The essence of the feature is the separation of the definition of a module procedure into two parts: the interface, which is defined in the module; and the body, which is defined in the submodule. Such a module procedure is known as a **separate module procedure**. A simple example is shown in Figure 16.1. The keyword module in the prefix of the function statement indicates in the interface block that this is the interface to a module procedure, and in the submodule that this is the implementation part of a module procedure. The submodule specifies the name of its parent. Both the interface and the submodule gain access to the type point by host association.

The interface specified in the submodule must be exactly the same as that specified in the interface block. For an external or dummy procedure, the interface is permitted to differ,

Modern Fortran Explained, 3rd Edition. M. Metcalf, J. Reid, M. Cohen, and R. Bader. Oxford University Press (2024). © M. Metcalf, J. Reid, M. Cohen, and R. Bader (2024). DOI 10.1093/oso/9780198876571.001.0016

Figure 16.1 A separate module procedure.

```
module points
   type :: point
      real :: x, y
   end type point
   interface
      real module function point_dist(a, b)
         type(point), intent(in) :: a, b
      end function point_dist
   end interface
end module points

submodule (points) points_a
contains
   real module function point_dist(a, b)
      type(point), intent(in) :: a, b
      point_dist = sqrt((a%x-b%x)**2+(a%y-b%y)**2)
   end function point_dist
end submodule points_a
```

---

for example, in respect of the names of the arguments, whether it is pure, and whether it is recursive (see Section 5.12); such variations are not permitted for a submodule since the intention is simply to separate the definition of the procedure into two parts. The name of the result variable is not part of the interface and so is permitted to be different in the two places; in this case, the name in the interface block is ignored. An import statement is not needed to access the host environment but is permitted if control is required over what is accessed.

An alternative is to give none of the interface information is the submodule. Instead, the whole interface is taken from the interface block in the module, including whether it is a function or a subroutine and the name of the result variable if it is a function. Here is an example:

```
submodule (points) points_a
contains
   module procedure point_dist
      point_dist = sqrt((a%x-b%x)**2+(a%y-b%y)**2)
   end procedure point_dist
end submodule points_a
```

Note the use of the keyword procedure, which avoids specifying whether it is a function or a subroutine.

16.3   Submodules of submodules

Submodules are themselves permitted to have submodules, which is useful for very large programs. The module or submodule of which a submodule is a direct subsidiary is called its **parent** and it is called a **child** of its parent. We do not expect the number of levels of submodules often to exceed two (that is, a module with submodules that themselves have submodules) but there is no limit and we refer to **ancestors** and **descendants** with the obvious meanings. Each module or submodule is the root of a tree whose other nodes are its descendants and have access to it by host association. No other submodules have such access, which is helpful for developing parts of large modules independently. Furthermore, there is no mechanism for accessing anything declared in a submodule from elsewhere – it is effectively private.

If a change is made to a submodule, only it and its descendants will need recompilation.

To indicate that a submodule has a submodule as its parent, the module name in the `submodule` statement is replaced by *module-name:parent-name*. For example,

```
submodule (points:points_a) points_b
```

declares that `points_b` has the submodule of Figure 16.1 as its parent. Note that this syntax allows two submodules to have the same name if they are descendants of different modules, which is useful when they are developed independently.

16.4 Submodule entities

A submodule can also contain entities of its own. These are not module entities and so are neither public nor private; they are, however, inaccessible outside of the defining submodule except to its descendants.

Typically, these will be variables, types, named constants, etc., for use in the implementation of some separate module procedure. As per the usual rules of host association, if any submodule entity has the same name as a module entity, the module entity is hidden.

A submodule can also contain procedures, which we will call **submodule procedures**. A submodule procedure is only accessible in the submodule and its descendants, and so can be invoked only there. To ensure this property holds for a submodule procedure with the `bind` attribute (see Section 20.9), such a procedure is not permitted to have a `bind` attribute with a `name=` specifier. This means that it has no binding label, so there is no additional mechanism for invoking the procedure.

Like a module procedure, a submodule procedure can also be separate; a separate submodule procedure has its interface declared in one submodule and its body in a descendant.

16.5 Submodules and use association

A submodule may access any module by use association. In particular, it is possible for a submodule of module `one` to access module `two` and a submodule of module `two` to access module `one`. A simple example is where a procedure of module `one` calls a procedure of module `two` and a procedure of module `two` calls a procedure of module `one`. Because circular dependencies between modules are not permitted, without submodules this would require that `one` and `two` were the same module, or that a third module `three` be used (containing those parts which were mutually dependent).

In the absence of any use association, a submodule has access by host association to any entity in its ancestor module. However, this is lost if the entity is accessed by the same name by use association. We therefore recommend employing renaming and `only` clauses in any use statement in a submodule.

16.6   The advantages of submodules

A major benefit of submodules is that if a change is made to one, only it and its descendants are affected. Thus, a large module may be divided into small submodule trees, improving modularity (and thus maintainability) and avoiding unnecessary recompilation cascades. We now summarize other benefits.

Entities declared in a submodule are private to that submodule and its descendants, which controls their name management and accidental use within a large module.

Separate concepts with circular dependencies can be placed in submodules of different modules in the common case where it is just the implementations that reference each other (because circular dependencies are not permitted between modules, this was impossible before).

Where a large task has been implemented as a set of modules, it may be appropriate to replace this by a single module and a collection of submodules. Entities that were public only because they are needed by other modules of the set can become private to the module or to a submodule and its descendants.

Once the implementation details of a module have been separated into submodules, the text of the module itself can be published to provide authoritative documentation of the interface without exposing any trade secrets contained in the implementation.

On many systems, each source file produces a single object file that must be loaded in its entirety into the executable program. Breaking the module into several files will allow the loading of only those procedures that are actually invoked into a user program. This makes modules more attractive for building large libraries.

Exercises

1. Further refine the solution for Exercise 4 from Chapter 15 by ensuring that the type extensions are defined in modules separate from that of their parent type. Apart from avoiding recompilation cascades, such a design would be necessary if, for example, the parent type's module is not available in source code.

17. Coarrays

17.1 Introduction

The coarray programming model is designed to provide a simple syntactic extension to support parallel programming from the point of view of both **work distribution** and **data distribution**.

Firstly, consider work distribution. The coarray extension adopts the single program multiple data (SPMD) programming model. A single program is replicated a fixed number of times, each replication having its own set of data objects. Each replication of the program is called an **image**. The number of images could be the same as, or more than, or less than the number of physical processors. A particular implementation may permit the number of images to be chosen at compile time, at link time, or at execution time. Each image executes asynchronously, and the normal rules of Fortran apply within each image.[1] The execution sequence can differ from image to image as specified by the programmer who, with the help of a unique image index, determines the actual path using normal Fortran control constructs and explicit synchronizations. For code between synchronizations, the compiler is free to use almost all its normal optimization techniques as if only one image were executing, but with access to additional memory on other images.

Secondly, consider data distribution. This requires the programmer to make a decision for a data object as to whether other images can directly access it:

i) Objects that are declared in the usual way (as for a serial program) are not directly accessible from other images. Such objects might have different sizes and types on different images.

ii) **Coarray** objects are declared with the **codimension** attribute, for example

```
real, dimension(1000), codimension[*] :: x, y
real :: z[*]     ! implicit codimension attribute
```

and are accessible both as a local object and from any other image. x and y are array coarrays and z is a scalar coarray. A coarray always has the same shape and type on each image. In the above example, each image has two real rank-one coarrays of size 1000 and a scalar coarray.

---

[1] Although this is not required, we expect implementations to arrange for each image to execute the same executable file on identical hardware.

Modern Fortran Explained, 3rd Edition. M. Metcalf, J. Reid, M. Cohen, and R. Bader. Oxford University Press (2024). © M. Metcalf, J. Reid, M. Cohen, and R. Bader (2024). DOI 10.1093/oso/9780198876571.001.0017

The coarray extension allows the programmer to specify the relationship between images in a syntax very much like normal Fortran array syntax; if an image executes the statement:

```
x(:) = y(:)[q]
```

the coarray y on image q is copied into coarray x on the executing image.

Subscripts within parentheses follow the normal Fortran rules within one image. **Cosubscripts** within square brackets provide an equally convenient notation for accessing an object on another image. Bounds for cosubscripts are known as **cobounds** and are given in square brackets in coarray declarations, except that the final cobound must be an asterisk. The asterisk indicates that the cobound is determined by the number of images, given that the coarray exists on them all. This allows the programmer to write code without knowing the number of images on which the code will eventually run.

The programmer uses coarray syntax only where it is needed. A reference to a coarray with no square brackets attached to it is a reference to the object in the memory of the executing image. Since it is desirable for most references to data objects in a parallel program to be local, coarray syntax should appear only in isolated parts of the source code. Coarray syntax acts as a visual flag to the programmer that communication between images will take place. It also acts as a flag to the compiler to generate code that avoids latency[2] whenever possible.

Because a coarray has the same shape on every image, and because allocations and deallocations of coarrays occur in synchrony across all images, coarrays may be implemented in such a way that each image can calculate the address of a coarray on another image. This is sometimes called **symmetric memory**. On a shared-memory machine, a coarray on an image and the corresponding coarrays on other images might be implemented as a sequence of objects with evenly spaced addresses. On a distributed-memory machine with one physical processor for each image, a coarray might be stored at the same address in each physical processor. If it is an array coarray, each image can calculate the address of an element on another image relative to the array start address on that other image.

Because coarrays are integrated into the language, remote references automatically gain the services of Fortran's basic data capabilities, including

- the type system;
- automatic conversions in assignments;
- information about structure layout; and
- object-oriented features, but with some restrictions.

17.2   Referencing images

Data objects on other images are referenced by cosubscripts enclosed in square brackets. Each valid set of cosubscripts maps to an **image index**, which is an integer between one and the number of images, in the same way as a valid set of array subscripts maps to a position in the array element order.

The number of images is returned by the intrinsic function num_images with no arguments. The intrinsic function this_image with no arguments returns the image index of the invoking image. The set of cosubscripts that corresponds to the invoking image for a coarray z are

---

[2]Delay while the image waits for data to be transferred to or from another image.

available as the rank-one array `this_image(z)`. The image index that corresponds to an array `sub` of valid cosubscripts for a coarray `z` is available as `image_index(z, sub)`.

For example, on image 5, `this_image()` has the value 5 and, for the array coarray declared as

```
real :: z(10, 20)[10, 0:9, 0:*]
```

`this_image(z)` has the value `[5, 0, 0]`, whilst on image 213, `this_image(z)` has the value `[3, 1, 2]`. On any image, the value of `image_index(z, [5, 0, 0])` is 5 and the value of `image_index(z, [3, 1, 2])` is 213.

17.3 The properties of coarrays

Each image has its own set of data objects, all of which may be accessed in the normal Fortran way. Some objects are declared with **codimensions** in square brackets, for example:

```
real, dimension(20), codimension[20,*] :: a ! An array coarray
real :: c[*], d[*]                           ! Scalar coarrays
character :: b(20)[20,0:*]
integer :: ib(10)[*]
type(interval) :: s[20,*]
```

A coarray may be allocatable, see Section 17.6. Unless the coarray is allocatable, it must have **explicit coshape**, that is, be declared with integer expressions for cobounds except that the final upper cobound must be an asterisk. The number of cosubscripts is called the **corank**. The sum of the rank and corank is limited to 15.

A coarray is not permitted to be a pointer. Furthermore, because an object of type `c_ptr` or `c_funptr` (Section 20.3) has the essence of a pointer, a coarray is not permitted to be of either of these types. However, a coarray may be of a derived type with pointer or allocatable components, see Section 17.7.[3]

A subobject of a coarray is regarded as a coarray if and only if it has no cosubscripts, no vector subscripts, no allocatable component selection, and no pointer component selection. For example, `a(1)`, `a(2:10)`, and `s%lower` are coarrays if `a` and `s` are the coarrays declared at the start of this section. This definition means that passing a coarray subobject to a dummy coarray does not involve copy-in copy-out (which would be infeasible given that the coarray exists on all images). Also, it ensures that the subobject has the property of symmetric memory. The term **whole coarray** is used for the whole of an object that is declared as a coarray or the whole of a coarray component of a structure.

The corank of a whole coarray is determined by its declaration. Its **cobounds** are specified within square brackets in its declaration or allocation. Any subobject of a whole coarray that is a coarray has the corank, cobounds, and coextents of the whole coarray. The cosize of a coarray is always equal to the number of images. Even though the final upper bound is specified as an asterisk, a coarray has a final coextent and a final upper cobound which depend on the number of images. The final upper cobound is the largest value that the final cobound can have in a valid reference (we discuss this further in Section 17.4). For example, when the number of images is 128, the coarray declared thus

---

[3]It is also permitted to have a component of type `c_ptr` or `c_funptr`, but these are nearly useless, see Appendix A.17.

```
real :: array(10,20)[10,-1:8,0:*]
```

has rank 2, corank 3, and shape $[10,20]$; its lower cobounds are 1, -1, 0 and its upper cobounds are 10, 8, 1.

A named constant is not permitted to be a coarray, because this would be useless. Each image would hold exactly the same value so there would be no reason to access its value on another image.

To ensure that each image initializes only its own data, cosubscripts are not permitted in data statements. For example:

```
real :: a(10)[*]
data a(1)    /0.0/ ! Permitted
data a(1)[2] /0.0/ ! Not permitted
```

17.4   Accessing coarrays

A coarray on another image may be addressed by using cosubscripts in square brackets following any subscripts in parentheses, for example:

```
a(5)[3,7] = ib(5)[3]
d[3] = c
a(:)[2,3] = c[1]
```

and for a coarray of derived type, the coindices for a subobject must appear before component selection, for example:

```
s[1,1]%lower = 0.0
```

We call any object whose designator includes cosubscripts a **coindexed object**. Only one image may be referenced at a time, so each cosubscript must be a scalar integer expression (section cosubscripts, such as $d[1:n]$, are not permitted). Section subscripts must be used when the coindexed object has nonzero rank. For example, $a[2,3]$ is not permitted as a shorthand for $a(:)[2,3]$. This is in order to make it clear that the rank is nonzero.

Any object reference without square brackets is always a reference to the object on the executing image. For example, in

```
real :: z(20)[20,*], zmax[*]
    ⋮
zmax = maxval(z)
```

the value of the largest element of the array coarray z on the executing image is placed in the scalar coarray zmax on the executing image.

For a reference with square brackets, the cosubscript list must map to a valid image index. For example, if there are 16 images and the coarray z is declared thus

```
real :: z(10)[5,*]
```

then a reference to $z(:)[1,4]$ is valid because it refers to image 16, but a reference to $z(:)[2,4]$ is invalid because it refers to image 17. Like array subscripts, it is the programmer's responsibility to ensure that cosubscripts are within cobounds and refer to a valid image.[4]

Square brackets attached to objects alert the reader to probable communication between images. However, communication may also take place during the execution of a procedure reference, and this could be via a defined operation or defined assignment.

That an image is selected in square brackets has no bearing on whether the statement is executed on that image. For example, the statement

```
z[6] = 1
```

is executed by every image that encounters it. If code is to be executed selectively, conditional execution is needed. An example is

```
if (this_image()==6) z = 1
```

A coindexed object is permitted in most contexts, such as intrinsic operations, intrinsic assignment, input/output lists, and as an actual argument corresponding to a non-coarray dummy argument. On a distributed-memory machine, passing it as an actual argument is likely to cause a local copy of it to be made before execution of the procedure starts (unless it has intent `out`) and the result to be copied back on return (unless it has intent `in` or the `value` attribute). The rules for argument association have been carefully constructed so that such copying is always allowed.

Pointers are not allowed to have targets on remote images, because this would break the requirement for remote access to be obvious. Therefore, the target of a pointer is not permitted to be a coindexed object:

```
p => a(n)[1]    ! Not allowed (compile-time constraint)
```

A coindexed object is not permitted as the *selector* in an `associate` or `select type` statement because that would disguise a reference to a remote image (the associated name is without square brackets). However, a coarray is permitted as the selector, in which case the associated entity is also a coarray and its cobounds are those of the selector.

So that the implementation does not need to query the dynamic type of an object on another image, no references are permitted to a polymorphic subobject of a coindexed object or to a coindexed object of a type that has a potential subobject component (Section 2.16) that is polymorphic and allocatable.

In order to allow for failed images (Section 17.24) and for wishing to address an image in another team (Chapter 18), the general form of an **image selector** is

[*cosubscript-list [, is-spec-list]]*

where *is-spec* is `stat`=*stat-variable*, `team`=*team-value*, or `team_number`=*expr*; *stat-variable* is a scalar non-coindexed integer variable that, like all such variables, should have a decimal exponent range of at least four to ensure that the error code is representable in it, *team-value* is a scalar expression of type `team_type`, and *expr* is a scalar integer expression.

17.5 The `sync all` statement

Each image executes on its own without regard to the execution of other images except when it encounters a special kind of statement called an **image control statement**. The

---

[4]Some compilers have an option of checking for this.

programmer inserts image control statements to ensure that, whenever one image alters the value of a coarray variable, no other image still wants the old value, and that whenever an image accesses the value of a coarray variable, it receives the wanted value – either the old value (before the update) or the new value (from the update). Because a coarray may be of a derived type with a pointer component (see Section 17.7), coarray syntax might be used to access a variable that has the `target` attribute but is not a coarray. It follows that image control statements are needed for these variables, too. In this section we describe the simplest of these image control statements.

The `sync all` statement provides a barrier where all images synchronize before executing further statements. All statements executed before the barrier on image *P* execute before any statement executes after the barrier on image *Q*. If the value of a variable is changed by an image before the barrier, the new value is available to all other images after the barrier. If a variable is referenced before the barrier and is changed by another image after the barrier, the old value is obtained.

Figure 17.1 shows a simple example of the use of `sync all`. Image 1 reads data and broadcasts it to other images. The first `sync all` ensures that image 1 does not interfere with any previous use of z by another image. The second `sync all` ensures that another image does not access z before the new value has been set by image 1.

Figure 17.1 Read on image 1 and broadcast to the others.

```
real :: z[*]
   ⋮
sync all
if (this_image()==1) then
   read (*, *) z
   do image = 2, num_images()
      z[image] = z
   end do
end if
sync all
   ⋮
```

The full form of the `sync all` statement is

sync all [([*sync-stat-list*])]

where *sync-stat* is stat=*stat* or errmsg=*erm*. If the stat= specifier is present, the integer variable *stat* is given the value zero after a successful execution or a positive value otherwise. If the errmsg= specifier is present and an error occurs, an explanatory message is assigned to the character variable *erm*.

For another example of the use of `sync all` consider solving a partial differential equation with one space variable. Suppose that the vector of unknowns has been separated into parts on the images that are all of the same size. Each part is iterated independently but needs approximations for the end values of the adjacent parts to estimate derivatives at its ends. Periodically, all the images get the updated end values of their neighbours by calling the subroutine in Figure 17.2. The images must be synchronized at the start to be sure that the values are not still being changed and at the end to be sure that they have been updated before they are referenced.

Figure 17.2 Update neighbouring vectors.

```
subroutine update(a)
   real :: a(0:)[*]
   integer :: me, n
   me = this_image()
   n = size(a)-2
   sync all
   if (me>1) a(0) = a(n)[me-1]
   if (me<num_images()) a(n+1) = a(1)[me+1]
   sync all
end subroutine update
```

Although usually the synchronization will be initiated by the same `sync all` statement on all images, this is not a requirement. The additional flexibility may be useful, for example, when different images are executing different code and need to exchange data.

All images are synchronized at program initiation as if by a `sync all` statement. This ensures that initialized coarrays will have their initial values on all images before any image commences executing its executable statements.

There is an implicit barrier whenever a coarray is allocated or deallocated, see Section 17.6. Similarly, all images synchronize during program termination, see Section 17.19. Other image control statements are described in Sections 17.12 to 17.15 and a complete list is found in Section 17.17.

17.6   Allocatable coarrays and coarray components

A coarray may be allocatable; it must have **deferred coshape**, that is, be declared with colons for all codimensions. It is given normal cobounds by the `allocate` statement, for example,

```
real, allocatable :: a(:)[:], s[:, :]
   ⋮
allocate (a(10)[*], s[-1:34,0:*])
```

The cobounds must always be included in the `allocate` statement and the upper bound for the final codimension must always be an asterisk. For example, the following are not permitted (compile-time constraints):

```
allocate (a(n))       ! Not allowed for a coarray (no cobounds)
allocate (a(n)[p])    ! Not allowed (cobound not *)
```

Also, the value of each bound, cobound, or length type parameter is required to be the same on all images. For example, the following is not permitted (run-time constraint):

```
allocate (a(this_image())[*])   ! Not allowed (varying local bound)
```

For a polymorphic coarray, if the dynamic types and type parameters are given on the `allocate` statement by typed or sourced allocation (Section 15.4), they must be the same on all images. Together, these restrictions ensure that the coarrays exist on every image and are consistent.

The `move_alloc` intrinsic subroutine may be applied to coarrays of the same corank, but no actual argument of `move_alloc` is permitted to be coindexed. This ensures that each image performs an independent action.

There is implicit barrier synchronization of all images in association with each `allocate` statement that involves one or more coarrays. Images do not commence executing subsequent statements until all images finish executing the same `allocate` statement (on the same line or lines of the source code). Similarly, for the `deallocate` and `move_alloc` statements for coarrays, all images synchronize at the beginning of the same statement and do not continue with the next statement until all images have finished executing the statement. The synchronization means that `move_alloc` is not pure when applied to coarrays.

When an image executes an `allocate` statement, interaction is needed between images only for synchronization. The image allocates its local coarray and records how the corresponding coarrays on other images are to be addressed. The compiler is not required to check that the bounds and cobounds are the same on all images, although it may do so (or have an option to do so). Nor is the compiler required to detect when deadlock has occurred; for example, when one image is executing an `allocate` statement while another is executing a `deallocate` statement.

If an unsaved allocatable coarray is local to a procedure or block construct (Section 8.12), it comes into existence when the procedure or block is entered. Allocatable objects are initially undefined, so there is no need for synchronization. If the coarray is still allocated when the procedure or block construct completes execution, implicit deallocation of the coarray and therefore synchronization of all images occurs. Here is a simple example:

```
subroutine work
   real, allocatable :: a[:]
                    ! a becomes undefined without synchronization
      ⋮
   allocate (a)   ! Synchronization
      ⋮
   end            ! Implicit synchronization
```

The allocation of a polymorphic coarray is not permitted to create a coarray that is of type `c_ptr`, `c_funptr`, or of a type with a coarray ultimate component.

Fortran allows the shapes or length parameters to disagree on the two sides of an intrinsic assignment to an allocatable variable (see Section 6.7); the system performs the appropriate reallocation. Such disagreement is not permitted for an allocatable coarray because it would imply synchronization.

For the same reason, intrinsic assignment to a polymorphic coarray is not permitted.

A component of a type may be a coarray, and if so must be allocatable. A coarray component is required to be allocatable to avoid the need for synchronization on entering a procedure or block that has a local object of a type with a coarray ultimate component. Just as for an allocatable coarray, when a coarray component is allocated, the bounds, cobounds, and length type parameters must be the same on all images and the images synchronize. If it is still allocated when the procedure or block is left, an implicit deallocation and synchronization occurs. A variable or component of a type that has a coarray ultimate component must not itself be a coarray. Were Fortran to allow a coarray of a type with coarray components, we could be confronted with references such as z[p]%x[q], where it is unclear whether z%x[q] is to be accessed from image p, or z[p]%x from image q. For simplicity, Fortran 2018 requires that a variable of a type having a coarray ultimate component be a non-pointer non-allocatable scalar. In Fortran 2023, it is allowed to be an array or allocatable, see Section 23.6.1.

If an object with a coarray ultimate component is declared without the save attribute in a procedure and the coarray is still allocated on return, there is an implicit deallocation and associated synchronization. Similarly, if such an object is declared within a block construct and the coarray is still allocated when the block completes execution, there is an implicit deallocation and associated synchronization.

To avoid the possibility of implicit reallocation in an intrinsic assignment for a scalar of a derived type with a coarray ultimate component, no disagreement of allocation status or shape is permitted for the coarray ultimate component.

To avoid the coarray properties of a polymorphic variable being dependent on its dynamic type, it is not permissible to add a coarray ultimate component by type extension unless the type already has one or more coarray ultimate components.

17.7 Coarrays with allocatable or pointer components

17.7.1 Introduction

A coarray is permitted to be of a derived type with allocatable or pointer components.

17.7.2 Data components

To share data structures with different sizes, length parameter values, or types between different images, we may declare a coarray of a derived type with a non-coarray component that is allocatable or a pointer. On each image, the component is allocated locally or is pointer assigned to a local target, so that it has the desired properties on that image (or is not allocated or pointer assigned if it is not needed on that image). It is straightforward to access such data on another image. For example, if coarray z has an allocatable array component alloc, in

```
x(:) = z[p]%alloc(:)
```

the cosubscript is associated with the scalar variable z, not with its component. In words, this statement means 'Go to image p, obtain the address and size of the component alloc, and copy the data itself to the local array x'.

If coarray z contains a data pointer component ptr, the appearance of z[q]%ptr in a context that refers to its target is a reference to the target of component ptr of z on image q. This target must reside on image q and must have been established by an allocate statement executed on image q or a pointer assignment executed on image q, for example

```
z%ptr => r ! Local association
```

A local pointer may be associated with a target component on the local image,

```
r => z%ptr ! Local association
```

but may not be associated with a target component on another image,

```
r => z[q]%ptr ! Not allowed (compile-time constraint)
```

If an association with a target component on another image would otherwise be implied, the pointer component becomes undefined. For example, this happens when the following derived-type intrinsic assignments are executed on an image other than q:

```
z[q] = z ! The pointer component of z[q] may become undefined
z = z[q] ! The pointer component of z may become undefined
```

It can also happen in a procedure invocation on an image other than q if z[q] is an actual argument or z[q]%ptr is associated with a pointer dummy argument.

Similarly, for a coarray of a derived type that has a pointer or allocatable component, allocate, deallocate, or move_alloc for the component on another image is not allowed:

```
type(something), allocatable :: t[:]
  :
allocate (t[*])       ! Allowed
allocate (t%ptr(n))   ! Allowed
allocate (t[q]%ptr(n)) ! Not allowed (compile-time constraint)
```

However, the nullify statement may be applied to a coindexed pointer.

In order to prevent any possibility of a remote allocation:

- in an intrinsic assignment to a coindexed object that is allocatable, the shapes and length type parameters are required to agree;

- intrinsic assignment to a polymorphic coindexed object or a coindexed object with an allocatable ultimate component is not permitted; and

- if an actual argument is a coindexed object with an allocatable ultimate component, the corresponding dummy argument must be allocatable, a pointer, or have the intent in or value attribute.

17.7.3 Procedure pointer components

A coarray is permitted to be of a type that has a procedure pointer component or a type-bound procedure. Invoking a procedure pointer component of a coindexed object, for example

```
call a[p]%proc(x) ! Not allowed
```

is not permitted since the remote procedure target might be meaningless on the executing image. However, invoking a type-bound procedure (Section 15.8) is allowed except for a subobject of a coindexed polymorphic object, for example

```
call a[p]%tbp1(x)      ! Allowed even if a is polymorphic.
call a[p]%comp%tbp2(x) ! Not allowed if comp is polymorphic.
```

The restrictions on polymorphic coarrays ensure that the dynamic type and hence the procedure is the same on all images.

17.8   Coarrays in procedures

A dummy argument of a procedure is permitted to be a coarray. It may be a scalar, or an array that is explicit shape, assumed size (Section 20.5), assumed shape, or allocatable, see Figure 17.3. Unless it is allocatable, its coshape has to be explicit.

Figure 17.3 Coarray dummy arguments.

```
subroutine subr(n, p, u, w, x, y, z, a)
    integer :: n, p
    real :: u[2, p/2, *]                ! Scalar
    real :: w(n)[p, *]                  ! Explicit shape
    real :: y(:, :)[2, *]               ! Assumed shape
    real, allocatable :: z(:)[:, :]     ! Allocatable array
    real, allocatable :: a[:]           ! Allocatable scalar
```

When the procedure is called, the corresponding actual argument must be a coarray. The association is with the coarray itself and not with a copy; the restrictions below ensure that copy-in copy-out is never needed. (Making a copy would require synchronization on entry and return to ensure that remote references within the procedure are not to a copy that does not exist yet or that no longer exists.) Furthermore, the interface is required to be explicit so that the compiler knows it is passing the coarray and not just the local variable. An example is shown in Figure 17.4.

Figure 17.4 Calling a procedure with coarray dummy arguments.

```
        real, allocatable :: a(:)[:], b(:,:)[:]
        ⋮
        allocate ( a(1)[*], b(m,n)[*] )
        call sub(a(:), b(1,:))
        ⋮
contains
    subroutine sub(x, y)
        real :: x(:)[*], y(:)[*]
        ⋮
    end subroutine sub
```

The restrictions that apply to a coarray dummy argument and its associated actual argument are:

- the actual argument must be a coarray (see Section 17.3 for the rules on whether a subobject is a coarray);
- to avoid any possibility of copy-in copy-out occurring, if the dummy argument is an array other than an assumed-shape array without the `contiguous` attribute (see Section 7.17.1), the actual argument must be simply contiguous (satisfies conditions

given Section 7.17.2, which ensure that the array is known at compile time to be contiguous);[5]

- the dummy argument must not have the `value` attribute (this also applies to a non-coarray dummy argument that has a coarray ultimate component);
- if the dummy argument is an allocatable coarray, the corresponding actual argument must be an allocatable coarray of the same rank and corank. Furthermore, its chain of argument associations, perhaps through many levels of procedure call, must terminate with the same actual coarray on every image;[6] and
- if the dummy argument is an allocatable coarray or has an ultimate component that is a coarray, it must not have intent `out`.[7]

The restrictions on actual arguments described in Section 5.7.5 do not apply to an action taken through a coindexed object. Suitable synchronization must be in place, just as for an action taken through any other coindexed object.

A procedure with a non-allocatable coarray dummy argument will often be called on all images at the same time with the same actual coarray, but this is not a requirement. For example, the images may be grouped into two sets and the images of one set may be calling the procedure with one coarray while the images of the other set are calling the procedure with another coarray or are executing different code.

Automatic coarrays are not permitted. For example, the following is invalid:

```
subroutine solve3(n)
integer :: n
real :: work(n)[*]   ! Not permitted
```

Were automatic coarrays permitted, it would be necessary to require synchronization, both after memory is allocated on entry and before memory is deallocated on return. Furthermore, it would mean that the procedure would need to be called on all images concurrently.

A function result is not permitted to be a coarray or to have an ultimate component that is a coarray. This is because the function would need to allocate temporary coarrays for its results, which would involve synchronization of all images for every invocation within a statement.

The rules for resolving generic procedure references have not been extended to allow overloading of array and coarray versions because this would be ambiguous for a coarray actual argument.

A pure or elemental procedure is not permitted to define a coindexed object or contain any image control statements (Section 17.11), since these involve side-effects (defining a coindexed object is similar to defining a variable from the host or a module). However, it may reference the value of a coindexed object.

Neither a final subroutine (Section 15.11.1) nor an elemental procedure is permitted to have a coarray dummy argument.

Each image associates its non-allocatable coarray dummy argument with an actual coarray, perhaps through many levels of procedure call, and defines the corank and cobounds afresh.

---

[5]A high quality implementation will be able to identify violations of this rule at compile time, but this is not enforced by the Fortran standard.

[6]This allows the coarray to be allocated or deallocated and retain the property of symmetric memory.

[7]This is because the implicit deallocation (Section 6.9) of the coarray would also require an implicit synchronization.

It uses these to interpret each reference to a coindexed object, taking no account of whether a remote image is executing the same procedure with the corresponding coarray.

Allocatable coarrays may be declared in a procedure. Unless it is allocatable or a dummy argument, an object that is a coarray or has a coarray ultimate component is required to have the save attribute.[8] Again, this is because an unsaved non-allocatable coarray would come into existence on procedure invocation, which would require synchronization, and this is inappropriate because procedures are not invoked in lockstep on every image. An allocatable coarray is not required to have the save attribute because its initial state is unallocated; however, note that in a recursive procedure such a coarray has separate instances at different levels of recursion.

17.9 Asynchronous attribute

An asynchronous coindexed object is not permitted to be an actual argument that corresponds to an asynchronous or volatile dummy argument unless the dummy argument has the value attribute. This is because passing a coindexed object as an actual argument is likely to be done by copy-in copy-out.

17.10 Interoperability

Because coarrays may be used without restriction as local objects, a coarray has exactly the same interoperability properties (Chapter 20) as a non-coarray object of the same type, type parameters, and rank, including being an actual argument in a procedure call to C. However, it cannot correspond to a coarray because C does not have the concept of a data object like a coarray. In this sense, coarrays are not interoperable. Interoperability of coarrays with UPC[9] might be considered in the future.

17.11 Execution segments

We have encountered barrier synchronization in Sections 17.5 and 17.6. Before describing the statements that provide more selective synchronizations, we introduce the concept of the execution segment that underpins the behaviour of programs that employ them. Such statements are called **image control statements**.

On each image, the sequence of statements executed before the first execution of an image control statement or between the execution of two image control statements is known as a **segment**. The segment executed immediately before the execution of an image control statement includes the evaluation of all expressions within the statement.

For example, in Figure 17.1, each image executes a segment before executing the first sync all statement, executes a segment between executing the two sync all statements, and executes a segment after executing the second sync all statement.

---

[8]Note that variables declared in the specification part of a main program, module, or submodule automatically have the save attribute.

[9]Unified Parallel C, an extension of C which is similar to coarrays in Fortran.

On each image P, the statement execution order determines the segment order, P_i, $i = 1, 2, \ldots$ Between images, the execution of corresponding image control statements on images P and Q at the end of segments P_i and Q_j may ensure that either P_i precedes Q_{j+1}, or Q_j precedes P_{i+1}, or both.

A consequence is that the set of all segments on all images is partially ordered: the segment P_i precedes segment Q_j if and only if there is a sequence of segments starting with P_i and ending with Q_j such that each segment of the sequence precedes the next either because they are on the same image or because of the execution of corresponding image control statements.

A pair of segments P_i and Q_j are called **unordered** if P_i neither precedes nor succeeds Q_j. For example, if the middle segment of Figure 17.1 is P_i on image 1 and Q_j on another image Q, P_{i-1} precedes Q_{j+1} and P_{i+1} succeeds Q_{j-1}, but P_i and Q_j are unordered.

There are restrictions on what is permitted in a segment that is unordered with respect to another segment. These provide the compiler with scope for optimization. A coarray may be defined and referenced during the execution of unordered segments by calls to atomic subroutines (Section 17.23). Apart from this,

- if a variable is defined in a segment on an image, it must not be referenced, defined, or become undefined in a segment on another image unless the segments are ordered;
- if the allocation of an allocatable subobject of a coarray or the pointer association of a pointer subobject of a coarray is changed in a segment on an image, that subobject shall not be referenced or defined in a segment on another image unless the segments are ordered; and
- if a procedure invocation on image P is in execution in segments $P_i, P_{i+1}, \ldots, P_k$ and defines a non-coarray dummy argument, the argument-associated entity shall not be referenced or defined on another image Q in a segment Q_j unless Q_j precedes P_i or succeeds P_k (because a copy of the actual argument may be passed to the procedure).

It follows that for code in a segment, the compiler is free to use almost all its normal optimization techniques as if only one image were present. For example, during the execution of a segment, a variable might be held in a register and altered there several times without copying the value back until the segment ends.

17.12   The `sync images` statement

For greater flexibility than is available with `sync all`, the `sync images` statement

```
sync images ( image-set[, sync-stat-list] )
```

performs a synchronization of the image that executes it with each of the other images in its image set. Here, *image-set* is either an integer expression indicating distinct image indices or an asterisk (*) indicating all images. The expression must be scalar or of rank one. The optional *sync-stat-list* is as for `sync all`.

Execution of a `sync images` statement on image P corresponds to the execution of a `sync images` statement on image Q if the number of times image P has executed a `sync images` statement with Q in its image set is the same as the number of times image Q has executed a `sync images` statement with P in its image set. The segments that executed before the `sync images` statement on either image precede the segments that execute after the corresponding `sync images` statement on the other image. Figure 17.5 shows an example that imposes the fixed order 1, 2, ... on images.

Figure 17.5 Using `sync images` to impose an order on images.

```
me = this_image ()
p = 1
if (me > 1) sync images (me-1)
do step = 1, num_images ()
   if (step == me) then
      if (me > 1) p = p + p[me-1]
      if (me < num_images ()) sync images (me+1)
   end if
end do
```

Execution of a `sync images (*)` statement is not equivalent to the execution of a `sync all` statement. A `sync all` statement causes all images to wait for each other, whereas `sync images` statements are not required to specify the same image set on all the images participating in the synchronization. In the example in Figure 17.6, image 1 will wait for each of the other images to reach the `sync images (1)` statement. The other images wait for image 1 to set up the data, but do not wait for each other.

Figure 17.6 Using `sync images` to make other images wait for image 1.

```
if (this_image () == 1) then
   ! Set up coarray data needed by all other images
   sync images (*)
else
   sync images (1)
   ! Use the data set up by image 1
end if
```

17.13   The `lock` and `unlock` statements

Locks provide a mechanism for controlling access to data that are referenced or defined by more than one image.

A lock is a scalar variable of the derived type `lock_type` that is defined in the intrinsic module `iso_fortran_env` (Section 17.16) that is a coarray or a subobject of a coarray. It has one of two states: **locked** or **unlocked**. The unlocked state is represented by a single value and this is the initial value. All other values represent the locked state. The only permitted way to change the value of a lock is by executing the `lock` or `unlock` statement. For example, if a lock is a dummy argument or a subobject of a dummy argument, the dummy argument must not have intent `out` because this would cause it to be reinitialized on entry to the procedure. If a lock variable is locked, it can be unlocked only by the image that locked it.

Figures 17.7 and 17.8 illustrate the use of `lock` and `unlock` statements to manage stacks. Each image has its own stack; any image can add a task to the stack on any image by calling `put_task`, which uses the lock on that image to make sure that no other image is trying to do the same thing at the same time and that the image is not trying to get a task from its own stack. Similarly, the image can get a task from its own stack by calling `get_task`, which uses its lock to avoid interference with another image.

Figure 17.7 Module for using `lock` and `unlock` to manage stacks.

```
module stack_manager
    use, intrinsic :: iso_fortran_env, only: lock_type
    type task
        integer :: label(2)
    end type
    integer, parameter :: lstack=100
    type(lock_type), private :: stack_lock[*]
    type(task), private     :: stack(lstack)[*]
    integer, private        :: stack_size[*]
    interface
        module subroutine get_task(job, ok)
            type(task), intent(out) :: job
            logical, intent(out)    :: ok
        end subroutine
        module subroutine put_task(job, image, ok)
            type(task), intent(in) :: job
            integer, intent(in)    :: image
            logical, intent(out)   :: ok
        end subroutine
    end interface
end module stack_manager
```

There is a form of the `lock` statement that avoids a wait when the lock variable is locked:

```
logical :: success
lock (stack_lock, acquired_lock=success)
```

If the variable is unlocked, it is locked and the value of `success` is set to true; otherwise, `success` is set to false and execution continues immediately.

An error condition occurs for a `lock` statement if the lock variable is already locked by the executing image, and for an `unlock` statement if the lock variable is not already locked by the executing image. As for the `allocate` and `deallocate` statements, the `stat=` specifier is available to avoid this causing error termination. The values `stat_locked`, `stat_locked_other_image`, and `stat_unlocked` from the intrinsic module `iso_fortran_env` are given to the `stat=` variable in a `lock` statement for a variable locked

---

Figure 17.8 Submodule for using `lock` and `unlock` to manage stacks.

```
submodule (stack_manager) tasks
contains
    module procedure get_task
        lock (stack_lock)
        ok = stack_size>0
        if (ok) then
            job = stack(stack_size)
            stack_size = stack_size - 1
        end if
        unlock (stack_lock)
    end procedure get_task
    module procedure put_task
        lock (stack_lock[image])
        ok = stack_size[image]<lstack
        if (ok) then
            stack_size[image] = stack_size[image] + 1
            stack(stack_size[image])[image] = job
        end if
        unlock (stack_lock[image])
    end procedure put_task
end submodule tasks
```

---

by the image, in a `lock` statement for a variable locked by another image, or in an `unlock` statement for a variable that is unlocked.

Any particular lock variable is successively locked and unlocked by a sequence of `lock` and `unlock` statements, each of which separates two segments on the executing image. If execution of such an `unlock` statement P_u on image P is immediately followed in this sequence by execution of a `lock` statement Q_l on image Q, the segment that precedes the execution of P_u on image P precedes the segment that follows the execution of Q_l on image Q.

The full syntax of the `lock` statement is

```
lock ( lock-variable[, lock-stat-list] )
```

where *lock-stat* is `acquired_lock=`*scalar-logical-variable* or *sync-stat*. The full syntax of the `unlock` statement is

```
unlock ( lock-variable[, sync-stat-list] )
```

In both cases, *sync-stat* is as for `sync all` (Section 17.5).

17.14 Critical sections

Exceptionally, it may be necessary to limit execution of a piece of code to one image at a time. Such code is called a **critical section**. There is a construct to delimit a critical section:

```
critical
   ! code that is executed on one image at a time
end critical
```

No image control statement may be executed during the execution of a critical construct, that is, the code executed must be a single segment. Branching (Appendix A.12) into or out of a critical construct is not permitted.

The critical statement takes the general form

   [construct-name :] critical *[([sync-stat-list])]*

with optional stat= and errmsg= specifiers in its *sync-stat-list* to detect the case of an image failing while executing a critical construct, see Section 17.25.2. The errmsg= specifier provides a message in the event of failure.

If image *Q* is the next to execute the construct after image *P*, the segment in the critical section on image *P* precedes the segment in the critical section on image *Q*.

As for other constructs, the critical construct may be named:

```
example: critical
   ! code that is executed on one image at a time
end critical example
```

17.15   Events

Events allow an action to be delayed until one or more actions have been performed on other images. An image records that it has performed an action by executing an event post statement involving a scalar variable of type event_type from the intrinsic module iso_fortran_env (Section 17.16), known as an **event variable**. An event variable must be a coarray or a component of a coarray; when posting an event, it is almost always coindexed, but only the owning image can wait on an event.

An image executes an event wait statement involving the event variable if it needs to delay its action for a posted action on another image or for $k \geq 1$ posted actions on other images. This event variable is not allowed to be a coindexed object. The action on the waiting image is said to be **matched** by each of the k posted actions. The count of the number of posted actions that have not yet been matched is held in the event variable. This is known as the **event count**.

Each event post execution has a matching event wait execution that involves the same event variable. A segment that precedes the event post execution precedes any segment that succeeds the matching event wait execution.

The value of an event variable includes its event count, which is of type integer and initially has the value zero. All changes to the count of an event variable and accesses to its value are atomic, as for actions on an atomic variable, see Section 17.23. The count is atomically incremented by one when an event post statement is executed for its event variable and is atomically decremented by the chosen threshold k when an event wait statement is executed for its event variable.

An event variable may be defined only by appearance in an `event post` or `event wait` statement. It may be referenced or defined in a segment that is unordered with respect to another segment in which it is defined.

The `event post` statement is an image control statement that takes the general form

    `event post ( ` *event-variable [, sync-stat-list] *`)`

where *sync-stat-list* is as for the `sync all` statement. Successful completion of the statement atomically increases its event count by one. If there is an error condition, the value of the event count is processor dependent.

The `event wait` statement is an image control statement that takes the general form

    `event wait ( ` *event-variable [, event-wait-spec-list] *`)`

where an *event-wait-spec* is an `until_count=` specifier, `stat=` specifier, or `errmsg=` specifier. The event variable must not be a coindexed object. An `until_count=` specifier has the form

    `until_count = ` *scalar-integer-expression*

where the value of the expression provides the threshold. The threshold is one if there is no `until_count=` specifier. The executing image waits until the event count is at or above the threshold, then atomically decreases the count by the threshold and resumes execution. If the threshold is k, the first k unmatched `event post` executions for the event variable are matched with this `event wait` execution. After an error condition, the value of the event count is processor dependent.

A snapshot of the value of an event count may be determined by the intrinsic subroutine

call event_query(event, count *[*, stat*]*)

    event is an event variable that is not coindexed. It has intent `in`.

    count is a scalar integer with decimal range at least that of default integer. It has intent `out`. It is atomically assigned the value of the count of `event`. Note that the count of `event` may have already changed by the time `event_query` returns.

    stat is a scalar integer with decimal range of at least four. It must not be a coindexed object. If present, it is assigned the value zero if no error occurs, and a positive value otherwise. If an error occurs and `stat` is absent, error termination is initiated.

17.16   Derived types for locks, events, and teams

The following derived types are defined in the intrinsic module `iso_fortran_env`:

    `lock_type`
    `event_type`
    `team_type`

Their use is explained in Section 17.13 for locks, Section 17.15 for events, and Section 18.1 for teams. These types are extensible, have no type parameters, and have private components, all of which are default-initialized. Furthermore,

- a variable of type `lock_type` or `event_type` must be a coarray;
- a variable of a type with a potential subobject component of type `lock_type` or `event_type` must be a coarray;
- a type without a potential subobject component of type `lock_type` or `event_type` must not be extended to have one;
- a `source=` variable in an `allocate` statement must not be of a type with a potential subobject component of type `lock_type` or `event_type`;
- a variable of type `team_type` must not be a coarray or a component of a coarray.

There exist constraints that prevent the use of entities of types `lock_type` and `event_type` in variable definition contexts except in the specific statements (described in the sections referenced above) that have been designed to operate on them. Assignment between entities of type `team_type` is permitted, but the variable becomes undefined if it resides on another image. The restrictions and effects for `team_type` are the same as for type `c_ptr`. This facilitates an implementation where the information stored in a variable is different on different images, for the purpose of optimizing communication.

17.17   The image control statements

The full list of **image control statements** is

i) Explicit image control statements:
  - `sync all`;
  - `sync images`;
  - `lock` or `unlock`;
  - `sync memory` (see Appendix A.16);
  - `sync team`, `form team`, `change team`, or `end team` (see Chapter 18);
  - `event post` or `event wait`;
  - `allocate`, `deallocate`, or `call move_alloc` involving a coarray;
  - `critical` or `end critical`;
  - `stop` or `end program`.

ii) Implicit image control statements:
  - `end` or `return` that involves an implicit deallocation of a coarray;
  - a statement that completes the execution of a `block` construct (see Section 8.12) and results in an implicit deallocation of a coarray.

17.18   Error termination

An image initiates **error termination** if it encounters an error condition that causes the computation to be flawed, making it desirable to stop all images as soon as is practicable. This causes all other images that have not already initiated error termination to do so. Within the performance limits of the processor's ability to send signals to other images, this is expected to terminate all images immediately. Error termination is initiated by the statement

error stop *[stop-code] [* quiet= *quiet-exp]*

where *stop-code* is an integer or default character scalar expression and *quiet-exp* is a logical scalar expression whose value controls whether a stop code or exception summary is output. When executed on one image, this will cause all other images that have not already initiated error termination to do so. It thus causes the whole calculation to stop as soon as is practicable. The meaning of *stop-code* is the same as for the stop statement, see Section 5.3.

Error termination can also be initiated by a Fortran statement that encounters an error condition, for example an allocate statement that does not specify a stat= variable and fails to acquire the requested memory.

17.19   Normal termination

An image initiates **normal termination** by starting to execute a stop or end program statement, or by starting to execute a statement in a procedure defined by a C companion processor that has the same effect. Such an image is called a **stopped image**.

When this happens, other images might still need to continue executing, and might require access to the stopped image's data. To assure both of these work properly, the stopped image must wait until all other images have also initiated normal termination, before it (and all other images) complete execution.

The example in Figure 17.9 illustrates the use of stop and error stop in a simulation. If the check for data consistency fails on an image, it executes error stop and all images are forced to terminate execution without delay. Normal termination occurs after the call to the procedure write_output; if an error stop were used in this context, completion of the output operations on all images would not be assured.

17.20   Input/output

Just as each image has its own variables, so it has its own input/output units. Whenever an input/output statement uses an integer expression to index a unit, it refers to the unit on the executing image.

The default unit for input (* in a read statement or input_unit in the intrinsic module iso_fortran_env) is preconnected on image 1 only.

The default unit for output (* in a write statement or output_unit in the intrinsic module iso_fortran_env) and the unit that is identified by error_unit in the intrinsic module iso_fortran_env are preconnected on each image. The files to which these are connected are regarded as separate, but it is expected that the processor will merge their records into a single stream or a stream for all output_unit files and a stream for all error_unit files. If records from these separate files are being merged into one stream, the merging process is processor dependent. Synchronization and flush statements might be sufficient to control the ordering of records, but this is not guaranteed. For example, the processor might delay merging the files until termination.

Figure 17.9 `stop` and `error stop` in a simulation.

```
type(simdata_t) :: simulation_data[*]
real :: eps
real, parameter :: crit = 1.0e-5
   :
simloop : do
   ! image-local processing
   call process(simulation_data, eps) ! eps > 0 is returned
   call co_max(eps) ! Find maximum value on the images
   if (eps < crit) exit simloop
   sync all
   ! data transfer between images
   call exchange_boundaries(simulation_data)
   sync all
   ! image-local checking
   call check_data_consistency(simulation_data, ier)
   if (ier /= 0) error stop 'data consistency lost - aborting run.'
end do simloop
call write_output(simulation_data)
stop 'successfully concluded simulation'
```

Any other preconnected unit is connected on the executing image only, and the file is completely separate from any preconnected file on another image.

The `open` statement connects a file to a unit on the executing image only. Whether a file with a given name is the same file on all images or varies from one image to the next is processor dependent.

A file is permitted to be connected to more than one image.

17.21   Intrinsic procedures

17.21.1   Introduction

There are six intrinsic functions for inquiries about coarray properties. Again, we use italic square brackets *[ ]* to indicate optional arguments. None of the functions are permitted in a constant expression. Most of the functions have an argument `coarray`. If this is specified, it must be a coarray. If it is allocatable, it must be allocated. To ensure that it is subscriptable, if its designator involves component selection, the final component must be a coarray – for example, if the coarray properties are required of `a%lower` where a is a coarray, these are those of a and a must be used as an actual argument.

17.21.2 Inquiry functions

image_index (coarray, sub) returns a default integer scalar.

If sub holds a valid sequence of cosubscripts for coarray, the result is the corresponding image index. Otherwise, the result is zero.

coarray is a coarray of any type.

sub is a rank-one integer array of size equal to the corank of coarray.

The intrinsic function image_index has two additional forms, see Section 18.8.2.

lcobound (coarray [, dim] [, kind]) returns the lower cobounds of a coarray in just the same way as lbound returns the lower bounds of an array.

ucobound (coarray [, dim] [, kind]) returns the upper cobounds of a coarray in just the same way as ubound returns the upper bounds of an array. The value returned for the final upper cobound is that of the final cosubscript in the cosubscript list for the coarray that selects the image whose index is equal to the number of images.

coshape (coarray [, kind]) returns a rank-one integer array whose size is the corank of coarray. It is of kind kind if it is present, and default kind otherwise. Element i has the value $1 + \text{ucobound}(\text{coarray}, i) - \text{lcobound}(\text{coarray}, i)$.

coarray is a coarray of any type.

dim is an integer scalar whose value is in the range $1 \leq \text{dim} \leq n$ where n is the corank of coarray. The corresponding actual argument must not be an optional dummy argument.

kind is a scalar integer constant expression.

17.21.3 Transformational functions

num_images () returns the number of images as a default integer scalar.

this_image () returns the index of the invoking image as a default integer scalar.

this_image (coarray) returns a default integer array of rank one and size equal to the corank of coarray; it holds the set of cosubscripts of coarray for data on the invoking image.

coarray is a coarray of any type.

this_image (coarray, dim) returns a default integer scalar holding cosubscript dim of coarray for data on the invoking image.

coarray is a coarray of any type.

dim is an integer scalar whose value is in the range $1 \leq \text{dim} \leq n$ where n is the corank of coarray.

The intrinsic function num_images has two additional forms, see Section 18.8.3, and the intrinsic function this_image has three additional forms, see Section 18.8.4.

17.22   Collective subroutines

Collective intrinsic subroutines provide a means to perform collective operations on all the active images (or all the active images of the current team, Chapter 18), such as summing the values of a variable across the images. It is to be expected that the execution will have been optimized by the system, for example by associating the images with the leaves of a binary tree and grouping the operations by tree level.

The same `call` statement must be executed on all the active images and it must occur in a context that would allow an image control statement. There is no automatic synchronization at the statement, but it is to be expected that the system applies some form of synchronization while executing the subroutine. To avoid the possibility of these synchronizations causing deadlock, the sequence of invocations must be the same on all the active images.

The collective subroutines have all or most of these arguments:

a is a scalar or an array that has the same shape on all the active images. It has intent `inout`. It must not be coindexed. It may be a coarray but this is not required. If it is a coarray, corresponding coarrays must be specified on all the active images.

result_image is an optional intent `in` integer scalar. If present, it must have the same value on all the active images. It specifies the image on which the result is placed in a; a becomes undefined on all other active images. If it is not present, the result is broadcast to all active images.

stat is an optional intent `out` integer scalar with decimal range of at least four. If it is not present and an error occurs, error termination is initiated. If present on one image, it must be present on all active images. It must not be coindexed. If the invocation is successful, it is given the value zero. If an error occurs, it is given a nonzero value and the argument a becomes undefined. If there is a stopped image, it is given the value `stat_stopped_image`. Otherwise, if there is a failed image, it is given the value `stat_failed_image`. Otherwise, it is given a processor-dependent different value.

errmsg is an optional intent `inout` scalar of type default character. It must not be coindexed. When present, it provides an explanatory message in the event of an error condition.

The subroutines are as follows:

`call co_broadcast(a, source_image [, stat] [,errmsg])`   copies the value of a on `source_image` to the corresponding argument on all the other active images as if by intrinsic assignment. The variable a must not be polymorphic. Its shape, type, and type parameter values must be the same on all active images.

> **source_image** is an intent `in` integer scalar that specifies the image from which values are broadcast. It must have the same value on all active images.

`call co_max(a [,result_image][,stat][,errmsg])`   computes the maximum value of a on all active images. The result is placed in a on `result_image` if this is present and otherwise in a on all active images. The variable a may be of type integer, real, or character, but must have the same shape, type, and type parameters on all active images. If it is an array, the result is computed element by element.

call co_min(a[, result_image][, stat], errmsg]) behaves like the subroutine co_max but is for minimum instead of maximum values.

call co_sum(a[, result_image][, stat][, errmsg]) computes the sum of the values of a on all active images. The variable a may be of any numeric type, but must have the same shape, type, and type parameters on all active images. If it is an array, the result is computed element by element. The order of summation is not specified so the result may be affected by rounding errors that can vary even between different runs on the same computer. The result is placed in a on result_image if this is present and otherwise exactly the same result is placed in a on all active images.

call co_reduce(a, operation[, result_image][, stat][, errmsg])
behaves like co_sum but reduces the coarray by a given function instead of by summation. The variable a may be of any type but must not be polymorphic and must not have an allocatable or pointer ultimate component.

> **operation** is a pure function with two scalar non-coarray dummy arguments and a scalar non-coarray result, all of the type and type parameters of a. Neither argument may be polymorphic, allocatable, optional, or a pointer. If one has the asynchronous, target, or value attribute, the other must too. It must be the same function on all active images and must implement an operation that is mathematically associative but need not be computationally associative (for example, floating-point addition is not computationally associative due to rounding).

17.23 Atomic subroutines

An **atomic subroutine** is an intrinsic subroutine that acts on a scalar variable atom of type integer(atomic_int_kind) or logical(atomic_logical_kind), whose kind values are defined in the intrinsic module iso_fortran_env. The variable atom must be a coarray or a coindexed object. Two executions of atomic subroutines with the same atom actual argument are permitted to occur in unordered segments; the effect is as if each action on the argument atom occurs instantaneously and does not overlap with the other. The programmer has no control over which occurs first and it may vary from run to run on the same computer and the same compiled code. If the order matters, the programmer must add image control statements, but there can be performance gains where the order does not matter.

Atomic operations make asynchronous progress. If a variable x on image p is defined by an atomic subroutine on image q, image r repeatedly references x[p] by an atomic subroutine in an unordered segment, and no other image defines x[p] in an unordered segment, image r will eventually receive the value assigned by image q, even if none of the images p, q, or r execute an image control statement until after the definition of x[p] by image q and the reception of that value by image r.

Eleven atomic subroutines are available. They all have the optional argument

stat is an optional integer scalar with a decimal exponent range of at least four. It must not be coindexed. It has intent `out`. If it is not present and an error occurs, error termination is initiated. If the invocation is successful, it is given the value zero. If an error occurs, it is given a nonzero value and any other argument with intent `inout` or `out` becomes undefined. If `atom` is on a failed image and there is no other cause for the error, the value given is `stat_failed_image`.

The following five atomic subroutines have arguments with the same properties:

call atomic_add (atom, value [, stat]) gives the variable `atom` the value `atom+int(value,atomic_int_kind)`.

call atomic_and (atom, value [, stat]) gives the variable `atom` the value `iand(atom,int(value,atomic_int_kind))`.

call atomic_define (atom, value [, stat]) gives the variable `atom` the value `value`.

call atomic_or (atom, value [, stat]) gives the variable `atom` the value `ior(atom,int(value,atomic_int_kind))`.

call atomic_xor (atom, value [, stat]) gives the variable `atom` the value `ieor(atom,int(value,atomic_int_kind))`.

> **atom** is a scalar coarray or coindexed object of type `integer(atomic_int_kind)`. It has intent `inout`.
>
> **value** is an integer scalar. It has intent `in`.

The following four atomic subroutines have arguments with the same properties:

call atomic_fetch_add (atom, value, old [, stat]) gives the variable `atom` the value `atom+value` and `old` is given the value `atom` had on entry.

call atomic_fetch_and (atom, value, old [, stat]) gives the variable `atom` the value `iand(atom,int(value,atomic_int_kind))` and `old` is given the value `atom` had on entry.

call atomic_fetch_or (atom, value, old [, stat]) gives the variable `atom` the value `ior(atom,int(value,atomic_int_kind))` and `old` is given the value `atom` had on entry.

call atomic_fetch_xor (atom, value, old [, stat]) gives the variable `atom` the value `ieor(atom,int(value,atomic_int_kind))` and `old` is given the value `atom` had on entry.

> **atom** is a scalar coarray or coindexed object of type `integer(atomic_int_kind)`. It has intent `inout`.

value is an integer scalar. It has intent in.

old is a scalar of the same type and kind as atom. It has intent out.

The following two atomic subroutines have the optional argument stat with unchanged properties, but are otherwise a little different.

call atomic_cas (atom, old, compare, new [, stat]) compares the value of atom with that of compare. If they are equal, it causes atom to be given the value of new. Otherwise, the value of atom is not changed. In either case, old is given the value atom had on entry.

> **atom** is a scalar coarray or coindexed object of type integer(atomic_int_kind) or integer(atomic_logical_kind). It has intent inout.
>
> **old** is a scalar of the same type and kind as atom. It has intent out.
>
> **compare** is a scalar of the same type and kind as atom. It has intent in.
>
> **new** is a scalar of the same type and kind as atom. It has intent in.

call atomic_ref (value, atom [, stat]) gets the value of atom atomically and assigns it to value.

> **value** is a scalar coarray or coindexed object of type integer(atomic_int_kind) or integer(atomic_logical_kind). It has intent out.
>
> **atom** is a scalar coarray or coindexed object of the same type and kind as value. It has intent in.

17.24 Image failure

The Fortran standard specifies semantics for continued execution of the overall program in the presence of individual images that have ceased to operate due to a technical issue with the execution environment. Such an image is called a **failed image**. An image that has failed will remain failed,[10] and all its data become inaccessible. If the programmer wishes execution to continue, it is incumbent on him or her to deal with the situation. It is not required that an implementation support this feature, and even if it does, not all failure scenarios are covered by the model.

To avoid error termination after an image failure, the use of implicit image control statements must be avoided, and a stat= variable must be specified on every explicit image control statement and on every call to a collective or atomic subroutine. For image control statements this will also assure that synchronization still happens on active images. An explicit check of the stat= variable at certain key points in the computation then can be used as a trigger for taking corrective action. For example, it may be possible to go back to a previous state of the computation and repeat it with fewer images, or with 'reserve' images brought in to replace failed ones.[11]

---

[10]If an image is unresponsive for a certain length of time, the system might assume that it had failed. If it subsequently becomes apparent that it has not failed, Fortran treats it as continuing to fail because it would be very hard for the programmer to allow for this situation.

[11]See Section 18.10 for how to use teams to deal with this scenario.

Image failure also has an impact on references and definitions that were intended to involve data on a failed image. Suppose image 2 has failed and image 1 executes the following statements:

```
x = a[2]
b[2] = x + y
```

Both statements will be successfully executed, but the result will differ from the non-failed case. The variable x will receive a processor-dependent value in the first statement, and the second statement will not perform any data transfer. The image selector syntax offers the possibility of adding a stat= specifier for diagnostic purposes:

```
x = a[2, stat=tstat_1]
b[2, stat=tstat_2] = x + y
```

This does not affect the results in x and b, but the value stat_failed_image will be received in the stat= variables of both statements.

If image 2 has failed while executing a segment that normally would execute the statement

```
a[1] = x
```

the variable a[1] becomes undefined. Images that reference a[1] in subsequent segments need to use knowledge of the communication pattern as well as to note the failure of image 2 and then take measures to guard against potential data corruption issues.

In order to facilitate testing of image failure recovery methods, the statement

```
fail image
```

can be used to cause an executing image to behave as if it had failed. No further statements are executed by the image.

17.25 Diagnosing the state of a parallel computation

17.25.1 Detecting failed and stopped images

Three intrinsic functions assist in the detection of failed and stopped images. There are two transformational functions:

failed_images ([kind]) returns a rank-one integer array holding the image indices of the images that are known to have failed, in numerically increasing order.

stopped_images ([kind]) returns a rank-one integer array holding the image indices of the images that are known to have stopped, in numerically increasing order.

> **kind** is an optional integer scalar constant expression that specifies the kind of the result; if absent, the result is of default kind.

and an elemental function:

image_status (image) returns the default integer value stat_failed_image if the image with index image has failed, stat_stopped_image if it has stopped, and zero otherwise

> **image** is a scalar integer whose value must be an image index.

These three intrinsic functions have additional forms when teams are in use, see Section 18.9.

17.25.2 Coping with stopped or failed images

When an image has initiated normal termination or has failed, the remaining active images continue execution. This has an impact on the execution of statements that involve multiple images, that is,

- coindexed definitions and references,
- image control statements, and
- atomic and collective procedures.

Therefore, we describe the facilities that enable active images to cope with the situation of stopped or failed images arising during a computation.

All the explicit image control statements have optional stat= and errmsg= specifiers. They have the same role for these statements as they do for allocate and deallocate (Sections 6.6 and 6.5). They must not be coindexed and no specifier's variable may depend on the value of another's.

A nonzero stat= value will be received only if one of the images involved in executing the image control statements has encountered an image failure, normal termination, or another error condition. The integer constants stat_stopped_image and stat_failed_image are possible return values, and other errors cause a processor-dependent positive value to be returned.[12] If more than one error occurs, the precedence for the stat= value is stat_stopped_image first, a processor-dependent error condition next, then stat_failed_image.

The value of stat_failed_image is positive if failed image handling is supported and negative otherwise.

Execution of an image control statement that does not specify a stat= variable and encounters an error (including involvement with a failed or stopped image) will initiate error termination.

When an image control statement that specifies a stat= variable is executed the effect of a stopped image on the synchronization behaviour depends on the image control statement that is executed. For allocate and deallocate statements that involve coarrays, the synchronization behaviour is the usual one for all active images because the obvious way to implement this is to perform independent allocations or deallocations on the images and then synchronize. For sync all, sync images and sync team, the synchronization effect reduces to that of sync memory (see Appendix A.16). For locks, events, and critical constructs, it is important that the programmer avoids initiating normal termination on an image

- without unlocking a lock that another image might try to lock,
- without executing an event post statement for which another image is waiting, or
- inside a critical construct that another image might try to execute,

because the other image will then wait for ever (deadlock).

The effect of a failed image on the synchronization behaviour also depends on the image control statement that is executed, but in a different manner. For allocate, deallocate, and the move_alloc intrinsic with coarray arguments, the full semantics apply for execution

---

[12]Some image control statements, for example, lock, also define further stat= values related to their semantics.

on all active images; all other image control statements, with some exceptions described in the following, retain their synchronization properties on all active images involved in their execution.

For `lock`, a value of `stat_failed_image` is returned if the lock variable resides on the failed image.   If the failed image held the lock at the time it ceased to operate, the lock variable becomes unlocked, the executing image obtains the lock, and the value `stat_unlocked_failed_image` from `iso_fortran_env` is returned. Similarly, if an image fails while executing a `critical` construct that specifies a `stat=` variable, the construct can be subsequently executed by an active image, and any such subsequent execution will receive the value `stat_failed_image`.

An `event wait` statement might hang indefinitely (or run into a timeout error) if a stopped or failed image is responsible for satisfying its request. Additional code using the intrinsics for detection of inactive images (Section 18.9) might be used to work around this.

The effect of specifying `stat=` for atomic subroutines is described in Section 17.23, for collective subroutines in Section 17.22, and for coindexed references and definitions in Section 17.24.

Exercises

1. Write the following versions of a program in which image 1 reads a real value from a file and copies it to the other images, then all images print their values:

    i) the first version with a coarray and using `sync all`;

    ii) the second version should make use of the knowledge that images other than 1 do not need to synchronize with each other;

    iii) the third version in addition should use the knowledge that image 1 does not need information from other images;

    iv) the fourth version should use a data object that is not a coarray.

2. Write a program in which there is an allocatable array coarray that is allocated of size three with different values on all images, and then each image > 1 prints the value from its left neighbour image. Why is no explicit image control statement needed?

3. The collective procedure `co_broadcast` has the property that all images need to agree on what source image to use. But this information might not be readily available. Implement a subroutine that, from a logical input value that is assumed to have the value true on exactly one image and false on all others, supplies the value of the corresponding image index to all images with only τ references to remote images, assuming that the number of images is 2^τ. Hint: Treat the images as in a circle and arrange that at the start of the ith loop the number of images that have received an update has doubled. If the updates are handled with care, this specific algorithm will also work correctly if the number of images is not a power of two.

4. Suppose we have a rectangular grid of size `nrow` by `ncol` with a real value at each point and `ncol==num_images()`. The first and last rows are regarded as neighbours and the the first and last columns are regarded as neighbours. If the values are distributed in the coarray `u(1:nrow)[*]`, write a subroutine with arguments `nrow`, `ncol`, and `u` that replaces each value by the sum of the values at its four neighbours minus four times its own value.

5. Suppose we have the coarrays `a(1:nx,1:ny)[*]` and `b(1:ny,1:nz)[*]`. Assuming that `max(nx,ny,nz)` $\leq$ `num_images()`, write code to copy the data in b to a with redistribution so that `a(i,j)[k] == b(j,k)[i]` for all valid values of the indices.

Does your code have any bottlenecks where the same image is being asked for data by many images? If so, modify it to avoid this.

6. Adapt your subroutine from Exercise 3 to apply to a set of images by adding an array argument holding the indices of the set and a scalar argument holding the position of the executing image in the set. In a main program, set up two sets and values, then call your subroutine simultaneously for your two sets.

18. Coarray teams

18.1 Teams

Teams are a concept that allows separate sets of images to execute independently. An important design objective was that, given code that has been developed and tested on all images, it should be possible to run the code on a team without making changes. This requires that if a team has n images, the image indices within the team run from 1 to n, `sync all` applies to the images of the team, and a coarray actual argument refers to the team coarray.

It was decided that teams should always be formed by partitioning an existing team into parts, starting with the team of all the images, which is known as the **initial team**. The team in which a statement is executed by an image is known as the **current team**.

Information about a team is held collectively on all the images of the team in a scalar variable of type `team_type` from the intrinsic module `iso_fortran_env`. This design means that two maps determine the image to which a coindexed object refers. The values of the coindices together with the cobounds of the coarray determine the image index within the current team and the team variable determines the corresponding actual image. The components of a team variable are private and may contain information to improve communication efficiency within the team.

To facilitate implementation where team information may differ between images, a team variable created on one image is not usable on another image. Therefore, a coarray must not be of type `team_type`, and a polymorphic coarray must not be allocated to be of type `team_type` or to have a subobject of type `team_type`. Furthermore, assigning a value of type `team_type` to a variable on another image, or vice versa, causes the variable to become undefined. (These restrictions and effects are the same as for type `c_ptr`.)

A set of new teams is formed by executing a `form team` statement on all images of the current team. The new team to which an image of the current team will belong is determined by its **team number**, which is a positive integer. All the images with the same team number will belong to the same new team. The team number is specified on the `form team` statement by an integer expression. For example, the code

```
use iso_fortran_env
type ( team_type ) odd_even
    ⋮
form team ( 1 + mod(this_image(),2), odd_even )
```

Modern Fortran Explained, 3rd Edition. M. Metcalf, J. Reid, M. Cohen, and R. Bader. Oxford University Press (2024). © M. Metcalf, J. Reid, M. Cohen, and R. Bader (2024). DOI 10.1093/oso/9780198876571.001.0018

forms two new teams consisting of the images of the current team that have odd or even image indices. We describe the `form team` statement in detail in Section 18.2.

Changes of team take place at the `change team` and `end team` statements, which mark the beginning and end of a new construct, the `change team` construct:

```
change team ( odd_even )
    : ! Statements executed with odd_even as the current team
end team
```

The values of `odd_even` must have been constructed by the prior execution of a `form team` statement on all the images of the current team. Both the `change team` and the `end team` statement are image control statements. The executing image and the other images of its new team synchronize at these statements. These images must all execute the same `change team` statement. While we expect it to be usual for the images of the other new teams to execute the construct, this is not required – they might continue to execute in the previously current team. We describe the `change team` construct in detail in Section 18.3.

The team that is current during the execution of a `form team` statement is known as the **parent** of each new team formed. We find it convenient to refer to each new team as a **child** of the current team and a **sibling** of the others. If the current team is not the initial team, the children have as parent a team that itself has a parent. Parents of parents can occur at any depth and are known as **ancestors**. Note that `change team` constructs can be nested to any depth and can be executed in a procedure called from within a `change team` construct.

18.2 The `form team` statement

The `form team` statement takes the general form

```
form team (team-number, team-variable [ , form-team-spec ] ... )
```

where *team-number* is a scalar integer expression whose value must be positive, *team-variable* is a scalar variable of type `team_type`, and each *form-team-spec* is a `new_index=`, `stat=`, or `errmsg=` specifier. The *team-variable* on each image of a child team will be defined with information about that child team. All the images of the current team that specify the same *team-number* value will be in the same child team. This value is the **team number** for the child team and can be used to identify it in statements executed on an image of a sibling team. The team number of any image in the initial team is equal to -1.

By default, the processor chooses which image indices are assigned to which images of each child team, and the choice may vary from processor to processor. However, they may be specified by including the *form-team-spec*

```
new_index=expr
```

in the statement, where *expr* is a scalar integer expression; this specifier provides the image index for the executing image in the child team. If a child team has k images, the values on those images must be a permutation of $1, 2, \ldots, k$.

The `form team` statement is an image control statement. The same statement must be executed by all active images of the current team, and they synchronize.

18.3   The `change team` construct

The `change team` statement takes the general form

  [construct-name : *]* `change team` (*team-value [, association]* ... *[, sync-stat-list]*)

where *team-value* is a scalar of type `team_type`. Its values on the images that execute the same `change team` statement must be as constructed by corresponding executions of a `form team` statement on those images. Its value must not be altered during the execution of the `change team` construct.

Each *association* has the form

  coarray-name [*coarray-spec*] => *selector*

and declares a new name and new cobounds for *selector*, which is a named **associating** coarray that is in scope at the `change team` statement. For example, if `big` has cobounds [1:k, 1:k] and the current team is subdivided into *k* teams of *k* images, the *association*

  `part[*] => big`

makes `part` an associating coarray with cobounds [1:k] on each new team. The appearance of an *association* does not prevent the coarray being referenced by its original name. The mapping to an image index is then unchanged but the image index will refer to the image of the current (new) team.

In all cases, the largest value that the final cobound can have in a valid reference is returned by the intrinsic `ucobound`.

The *sync-stat-list* is as for the `sync all` statement (Section 17.5). If an image of the new team has failed but no other error occurs, `stat_failed_image` is assigned to the `stat=` variable and all the active images of the new team synchronize on entering the construct, the same as if the execution were successful.

The `end team` statement takes the general form

  `end team` *[([sync-stat-list])] [construct-name]*

The reason for the appearance of *sync-stat-list* here is to detect the possibility of image failure in the current team during execution of the construct. The `end team` statement is an image control statement and the images of the team that were current inside the construct synchronize here. After the execution of `end team` completes, the current team reverts to the one current before the execution of the matching `change team` statement.

18.4   Coarrays allocated in teams

Without teams, coarrays are always allocated and deallocated in synchrony across all images, which allows each image to calculate the address of a coarray element on another image (symmetric memory). With teams, synchronization is across the team, of course. Symmetric memory is maintained within teams by requiring that

- any allocatable coarray that is allocated before entry to a `change team` construct remains allocated during the execution of the construct, and
- any allocatable coarray that becomes allocated within a `change team` construct and is still allocated when the construct is left is automatically deallocated, even if it has the `save` attribute.

This allows each image to hold its allocated allocatable coarrays in a stack with those allocated in the initial team at the bottom, those allocated in the team that is the child of the initial team next, those allocated in the team that is the grandchild of the initial team next, etc. Of course, there is no requirement for exactly this form of memory management to be used.

18.5 The `sync team` statement

The `sync team` statement has been introduced to allow synchronization within an ancestor team without leaving a `change team` construct, or within a child team to which the executing image belongs without entering a `change team` construct. It has the general form

```
sync team ( team-value [ , sync-stat-list ] )
```

where *team-value* is of type `team_type` and identifies a child team, the current team, or an ancestor team. Successful execution synchronizes all the images of the specified team in the same way as `sync all` does for the current team.

The *sync-stat-list* is as for the `sync all` statement.

18.6 Image selectors and teams

Consider a coarray `old` that is in scope at a `change team` statement. If this is accessed directly within the `change team` construct, cosubscripts map to an image index within the current team. However, it is likely that data from other teams will need to be accessed from time to time, subject to suitable synchronization. Significant overheads are likely to be associated with leaving the construct, performing the data exchange, and changing teams again. Unless the current team is the initial team, an image of an ancestor team may be accessed thus

```
old[ cosubscript-list, team=team-value ]
```

and an image of a sibling team may be accessed thus

```
old[ cosubscript-list, team_number=team-number ]
```

where *team-number* is a scalar integer expression whose value is a team number of a team formed by the execution of the `form team` statement that created the current team. The `team=` and `team_number=` specifiers are mutually exclusive, but the `stat=` specifier may appear with or without either of them. These statements are not useful if the current team is the initial team, but are permitted if they select the initial team.

In both these cases, the coindices and the cobounds determine the image index within the team and the team variable determines the image to which the image index refers.

18.7 Procedure calls and teams

When a procedure with a non-allocatable coarray dummy argument is called, the current team does not change but the corank and cobounds are declared afresh and the mapping from coindices to image indices changes. This means that referencing an ancestor or sibling team in an image selector is not appropriate. This is also the case after the allocation of an allocatable coarray or the association of a coarray in a change team statement. This leads us to the concept of a coarray having a team in which it is **established** or not being established, which determines whether the coarray can be accessed on a team. Specifically,

- a non-allocatable coarray with the save attribute is established in the initial team;
- an allocatable coarray is established in the team in which it was allocated and is not established as long as it is unallocated;
- an associating coarray in a change team construct is established in the team that is current in the change team construct;
- a non-allocatable coarray that is an associating entity in an associate, select rank, or select type construct is established in the team in which the associate, select rank, or select type statement is executed; and
- a non-allocatable coarray that is a dummy argument or host associated with a dummy argument is established in the team in which the procedure was invoked; such a dummy coarray is not established in any ancestor team.

In an image selector, the coarray is required to be established in the team selected.

18.8   Intrinsic functions

All the intrinsic procedures of Section 17.21 are available when executing in a team and refer to the coarray in that team of images. Several of them have versions for specifying the team, and we describe these here. We also describe two functions that are specifically concerned with teams.

18.8.1   Intrinsic functions get_team and team_number

get_team (*[level]*) returns a value of type team_type; level is an optional integer scalar with value one of the constants initial_team, parent_team, and current_team in the intrinsic module iso_fortran_env. The function returns a team value that identifies the current team if level is not present or the team indicated by level if it is present.

team_number (*[team]*) returns a value of type default integer. It is the team number (Section 18.2) of the specified team within its set of sibling teams, or −1 for the initial team; team is an optional scalar value of type team_type that specifies the current or an ancestor team. Absence specifies the current team.

Referencing get_team is the only way to obtain the team value of the initial team; this will be needed in order to refer to it in a sync_team statement or image selector when executing in a child team.

Referencing team_number allows the executing image to determine in which team it lies and execute appropriate code, as illustrated in Figure 18.1, using the team odd_even of Section 18.1.

Figure 18.1 Use of `team_number`.

```
change team (odd_even)
   select case (team_number())
      case (1)
         ⋮  ! Code for images in team 1.
      case (2)
         ⋮  ! Code for images in team 2.
   end select
   ⋮
end team
```

18.8.2   Intrinsic function `image_index`

The intrinsic function `image_index` has two additional forms:

image_index (coarray, sub, team) returns a default integer scalar. If `sub` holds a valid sequence of cosubscripts for `coarray` in the team `team`, the result is the corresponding image index. Otherwise, the result is zero.

> **coarray** is a coarray that can have cosubscripts and is of any type. It must be established in the team `team`. If its designator involves component selection, the final component must be a coarray.
>
> **sub** is a rank-one integer array of size equal to the corank of `coarray` in the team `team`.
>
> **team** is a scalar value of type `team_type` that specifies the current or an ancestor team.

image_index (coarray, sub, team_number) returns a default integer scalar. If `sub` holds a valid sequence of cosubscripts for `coarray` in the team `team_number`, the result is the corresponding image index. Otherwise, the result is zero.

> **coarray** is a coarray that can have cosubscripts and is of any type. It must be established in an ancestor of the current team unless the current team is the initial team. If its designator involves component selection, the final component must be a coarray.
>
> **sub** is a rank-one integer array of size equal to the corank of `coarray` in the sibling team specified by `team_number`.
>
> **team_number** is an integer scalar value identifying a sibling team of the current team or the initial team if the value is -1.

For a given set of cosubscripts, the value of the function may change on entering a `change team` construct. For example, `image_index (coarray, [30])` might be 30 in the parent team and 0 in the child team if this has only 15 images. Therefore, `image_index` cannot remain an inquiry function and becomes transformational.

18.8.3   Intrinsic function `num_images`

The intrinsic function `num_images` has two additional forms:

num_images (team)   returns the number of images in the team `team` as a default integer scalar.

> **team**   is a scalar value of type `team_type` that identifies the current or an ancestor team.

num_images (team_number)   returns the number of images in the team identified by `team_number` as a default integer scalar.

> **team_number**   is an integer scalar value identifying a sibling team of the current team or the initial team if the value is -1.

18.8.4   Intrinsic function `this_image`

The intrinsic function `this_image` has three additional forms:

this_image ([team])   returns the image index of the executing image in the team `team`, or the current team if `team` is absent, as a default integer scalar.

this_image (coarray [, team])   returns a default integer rank-one array holding the sequence of cosubscript values for `coarray` that would specify the executing image in the team specified by `team`, or the current team if `team` is absent.

this_image (coarray, dim [, team])   returns the value of cosubscript `dim` in the sequence of cosubscript values for `coarray` that would specify the executing image in the team specified by `team`, or the current team if `team` is absent, as a default integer scalar.

> **coarray**   is a coarray of any type. If allocatable, it must be allocated. If it is of type `team_type`, the argument `team` must be present. If its designator involves component selection, the final component must be a coarray.

> **dim**   is an integer scalar value in the range $1 \le \text{dim} \le n$ where n is the corank of `coarray` in the specified team.

> **team**   is a scalar value of type `team_type` that identifies the current or an ancestor team.

18.9   Detecting failed and stopped images

Three intrinsic functions assist in the detection of failed and stopped images. There are two transformational functions:

```
failed_images ( [team ] [, kind ])
stopped_images ( [team ] [, kind ])
```

These functions return a rank-one integer array holding the image indices of the images in the specified team that are known to have failed or stopped respectively, in numerically increasing order.

team is an optional scalar value of type team_type that identifies the current or an ancestor team; absence specifies the current team.

kind is an optional integer scalar constant expression that specifies the kind of the result; if absent, the result is of default kind.

and an elemental function:

image_status (image [,team]) returns a value of type default integer that is stat_failed_image if the image with index image has failed, stat_stopped_image if it has stopped, and zero otherwise.

image is a scalar integer whose value must be the image index of an image in the specified team.

team is an optional scalar value of type team_type that identifies the current or an ancestor team; absence specifies the current team.

18.10 Recovering from image failure

Teams may be used to enable recovery from image failure by setting up a team of images to execute the code while leaving other images idle but ready to replace images that fail. We illustrate this by the module in Figure 18.2 containing a subroutine that forms a team of executing images of a required size and the program in Figure 18.3 that uses it.

The subroutine executes a do loop on image 1 to choose a set of executing images of the required size for the team. These images are given the team number 1. The remaining executing images (the spares) are given the team number 2. This allows a form team statement to form the team simulation_team of the required size. The subroutine cannot working if image 1 has failed or if there are insufficient executing images; in both these cases an error stop statement is executed.

The code in Figure 18.3 makes use of the module in Figure 18.2. It begins by choosing how many images to use for the computation, holding this in images_used. The rest will be spares (or failed). It has an outer do loop that calls the module procedure form_work_team to find images_used executing images and set up a team for them, then executes within this team. The computation is checkpointed from time to time and if any image of the team fails, the outer do loop is re-executed to allow a new team of size images_used to restart from the most recent checkpointed data.

Figure 18.2 Module to form working team.

```
module recover
    use, intrinsic :: iso_fortran_env, only:team_type, stat_failed_image
contains
    subroutine form_work_team(images_used,simulation_team,status)
        implicit none
        integer :: images_used ! Number of images used
        type (team_type) :: simulation_team
        integer :: status ! stat value
        integer, save :: team_no [*] ! 1 if in working team; 2 otherwise.
        integer :: i, j, k ! Temporaries
        sync all (stat = status)
        if (status/=0 .and. status/=stat_failed_image) return
        if (image_status(1)==stat_failed_image) error stop &
            "Failure of image 1 in form_work_team"
        if (this_image () == 1) then
          j = 0
          do k = 1, num_images ()
            if (image_status (k) == 0) then
              j = j+1
              if (j<=images_used) then
                team_no [k] = 1 ! Image in work team
              else
                team_no [k] = 2 ! Spare image
              end if
            end if
          end do
          if (j<images_used) error stop &
                "Failure of too many images in form_work_team"
        end if
        sync all (stat = status)
        if (status/=0 .and. status/=stat_failed_image) return
        form team (team_no, simulation_team, stat=status)
    end subroutine
end module
```

Figure 18.3 Using spare images.

```
program possibly_recoverable_simulation
  use recover
  implicit none
  integer :: images_used ! Number of images used
  integer :: status ! stat= value
  type (team_type) :: simulation_team
  logical :: done [*] ! True if computation finished on the image.
  images_used = min(int(num_images()*0.99), num_images()-1)
  images_used = num_images()-2 ! Extra
  outer : do
    call form_work_team(images_used,simulation_team,status)
    if (status/=0 .and. status/=stat_failed_image) exit outer
    simulation : change team (simulation_team, stat=status)
      if (status == stat_failed_image) exit simulation
      if (team_number() == 1) then
        ! Each image reads checkpoint data for itself, unless at the
        ! computation start.
        iter : do
          call simulation_procedure (status, done)
          ! Each image performs its part of the simulation procedure,
          ! synchronizing as necessary, storing checkpoint data from
          ! time to time, indicating failures in the variable status,
          ! and sets done to .true. when the simulation has completed.
          if (status == stat_failed_image) exit simulation
          if (done) exit iter
        end do iter
      end if
    end team (stat=status) simulation
    sync all (stat=status)
    if (team_number() == 2) done = done[1]
    if (done) exit outer
  end do outer
  if (status/=0 .and. status/=stat_failed_image) &
    print *,'unexpected failure ',status,' in program.'
end program possibly_recoverable_simulation
```

Exercises

1. Redo Exercise 6 from Chapter 17, but make use of teams and reuse the version of the subroutine developed for Exercise 3. In which team is the coarray declared in the procedure established?

2. Write a procedure that circularly shifts coarray data between images:

    x(:)[i]   →   x(:)[i+1]   for i < num_images()
    x(:)[num_images()]   →   x(:)[1]

    Then, using a square *nc* × *nc* subset of images, create a rank-two matrix of coshape [nc, *] and apply the shift to the diagonal blocks a(:,:)[i, i]. In which team is the dummy argument of the procedure established?

19. Floating-point exception handling

19.1   Introduction

Exception handling is required for the development of robust and efficient numerical software, a principal application of Fortran.   Indeed, the existence of such a facility makes it possible to develop more efficient software than would otherwise be possible. Most computers nowadays have hardware based on the IEEE standard for floating-point arithmetic,[1] which is also an ISO standard.[2]   Therefore, the Fortran exception handling features are based on the ability to test and set the flags for the five floating-point exceptions that the IEEE standard specifies. However, non-IEEE computers have not been ignored; they may provide support for some of the features and the programmer is able to find out what is supported or state that certain features are essential.

Few (if any) computers support every detail of the IEEE standard.   This is because considerable economies in construction and increases in execution performance are available by omitting support for features deemed to be necessary to few programmers. It was therefore decided to include inquiry facilities for the extent of support of the standard, and for the programmer to be able to state which features are essential in his or her program.

The mechanism finally chosen by the committees is based on a set of procedures for setting and testing the flags and inquiring about the features, collected in an intrinsic module called `ieee_exceptions`.

Given that procedures were being provided for the IEEE flags, it seemed sensible to provide procedures for other aspects of the IEEE standard. These are collected in a separate intrinsic module, `ieee_arithmetic`, which contains a `use` statement for `ieee_exceptions`.

To provide control over which features are essential, there is a third intrinsic module, `ieee_features`, containing named constants corresponding to the features. If a named constant is accessible in a scoping unit, the corresponding feature must be available there.

19.2   The IEEE standard

In this section we explain those aspects of the IEEE standard that the reader needs to know in order to understand the features of this chapter. We do not attempt to give a complete description of the standard.

---

[1] IEEE 754-2008, Standard for Floating-Point Arithmetic.

[2] ISO/IEC/IEEE 60559:2011, Information Technology – Microprocessor Systems – Floating-Point Arithmetic.

Modern Fortran Explained, 3rd Edition. M. Metcalf, J. Reid, M. Cohen, and R. Bader. Oxford University Press (2024). © M. Metcalf, J. Reid, M. Cohen, and R. Bader (2024). DOI 10.1093/oso/9780198876571.001.0019

Five floating-point data formats are specified, three for binary arithmetic and two for decimal arithmetic. They are supersets of the Fortran model (see Section 9.9.1), repeated here:

$$x = 0 \quad \text{and} \quad x = s \times b^e \times \sum_{k=1}^{p} f_k \times b^{-k},$$

where s is ± 1, p and b are integers exceeding one, e is an integer in a range $e_{\min} \le e \le e_{\max}$, and each f_k is an integer in the range $0 \le f_k < b$ except that f_1 is also nonzero. The IEEE formats are binary with $b = 2$ or decimal with $b = 10$. The precisions are $p = 24$, $p = 53$, or $p = 113$ for binary and $p = 16$ or $p = 34$ for decimal. The exponent ranges are $-125 \le e \le 128$, $-1021 \le e \le 1024$, and $-16381 \le e \le 16384$ for binary, and $-382 \le e \le 385$ and $-6142 \le e \le 6145$ for decimal.

In addition, there are numbers with $e = e_{\min}$ and $f_1 = 0$, which are known as **subnormal**[3] numbers; note that they all have absolute values less than that returned by the intrinsic `tiny` since it considers only numbers within the Fortran model. Also, zero has a sign and both 0 and -0 have inverses, ∞ and $-\infty$. Within Fortran, -0 is treated as the same as a zero in all intrinsic operations and comparisons, but it can be detected by the `sign` function and is respected on formatted output.

The IEEE standard also specifies that some of the binary patterns that do not fit the model be used for the results of exceptional operations, such as 0/0. Such a number is known as a **NaN** (Not a Number). A NaN may be **signaling** or **quiet**. Whenever a signaling NaN appears as an operand, the invalid exception signals and the result is a quiet NaN. Quiet NaNs propagate through almost every arithmetic operation without signaling an exception.

The standard specifies five rounding modes:

roundTiesToEven rounds the exact result to the nearest representable value; if there are two such numbers, to the one with an even least significant digit; this corresponds to the Fortran rounding mode `ieee_nearest`.

roundTiesToAway rounds the exact result to the nearest representable value; if there are two such numbers, to the one with larger magnitude; this corresponds to the Fortran rounding mode `ieee_away`.

roundTowardZero rounds the exact result towards zero to the next representable value; this corresponds to the Fortran rounding mode `ieee_to_zero`.

roundTowardPositive rounds the exact result towards $+\infty$ to the next representable value; this corresponds to the Fortran rounding mode `ieee_up`.

roundTowardNegative rounds the exact result towards $-\infty$ to the next representable value; this corresponds to the Fortran rounding mode `ieee_down`.

Some computers perform division by inverting the denominator and then multiplying by the numerator. The additional round-off that this involves means that such an implementation does not conform with the IEEE standard. The IEEE standard also specifies that its *squareRoot* operation returns the properly rounded value of the exact square root and returns -0 for *squareRoot*(-0). The Fortran facilities include inquiry functions for IEEE division and *squareRoot*.

---

[3]Formerly called 'denormalized'.

The presence of −0, ∞, −∞, and the NaNs allows IEEE arithmetic to be closed, that is, every operation has a result. This is very helpful for optimization on modern hardware since several operations, none needing the result of any of the others, may actually be progressing in parallel. If an exception occurs, execution continues with the corresponding flag signaling, and the flag remains signaling until explicitly set quiet by the program. The flags are therefore called **sticky**.

There are five flags:

overflow occurs if the exact result of an operation with two normal values is too large for the data format. The stored result is ∞, huge(x), -huge(x), or −∞, according to the rounding mode in operation, always with the correct sign.

divideByZero occurs if a finite nonzero value is divided by zero. The stored result is ∞ or −∞ with the correct sign.

invalid occurs if the operation is invalid, for example, ∞ × 0, 0/0, or when an operand is a signaling NaN.

underflow occurs if the result of an operation with two finite nonzero values cannot be represented exactly and is too small to represent with full precision. The stored result is the best available, depending on the rounding mode in operation.

inexact occurs if the exact result of an operation cannot be represented in the data format without rounding.

The IEEE standard specifies the possibility of exceptions being trapped by user-written handlers, but this inhibits optimization and is not supported by Fortran. Instead, Fortran supports the possibility of halting program execution after an exception signals. For the sake of optimization, such halting need not occur immediately.

19.3   Access to the features

To access the features of this chapter, the user needs to employ use statements for one or more of the intrinsic modules ieee_exceptions, ieee_arithmetic (which contains a use statement for ieee_exceptions), and ieee_features. If the processor does not support a module accessed in a use statement, the compilation, of course, fails.

The module ieee_arithmetic contains the following transformational function that is permitted in a constant expression (Section 8.4) and can be used to declare an IEEE real:

ieee_selected_real_kind (*[p]* *[,r]* *[,* **radix** *]***)** is similar to the function selected_real_kind (Section 9.9.4) except that the result is the kind value of a real x for which ieee_support_datatype(x) (Section 19.7.3) is true.

If a scoping unit does not access ieee_exceptions or ieee_arithmetic, the level of support is processor dependent, and need not include support for any exceptions. If a flag is signaling on entry to such a scoping unit, the processor ensures that it is signaling on exit. If a flag is quiet on entry to such a scoping unit, whether it is signaling on exit is processor dependent.

The module ieee_features contains the derived type

    ieee_features_type

for identifying a particular feature. The only possible values objects of this type may take are those of named constants defined in the module, each corresponding to an IEEE feature. If a scoping unit has access to one of these constants, the compiler must support the feature in the scoping unit or reject the program. For example, some hardware is much faster if subnormal numbers are not supported and instead all underflowed values are flushed to zero. In such a case, the statement

```
use, intrinsic :: ieee_features, only: ieee_subnormal
```

will ensure that the scoping unit is compiled with (slower) code supporting subnormal numbers. This form of the `use` statement is safer because it ensures that should there be another module with the same name, the intrinsic one is used. It is described fully in Section 9.23.

The module is unusual in that all a program ever does is to access it with `use` statements, which affect the way the code is compiled in the scoping units with access to one or more of the module's constants. There is no purpose in declaring data of type `ieee_features_type`, though it is permitted; the components of the type are private, no operation is defined for it, and only intrinsic assignment is available for it. In a scoping unit containing a `use` statement the effect is that of a compiler directive, but the other properties of `use` make the feature more powerful than would be possible with a directive.

The complete set of named constants in the module and the effect of their accessibility is:

ieee_datatype The scoping unit must provide IEEE arithmetic for at least one kind of real.

ieee_denormal Same as **ieee_subnormal**.

ieee_divide The scoping unit must support IEEE divide for at least one kind of real.

ieee_halting The scoping unit must support control of halting for each flag supported.

ieee_inexact_flag The scoping unit must support the inexact exception for at least one kind of real.

ieee_inf The scoping unit must support ∞ and $-\infty$ for at least one kind of real.

ieee_invalid_flag The scoping unit must support the invalid exception for at least one kind of real.

ieee_nan The scoping unit must support NaNs for at least one kind of real.

ieee_rounding The scoping unit must support control of the rounding mode for the four rounding modes `ieee_nearest`, `ieee_to_zero`, `ieee_up`, and `ieee_down` for at least one kind of real. Support for `ieee_away` is required if at least one kind of real with radix 10 is supported.

ieee_sqrt The scoping unit must support IEEE *squareRoot* for at least one kind of real.

ieee_subnormal The scoping unit must support subnormal numbers for at least one kind of real.

ieee_underflow_flag The scoping unit must support the underflow exception for at least one kind of real.

Execution may be slowed on some processors by the support of some features. If `ieee_exceptions` is accessed but `ieee_features` is not accessed, the vendor is free to choose which subset to support. The processor's fullest support is provided when all of `ieee_features` is accessed:

```
use, intrinsic :: ieee_arithmetic
use, intrinsic :: ieee_features
```

but execution may then be slowed by the presence of a feature that is not needed. In all cases, inquiries about the extent of support may be made by calling the functions of Sections 19.6.3 and 19.7.3.

19.4   The Fortran flags

There are five Fortran exception flags, corresponding to the five IEEE flags. Each has a value that is either quiet or signaling. The value may be determined by the function ieee_get_flag (Section 19.6.4). Its initial value is quiet and it signals when the associated exception occurs in a real or complex operation. Its status may also be changed by the subroutine ieee_set_flag (Section 19.6.4) or the subroutine ieee_set_status (Section 19.6.4). Once signaling, it remains signaling unless set quiet by an invocation of the subroutine ieee_set_flag or the subroutine ieee_set_status. For invocation of an elemental procedure, it is as if the procedure were invoked once for each set of corresponding elements; if any of the invocations return with a flag signaling, it will be signaling in the caller on completion of the call.

If a flag is signaling on entry to a procedure, the processor will set it to quiet on entry and restore it to signaling on return. This allows exception handling within the procedure to be independent of the state of the flags on entry, while retaining their 'sticky' properties: within a scoping unit, a signaling flag remains signaling until explicitly set quiet. Evaluation of a specification expression may cause an exception to signal.

If a scoping unit has access to ieee_exceptions and references an intrinsic procedure that executes normally, the values of the overflow, divide-by-zero, and invalid flags are as on entry to the intrinsic procedure, even if one or more signals during the calculation. If a real or complex result is too large for the intrinsic procedure to handle, overflow may signal. If a real or complex result is a NaN because of an invalid operation (for example, log(-1.0)), invalid may signal. Similar rules apply to format processing and to intrinsic operations: no signaling flag shall be set quiet and no quiet flag shall be set signaling because of an intermediate calculation that does not affect the result.

An implementation may provide alternative versions of an intrinsic procedure; for example, one might be rather slow but be suitable for a call from a scoping unit with access to ieee_exceptions, while an alternative faster one might be suitable for other cases.

If it is known that an intrinsic procedure will never need to signal an exception, there is no requirement for it to be handled – after all, there is no way that the programmer will be able to tell the difference. The same principle applies to a sequence of in-line code with no invocations of the procedures of Section 19.6.4 for getting and setting the flags and modes. If the code, as written, includes an operation that would signal a flag, but after execution of the sequence no value of a variable depends on that operation, whether the exception signals is processor dependent. Thus, an implementation is permitted to optimize such an operation away. For example, when y has the value zero, whether the code

```
x = 1.0/y
x = 3.0
```

signals divide-by-zero is processor dependent. Another example is:

```
real, parameter :: x=0.0, y=6.0
    :
if (1.0/x == y) print *,'Hello world'
```

where the processor is permitted to discard the `if` statement since the logical expression can never be true and no value of a variable depends on it.

An exception does not signal if this could arise only during execution of code not required or permitted by the standard. For example, the statement

```
if (f(x) > 0.0) y = 1.0/z
```

must not signal divide-by-zero when both `f(x)` and `z` are zero, and the statement

```
where(a > 0.0) a = 1.0/a
```

must not signal divide-by-zero. On the other hand, when x has the value 1.0 and y has the value 0.0, the expression

```
x > 0.00001 .or. x/y > 0.00001
```

is permitted to cause the signaling of divide-by-zero.

The processor need not support the invalid, underflow, and inexact exceptions. If an exception is not supported, its flag is always quiet. The function `ieee_support_flag` (Section 19.6.3) may be used to inquire whether a particular flag is supported. If invalid is supported, it signals in the case of conversion to an integer (by assignment or an intrinsic procedure) if the result is too large to be representable.

19.5   The floating-pointing modes

Some processors allow

- control during program execution of whether to abort or continue execution after an exception has occurred,

- alteration of the rounding mode during execution, and

- alteration of the underflow mode during execution, that is, whether small values are represented as subnormal values or are set to zero.

These are known as the floating-pointing modes. The functions `ieee_support_halting`, (Section 19.6.3), and `ieee_support_rounding` and `ieee_support_underflow_control` (Section 19.7.3) may be used to inquire whether they are available.

Halting after an exception has occurred is not precise and may occur any time after the exception has occurred. The underflow mode is said to be **gradual** if subnormal values are employed. The processor might be able execute faster without subnormal values.

The modes may be set individually by `ieee_set_halting_mode` (Section 19.6.4), and `ieee_set_rounding_mode` and `ieee_set_underflow_mode` (Section 19.7.5). Also (see Section 19.6.4), they may be stored collectively by `ieee_get_modes` or `ieee_get_status` and set collectively using stored values by `ieee_set_modes` or `ieee_get_status`. In a procedure other than these, the processor does not change the modes on entry and on return ensures that they are the same as they were on entry. The initial settings of the modes are processor dependent. In a procedure other than one of these procedures, the processor does not change any mode on entry and on return ensures that the modes are the same as they were on entry.

19.6   The `ieee_exceptions` module

19.6.1   Introduction

When the module `ieee_exceptions` is accessible, the overflow and divide-by-zero flags are supported in the scoping unit for all available kinds of IEEE real and complex data. This minimal level of support has been designed to be possible also on a non-IEEE computer. Which other exceptions are supported may be determined by the function `ieee_support_flag`, see Section 19.6.3. Whether control of halting is supported may be determined by the function `ieee_support_halting`, see Section 19.6.3. The extent of support of the other exceptions may be influenced by the accessibility of the named constants `ieee_inexact_flag`, `ieee_invalid_flag`, and `ieee_underflow_flag` of the module `ieee_features`, see Section 19.3.

The module contains two derived types (Section 19.6.2), named constants of these types (Section 19.6.2), and a collection of generic procedures (Sections 19.6.3 and 19.6.4). None of the procedures is permitted as an actual argument.

19.6.2   Derived types for floating-point flags, modes, and status

The module `ieee_exceptions` contains three derived types:

`ieee_flag_type` for identifying a particular exception flag. The only values that can be taken by objects of this type are those of named constants defined in the module,

```
ieee_overflow    ieee_divide_by_zero   ieee_invalid
ieee_underflow   ieee_inexact
```

and these are used in the module to define the named array constants:

```
type(ieee_flag_type), parameter ::                        &
ieee_usual(3) =                                           &
    (/ieee_overflow, ieee_divide_by_zero, ieee_invalid/), &
ieee_all(5) = (/ieee_usual, ieee_underflow, ieee_inexact/)
```

These array constants are convenient for inquiring about the state of several flags at once by using elemental procedures. Besides convenience, such elemental calls may be more efficient than a sequence of calls for single flags.

`ieee_modes_type` for saving all the current floating-point modes, that is, the values of the rounding mode, underflow mode, and halting mode.

ieee_status_type for saving the current floating-point status, that is, the values of all the exception flags supported and all the floating-point modes.

The components of these types are private. Intrinsic assignment is available for them and they may be argument associated, but no intrinsic operations are defined for them.

19.6.3   Inquiring about support of IEEE exceptions

The module ieee_exceptions contains two functions for inquiring about the support of IEEE exceptions, both of which are pure. Their argument flag must be a scalar of type type(ieee_flag_type). The inquiries are about the support for kinds of reals and the same level of support is provided for the corresponding kinds of complex type.

ieee_support_flag(flag) or **ieee_support_flag(flag, x)** returns .true. if the processor supports the exception flag for all reals or for reals of the same kind type parameter as the real argument x, respectively; otherwise, it returns .false..

ieee_support_halting (flag) returns .true. if the processor supports the ability to change the mode by call ieee_set_halting_mode(flag, halting); otherwise, it returns .false..

19.6.4   Subroutines for getting and setting the flags and modes

The module ieee_exceptions contains the following eight subroutines for getting and setting the flags and halting modes:

call ieee_get_flag (flag, flag_value) where:

> **flag** is of type type(ieee_flag_type). It specifies a flag.

> **flag_value** is of type logical and has intent out. It is given the value true if the exception specified by flag is signaling and false otherwise.

call ieee_get_halting_mode (flag, halting) where:

> **flag** is of type type(ieee_flag_type). It specifies a flag.

> **halting** is of type logical and has intent out. If the exception specified by flag will cause halting, halting is given the value true; otherwise, it is given the value false.

call ieee_get_modes (modes) where modes is a scalar of type ieee_modes_type that has intent out and is assigned the current value of the floating-point modes.

call ieee_get_status (status_value) where status_value is scalar and of type ieee_status_type, and has intent out. It returns the whole floating-point status.

call ieee_set_flag (flag, flag_value) where:

flag is of type type(ieee_flag_type). It may be scalar or array valued. If it is an array, no two elements may have the same value.

flag_value is of type logical. It must be conformable with flag. Each flag specified by flag is set to be signaling if the corresponding flag_value is true, and to be quiet if it is false.

call ieee_set_halting_mode (flag, halting) which may be called only if the value returned by ieee_support_halting(flag) is true:

flag is of type type(ieee_flag_type). It may be scalar or array valued. If it is an array, no two elements may have the same value.

halting is of type logical. It must be conformable with flag. Each exception specified by flag will cause halting if the corresponding value of halting is true and will not cause halting if the value is false.

call ieee_set_modes (modes) where modes is a scalar of type ieee_modes_type that has intent in and has a value obtained by a previous call of ieee_get_modes. The floating-point modes are restored to the values they had then.

call ieee_set_status (status_value) where status_value is scalar of type ieee_status_type that has intent in and has a value obtained by a previous call of ieee_get_status. The whole floating-point status is reset to what it was then.

The subroutines ieee_get_status and ieee_set_status have been included for convenience and efficiency when a subsidiary calculation is to be performed and one wishes to resume the main calculation with exactly the same environment, as shown in Figure 19.1. There are no facilities for finding directly the value held within such a variable of a particular flag or mode.

Figure 19.1 Performing a subsidiary calculation with an independent set of flags.

```
use, intrinsic          :: ieee_exceptions
type(ieee_status_type) :: status_value
  :
call ieee_get_status(status_value)    ! Get the flags
call ieee_set_flag(ieee_all,.false.)  ! Set the flags quiet
  : ! Calculation involving exception handling
call ieee_set_status(status_value)    ! Restore the flags
```

19.7   The **ieee_arithmetic** module

19.7.1   Introduction

The module ieee_arithmetic behaves as if it contained a use statement for the module ieee_exceptions, so all the features of ieee_exceptions are also features of ieee_arithmetic.

The module contains two derived types (Section 19.7.2), named constants of these types (Section 19.7.2), and a collection of generic procedures (Sections 19.7.3, 19.7.4, and 19.7.5). None of the procedures is permitted as an actual argument.

If x_1 and x_2 are of IEEE numeric types and the type of $x_1 + x_2$ is real, comparisons are made as follows:

$x_1 == x_2$ compareQuietEqual

$x_1 >= x_2$ compareSignalingGreaterEqual

$x_1 > x_2$ compareSignalingGreater

$x_1 <= x_2$ compareSignalingLessEqual

$x_1 < x_2$ compareSignalingLess

$x_1 /= x_2$ compareQuietNotEqual

If x_1 and x_2 are of IEEE numeric types and the type of $x_1 + x_2$ is complex, comparisons are made as follows:

$x_1 == x_2$ compareQuietEqual

$x_1 /= x_2$ compareQuietNotEqual

19.7.2   Derived types

The module `ieee_arithmetic` contains two derived types:

ieee_class_type for identifying a class of floating-point values.   The only values objects of this type may take are those of the named constants defined in the module,

```
ieee_signaling_nan        ieee_quiet_nan
ieee_negative_inf         ieee_negative_normal
ieee_negative_denormal    ieee_negative_zero
ieee_positive_zero        ieee_positive_denormal
ieee_positive_normal      ieee_positive_inf
ieee_positive_subnormal   ieee_negative_subnormal
```

with obvious meanings, and `ieee_other_value` for any cases that cannot be so identified, for example if an unformatted file were written with gradual underflow enabled and read with it disabled.   The values `ieee_negative_denormal` and `ieee_negative_subnormal` are the same, as are `ieee_positive_denormal` and `ieee_positive_subnormal`.

ieee_round_type for identifying a particular rounding mode. The only possible values objects of this type may take are those of the named constants defined in the module,

```
ieee_nearest   ieee_away   ieee_to_zero   ieee_up   ieee_down
```

for the IEEE modes and `ieee_other` for any other mode.

The components of both types are private. The only operations defined for them are == and /= for comparing values of one of the types; they return a value of type default logical. They may appear as arguments and in intrinsic assignments.

19.7.3   Inquiring about IEEE arithmetic

The module `ieee_arithmetic` contains the following functions, all of which are pure. The inquiries are about the support of reals, and the same level of support is provided for the corresponding kinds of complex type. The argument x may be a scalar or an array. All but `ieee_support_rounding` are inquiry functions.

`ieee_support_datatype ()` or **`ieee_support_datatype (x)`** returns `.true.` if the processor supports IEEE arithmetic for all reals or for reals of the same kind type parameter as the real argument x, respectively. Otherwise, it returns `.false.`. Complete conformance with the IEEE standard is not required for `.true.` to be returned, but the normalized numbers must be exactly those of an IEEE format; the binary arithmetic operators +, -, and * must be implemented with at least one of the IEEE rounding modes when the operands and result specified by the IEEE standard are normal; and the functions `abs`, `ieee_copy_sign`, `ieee_logb`, `ieee_rem`, and `ieee_unordered` must implement the corresponding IEEE functions.

`ieee_support_denormal ()` or **`ieee_support_denormal (x)`** returns `.true.` if the processor supports the IEEE denormal numbers for all reals or for reals of the same kind type parameter as the real argument x respectively. Otherwise, it returns `.false.`.

`ieee_support_divide ()` or **`ieee_support_divide (x)`** returns `.true.` if the processor supports divide with the accuracy specified by the IEEE standard for all reals or for reals of the same kind type parameter as the real argument x respectively. Otherwise, it returns `.false.`.

`ieee_support_inf ()` or **`ieee_support_inf (x)`** returns `.true.` if the processor supports the IEEE infinity facility for all reals or for reals of the same kind type parameter as the real argument x, respectively. Otherwise, it returns `.false.`.

`ieee_support_io ()` or **`ieee_support_io (x)`** returns `.true.` if the results of formatted input/output satisfy the requirements of the IEEE standard for the input/output rounding modes `up`, `down`, `zero`, and `nearest` for all reals or for reals of the same kind type parameter as the real argument x, respectively. Otherwise, it returns `.false.`.

`ieee_support_nan ()` or **`ieee_support_nan (x)`** returns `.true.` if the processor supports the IEEE NaN facility for all reals or for reals of the same kind type parameter as the real argument x, respectively. Otherwise, it returns `.false.`.

`ieee_support_rounding (round_value)` or
`ieee_support_rounding (round_value, x)` for a `round_value` of the type `ieee_round_type` is a transformational function that returns `.true.` if the processor supports that rounding mode for all reals or for reals of the same kind type parameter as the argument x, respectively. Otherwise, it returns `.false.`. Here, support includes the ability to change the mode by the invocation

```
call ieee_set_rounding_mode (round_value)
```

ieee_support_sqrt () or **ieee_support_sqrt (x)** returns .true. if sqrt implements IEEE square root for all reals or for reals of the same kind type parameter as the argument x, respectively. Otherwise, it returns .false..

ieee_support_subnormal same as **ieee_support_denormal**.

ieee_support_standard () or **ieee_support_standard (x)** returns .true. if the processor supports all the functions

```
ieee_support_datatype   ieee_support_divide   ieee_support_flag
ieee_support_halting    ieee_support_inf      ieee_support_nan
ieee_support_rounding   ieee_support_sqrt     ieee_support_subnormal
```

for valid arguments flag and round_value for all reals or for reals of the same kind type parameter as the real argument x, respectively. Otherwise, it returns .false..

ieee_support_underflow_control () or
ieee_support_underflow_control (x) returns .true. if the processor supports control of the underflow mode for all reals or for reals of the same kind type parameter as the real argument x, respectively. Otherwise, it returns .false..

19.7.4   Elemental functions

The module ieee_arithmetic contains the following elemental functions for the reals a, b, c, x, and y for which the values of ieee_support_datatype are true. If the value of any of these is an infinity or a NaN, the behaviour is consistent with the general rules of the IEEE standard for arithmetic operations. For example, the result for an infinity is constructed as the limiting case of the result with a value of arbitrarily large magnitude, when such a limit exists.

ieee_class (x) is of type type(ieee_class_type) and returns the IEEE class of the real argument x. The possible values are explained in Section 19.7.2.

ieee_copy_sign (x, y) returns a real with the same type parameter as x, holding the value of x with the sign of y. This is true even for the IEEE special values, such as NaN and ∞ (on processors supporting such values).

ieee_fma (a, b, c) where a, b, and c are real with the same kind type parameter returns the mathematical value of $a \times b + c$ (fused multiply-add operation) rounded according to the rounding mode; ieee_overflow, ieee_underflow, and ieee_inexact are signaled according to the final step in the calculation and not by any intermediate calculation. The kind of the result is the same as that of the arguments.

ieee_int (a, round [, kind]) where round is of type ieee_round_type returns the value of a converted to an integer according to the rounding mode specified by round. If kind is absent, the result is default integer; otherwise the result is of type integer with kind type parameter `kind`, which must be a scalar integer constant expression. If the rounded value is not representable in the result kind, `ieee_invalid` is signaled and the result is processor dependent. The processor is required to consistently signal or consistently not signal `ieee_inexact` when the result is not exactly equal to a.

ieee_is_finite (x) returns the value `.true.` if `ieee_class(x)` has one of the values

```
ieee_negative_normal      ieee_negative_subnormal
ieee_negative_zero        ieee_positive_zero
ieee_positive_subnormal   ieee_positive_normal
```

and `.false.` otherwise.

ieee_is_nan (x) returns the value `.true.` if the value of x is an IEEE NaN and `.false.` otherwise.

ieee_is_negative (x) returns the value `.true.` if `ieee_class(x)` has one of the values

```
ieee_negative_normal  ieee_negative_subnormal
ieee_negative_zero    ieee_negative_inf
```

and `.false.` otherwise.

ieee_is_normal (x) returns the value `.true.` if `ieee_class (x)` has one of the values

```
ieee_negative_normal   ieee_negative_zero
ieee_positive_zero     ieee_positive_normal
```

and `.false.` otherwise.

ieee_logb (x) returns a real with the same type parameter as x. If x is neither zero, infinity, nor NaN, the value of the result is the unbiased exponent of x, that is, `exponent(x)-1`. If x == 0, the result is $-\infty$ if `ieee_support_inf(x)` is true and `-huge(x)` otherwise; `ieee_divide_by_zero` signals. If `ieee_support_inf(x)` is true and x is infinite, the result is +infinity. If `ieee_support_nan(x)` is true and x is a NaN, the result is a NaN.

ieee_max_num (x, y) returns x if x > y or y if x < y. If one of x and y is a quiet NaN, the result is the other. If one or both of x and y are signaling NaNs, `ieee_invalid` signals and the result is a NaN. Otherwise, either x and y is returned and which is returned is processor dependent.

ieee_max_num_mag (x, y) returns x if `abs(x) > abs(y)` or y if `abs(y) > abs(x)`. Otherwise, the result is `ieee_max_num(x, y)`.

ieee_min_num (x, y) returns x if x < y or y if x > y. Otherwise, the result is `ieee_max_num (x, y)`.

ieee_min_num_mag (x, y) returns x if `abs(x) < abs(y)` or y if `abs(y) < abs(x)`. Otherwise, the result is `ieee_max_num(x, y)`.

ieee_next_after (x, y) returns a real with the same type parameter as x. If x == y, the result is x, without an exception ever signaling. Otherwise, the result is the neighbour of x in the direction of y. The neighbours of zero (of either sign) are both nonzero. Overflow is signaled when x is finite but ieee_next_after(x, y) is infinite; underflow is signaled when ieee_next_after(x, y) is subnormal; in both cases, ieee_inexact signals.

ieee_next_down (x) where x is real returns the greatest value in the representation method of x that compares less than x, except that when x is equal to $-\infty$ the result has the value $-\infty$, and when x is a NaN the result is a NaN (IEEE operation of nextDown). If x is a signaling NaN, ieee_invalid signals; otherwise, no exception is signaled. If ieee_support_inf(x) has the value false, this function must not be invoked when x has the value -huge(x).

ieee_next_up (x) where x is real returns the least value in the representation method of x that compares greater than x, except that when x is equal to $+\infty$ the result has the value $+\infty$, and when x is a NaN the result is a NaN (IEEE operation of nextUp). If x is a signaling NaN, ieee_invalid signals; otherwise, no exception is signaled. If ieee_support_inf(x) has the value false, this function must not be invoked when x has the value huge(x).

ieee_quiet_eq (a, b) returns true if a compares with b as equal. If a or b is a NaN, the result will be false. If a or b is a signaling NaN, ieee_invalid signals; otherwise, no exception is signaled. This is the IEEE compareQuietEqual operation.

ieee_quiet_ge (a, b) returns true if a compares greater than or equal to b. If a or b is a NaN, the result will be false. If a or b is a signaling NaN, ieee_invalid signals; otherwise, no exception is signaled. This is the IEEE compareQuietGreaterEqual operation.

ieee_quiet_gt (a, b) returns true if a compares greater than b. If a or b is a NaN, the result will be false. If a or b is a signaling NaN, ieee_invalid signals; otherwise, no exception is signaled. This is the IEEE compareQuietGreater operation.

ieee_quiet_ne (a, b) returns true if a compares with b as not equal. If a or b is a NaN, the result will be false. If a or b is a signaling NaN, ieee_invalid signals; otherwise, no exception is signaled. This is the IEEE compareQuietNotEqual operation.

ieee_quiet_le (a, b) returns true if a compares less than or equal to b. If a or b is a NaN, the result will be false. If a or b is a signaling NaN, ieee_invalid signals; otherwise, no exception is signaled. This is the IEEE compareQuietLessEqual operation.

ieee_quiet_lt (a, b) returns true if a compares less than b. If a or b is a NaN, the result will be false. If a or b is a signaling NaN, ieee_invalid signals; otherwise, no exception is signaled. This is the IEEE compareQuietLess operation.

ieee_real (a [,kind]) converts the value of a, which must be of type integer or real, to type real rounded according to the current rounding mode (as specified in the IEEE standard by the convertFromInt and convertFormat operations). If kind is absent, the result is default real; otherwise the result is of type real with kind type parameter kind, which must be a scalar integer constant expression. The kind of the result, and of a if a is of type real, must be an IEEE real kind.

ieee_rem (x, y) returns a real with the type parameter of whichever argument has the greater precision and value exactly $x - y * n$, where n is the integer nearest to the exact value x/y; whenever $|n - x/y| = 1/2$, n is even. If the result value is zero, the sign is that of x. The two arguments are required to have the same radix.

ieee_rint (x [,round]) where round is of type ieee_round_type, returns the value of x rounded to an integer according to the mode specified by round; if round is absent, the current rounding mode is used. If the result value is zero, the sign is that of x.

ieee_scalb (x, i) where i is of type integer returns a real with the same type parameter as x whose value is $2^i x$ if this is within the range of normal numbers. If x is finite and $2^i x$ is too large, ieee_overflow signals; if ieee_support_inf(x) is true, the result value is infinity with the sign of x; otherwise, it is sign(huge(x),x). If $2^i x$ is too small and cannot be represented exactly, ieee_underflow signals; the result is the representable number having a magnitude nearest to $|2^i x|$ and the same sign as x. If x is infinite, the result is the same as x; no exception signals.

ieee_signaling_eq (a, b) returns true if a compares with b as equal. If a or b is a NaN, the result will be false and ieee_invalid signals; otherwise, no exception is signaled. This is the IEEE compareSignalingEqual operation.

ieee_signaling_ge (a, b) returns true if a compares greater than or equal to b. If a or b is a NaN, the result will be false and ieee_invalid signals; otherwise, no exception is signaled. This is the IEEE compareSignalingGreaterEqual operation.

ieee_signaling_gt (a, b) returns true if a compares greater than b. If a or b is a NaN, the result will be false and ieee_invalid signals; otherwise, no exception is signaled. This is the IEEE compareSignalingGreater operation.

ieee_signaling_ne (a, b) returns true if a compares with b as not equal. If a or b is a NaN, the result will be false and ieee_invalid signals; otherwise, no exception is signaled. This is the IEEE compareSignalingNotEqual operation.

ieee_signaling_le (a, b) returns true if a compares less than or equal to b. If a or b is a NaN, the result will be false and ieee_invalid signals; otherwise, no exception is signaled. This is the IEEE compareSignalingLessEqual operation.

ieee_signaling_lt (a, b) returns true if a compares less than b. If a or b is a NaN, the result will be false and ieee_invalid signals; otherwise, no exception is signaled. This is the IEEE compareSignalingLess operation.

ieee_signbit (x) where x is type real and an IEEE kind returns a default logical value that is true if and only if the sign bit of x is nonzero (as specified in the IEEE standard by the isSignMinus operation). Even when x is a signaling NaN, no exception is signaled by this function.

ieee_unordered (x, y) returns .true. if x or y is a NaN or both are, and .false. otherwise. If x or y is a signaling NaN, ieee_invalid may signal.

ieee_value (x, class) returns a real with the same type parameter as x and a value specified by class. The argument class is of type type(ieee_class_type) and may have value

- ieee_signaling_nan or ieee_quiet_nan if ieee_support_nan(x) is true;
- ieee_negative_inf or ieee_positive_inf if ieee_support_inf(x) is true;
- ieee_negative_subnormal or ieee_positive_subnormal if the value of ieee_support_subnormal(x) is true; or
- ieee_negative_normal, ieee_negative_zero, ieee_positive_zero, or ieee_positive_normal.

Although in most cases the value is processor dependent, it does not vary between invocations for any particular kind type parameter of x and value of class.

19.7.5   Non-elemental subroutines

The IEEE standard has two modes for rounding to the nearest representable number, which differ in the way they handle ties: in the case of a tie, roundTiesToEven uses the value with an even least significant digit and roundTiesToAway uses the value further from zero. The value ieee_nearest of the type ieee_round_type corresponds to roundTiesToEven and the value ieee_away corresponds to roundTiesToAway.

The subroutines ieee_get_rounding_mode and ieee_set_rounding_mode have an optional argument radix to allow the decimal rounding mode to be inquired about and set independently of the binary rounding mode.

The module ieee_arithmetic altogether contains the following non-elemental subroutines:

call ieee_get_rounding_mode (round_value [, radix]) where:

> **round_value** is scalar, of type type(ieee_round_type), with intent out. It is assigned the floating-point rounding mode for the specified radix; the value will be ieee_nearest, ieee_to_zero, ieee_up, ieee_down, or ieee_away if one of the IEEE modes is in operation, and ieee_other otherwise.

> **radix** is a scalar integer with intent in. It must have the value 2 or 10; if absent, the specified radix is 2.

call ieee_get_underflow_mode (gradual) where:

gradual is scalar, of type logical, and has intent `out`. It returns `.true.` if gradual underflow is in effect, and `.false.` otherwise.

call ieee_set_rounding_mode (round_value [,radix]) where:

round_value is scalar, of type `type(ieee_round_type)`, with intent `in`. It specifies the mode to be set for the specified radix.

radix is a scalar integer with intent `in`. It must have the value 2 or 10; if absent, the specified radix is 2.

The subroutine must not be called unless the value of `ieee_support_rounding` `(round_value,x)` is true for some x with the specified radix such that the value of `ieee_support_datatype(x)` is true.

call ieee_set_underflow_mode (gradual) where:

gradual is scalar, of type logical. If its value is `.true.`, gradual underflow comes into effect; otherwise gradual underflow ceases to be in effect.

The subroutine must not be called unless `ieee_support_underflow_control(x)` is true for some x.

The example in Figure 19.2 shows the use of these subroutines to store the rounding mode, perform a calculation with round to nearest, and restore the rounding mode.

Figure 19.2 Store the rounding mode, perform a calculation with another mode, and restore the previous mode.

```
use, intrinsic :: ieee_arithmetic
type(ieee_round_type) round_value
   :
call ieee_get_rounding_mode(round_value) ! Store the rounding mode
call ieee_set_rounding_mode(ieee_nearest)
   : ! Calculation with round to nearest
call ieee_set_rounding_mode(round_value) ! Restore the rounding mode
```

19.8   Examples

19.8.1   Dot product

Our first example, Figure 19.3, is of a module for the dot product of two real arrays of rank one. It contains a logical scalar `dot_error`, which acts as an error flag. If the sizes of the arrays are different, an immediate return occurs with `dot_error` true. If overflow occurs during the actual calculation, the overflow flag will signal and `dot_error` is set true. If all is well, its value is unchanged.

Figure 19.3 Module for the dot product of two real rank-one arrays.

```
module dot
  ! The caller must ensure that exceptions do not cause halting.
    use, intrinsic :: ieee_exceptions
    implicit none
    private        :: mult
    logical        :: dot_error
    interface operator(.dot.)
       module procedure mult
    end interface
contains
    real function mult(a, b)
       real, intent(in) :: a(:), b(:)
       integer          :: i
       logical          :: overflow
       dot_error = .false.
       mult = 0.0
       if (size(a)/=size(b)) then
          dot_error = .true.
          return
       end if
  ! The processor ensures that ieee_overflow is quiet
       do i = 1, size(a)
          mult = mult + a(i)*b(i)
       end do
       call ieee_get_flag(ieee_overflow, overflow)
       dot_error = overflow
    end function mult
end module dot
```

19.8.2   Calling alternative procedures

Suppose the function `fast_inv` is code for matrix inversion that 'lives dangerously' and may cause a condition to signal. The alternative function `slow_inv` is far less likely to cause a condition to signal, but is much slower. The code in Figure 19.4 tries `fast_inv` and, if necessary, makes another try with `slow_inv`. If this still fails, a message is printed and the program stops. Note, also, that it is important to set the flags quiet before the second try. The state of all the flags is stored and restored.

Figure 19.4 Try a fast algorithm and, if necessary, try again with a slower but more reliable algorithm.

```
use, intrinsic :: ieee_exceptions
use, intrinsic :: ieee_features, only: ieee_invalid_flag
! The other exceptions of ieee_usual (ieee_overflow and
! ieee_divide_by_zero) are always available with ieee_exceptions
type(ieee_status_type) :: status_value
logical, dimension(3)  :: flag_value
:

call ieee_get_status(status_value)
call ieee_set_halting_mode(ieee_usual,.false.) ! Needed in case the
!                  default on the processor is to halt on exceptions.
call ieee_set_flag(ieee_usual,.false.)          ! Elemental
! First try the "fast" algorithm for inverting a matrix:
matrix1 = fast_inv(matrix) ! This must not alter matrix.
call ieee_get_flag(ieee_usual, flag_value)      ! Elemental
if (any(flag_value)) then
! "Fast" algorithm failed; try "slow" one:
   call ieee_set_flag(ieee_usual,.false.)
   matrix1 = slow_inv(matrix)
   call ieee_get_flag(ieee_usual, flag_value)
   if (any(flag_value)) then
      write (*, *) 'Cannot invert matrix'
      stop
   end if
end if
call ieee_set_status(status_value)
```

19.8.3 Calling alternative in-line code

The example in Figure 19.5 is similar to the inner part of the previous one, but here the code for matrix inversion is in line, we know that only overflow can signal, and the transfer is made more precise by adding extra tests of the flag.

19.8.4 Reliable hypotenuse function

The most important use of a floating-point exception handling facility is to make possible the development of much more efficient software than is otherwise possible. The code in Figure 19.6 for the 'hypotenuse' function, $\sqrt{x^2 + y^2}$, illustrates the use of the facility in developing efficient software.

An attempt is made to evaluate this function directly in the fastest possible way. This will work almost every time, but if an exception occurs during this fast computation, a safe but

Figure 19.5 As for Figure 19.4 but with in-line code.

```
use, intrinsic :: ieee_exceptions
logical        :: flag_value
  ⋮

call ieee_set_halting_mode(ieee_overflow,.false.)
call ieee_set_flag(ieee_overflow,.false.)
! First try a fast algorithm for inverting a matrix.
do k = 1, n
    ⋮

   call ieee_get_flag(ieee_overflow, flag_value)
   if (flag_value) exit
end do
if (flag_value) then
! Alternative code which knows that k-1 steps have
! executed normally.
  ⋮

end if
```

slower way evaluates the function. This slower evaluation may involve scaling and unscaling, and in (very rare) extreme cases this unscaling can cause overflow (after all, the true result might overflow if x and y are both near the overflow limit). If the overflow or underflow flag is signaling on entry, it is reset on return by the processor so that earlier exceptions are not lost.

19.8.5   Access to IEEE arithmetic values

The program in Figure 19.7 illustrates how the ieee_arithmetic module can be used to test for special IEEE values. It repeatedly doubles a and halves b, testing for overflowed, subnormal, and zero values. It uses ieee_set_halting_mode to prevent halting. The beginning and end of a sample output are shown. Note the warning messages; the processor is required to produce some such output if any exceptions are signaling at termination.

Figure 19.6 A reliable hypotenuse function.

```fortran
real function hypot(x, y)

! In rare circumstances this may lead to the signaling of
! ieee_overflow.
!

  use, intrinsic :: ieee_exceptions
  use, intrinsic :: ieee_features, only: ieee_underflow_flag
! ieee_overflow is always available with ieee_exceptions

  implicit none
  real               :: x, y
  real               :: scaled_x, scaled_y, scaled_result
  logical, dimension(2) :: flags
  type(ieee_flag_type), parameter, dimension(2) ::          &
        outside_range = (/ ieee_overflow, ieee_underflow /)
  intrinsic :: sqrt, abs, exponent, max, digits, scale
! The processor clears the flags on entry
  call ieee_set_halting_mode(outside_range, .false.) ! Needed in
!    case the default on the processor is to halt on exceptions.
! Try a fast algorithm first
  hypot = sqrt( x**2 + y**2 )
  call ieee_get_flag(outside_range, flags)
  if ( any(flags) ) then
    call ieee_set_flag(outside_range, .false.)
    if ( x==0.0 .or. y==0.0 ) then
      hypot = abs(x) + abs(y)
    else if ( 2*abs(exponent(x)-exponent(y)) > digits(x)+1 ) then
      hypot = max( abs(x), abs(y) )! We can ignore one of x and y
    else     ! Scale so that abs(x) is near 1
      scaled_x = scale( x, -exponent(x) )
      scaled_y = scale( y, -exponent(x) )
      scaled_result = sqrt( scaled_x**2 + scaled_y**2 )
      hypot = scale(scaled_result, exponent(x)) ! May cause
    end if                                 ! overflow
  end if
! The processor resets any flag that was signaling on entry
end function hypot
```

Figure 19.7 Test for overflowed, subnormal, and zero values.

```
program test
   use ieee_arithmetic; use ieee_features
   real    :: a=1.0, b=1.0
   integer :: i
   call ieee_set_halting_mode(ieee_overflow, .false.)
   do i = 1,1000
      a = a*2.0
      b = b/2.0
      if (.not. ieee_is_finite(a)) then
         write (*, *) '2.0**', i, ' is infinite'
         a = 0.0
      end if
      if (.not. ieee_is_normal(b)) &
         write (*, *) '0.5**', i, ' is subnormal'
      if (b==0.0) exit
   end do
   write (*, *) '0.5**', i, ' is zero'
end program test
```

```
 0.5** 127  is subnormal
 2.0** 128  is infinite
 0.5** 128  is subnormal
 0.5** 129  is subnormal
   :
 0.5** 148  is subnormal
 0.5** 149  is subnormal
 0.5** 150  is zero
Warning: Floating overflow occurred during execution
Warning: Floating underflow occurred during execution
```

20. Basic interoperability with C

20.1 Introduction

Fortran provides a standardized mechanism for interoperating with C. Clearly, any entity involved must be such that equivalent declarations of it may be made in the two languages. This is enforced within the Fortran program by requiring all such entities to be *interoperable*. We will explain in turn what this requires for types, variables, and procedures. They are all requirements on the syntax so that the compiler knows at compile time whether an entity is interoperable. We continue with examining interoperability for global data and then discuss some examples. We conclude with a syntax for defining sets of integer constants that is useful in this context.

20.2 Interoperability of intrinsic types

There is an intrinsic module named `iso_c_binding` that contains named constants of type default integer holding kind type parameter values for intrinsic types. Their names are shown in Table 20.1, together with the corresponding C types. The processor is required to support only `int`. Lack of support is indicated with a negative value of the constant. If the value is positive, it indicates that the Fortran type and kind type parameter interoperate with the corresponding C type.

The negative values are as follows. For the integer types, the value is -1 if there is such a C type but no interoperating Fortran kind or -2 if there is no such C type. For the real types, the value is -1 if the C type does not have a precision equal to the precision of any of the Fortran real kinds, -2 if the C type does not have a range equal to the range of any of the Fortran real kinds, -3 if the C type has neither the precision nor range of any of the Fortran real kinds, and -4 if there is no interoperating Fortran kind for other reasons.

The values of `c_float_complex`, `c_double_complex`, and `c_long_double_complex` are the same as those of `c_float`, `c_double`, and `c_long_double`, respectively.

For logical, the value of `c_bool` is -1 if there is no Fortran kind corresponding to the C type `_Bool`.

For character, the value of `c_char` is -1 if there is no Fortran kind corresponding to the C type `char`. For character type, interoperability also requires that the length type parameter be omitted or be specified by a constant expression whose value is 1.

Modern Fortran Explained, 3rd Edition. M. Metcalf, J. Reid, M. Cohen, and R. Bader. Oxford University Press (2024). © M. Metcalf, J. Reid, M. Cohen, and R. Bader (2024). DOI 10.1093/oso/9780198876571.001.0020

Table 20.1. Named constants for interoperable kinds of intrinsic Fortran types.

Type	Named constant	C type or types
integer	c_int	int
	c_short	short int
	c_long	long int
	c_long_long	long long int
	c_signed_char	signed char, unsigned char
	c_size_t	size_t
	c_int8_t	int8_t
	c_int16_t	int16_t
	c_int32_t	int32_t
	c_int64_t	int64_t
	c_int_least8_t	int_least8_t
	c_int_least16_t	int_least16_t
	c_int_least32_t	int_least32_t
	c_int_least64_t	int_least64_t
	c_int_fast8_t	int_fast8_t
	c_int_fast16_t	int_fast16_t
	c_int_fast32_t	int_fast32_t
	c_int_fast64_t	int_fast64_t
	c_intmax_t	intmax_t
	c_intptr_t	intptr_t
	c_ptrdiff_t	ptrdiff_t
real	c_float	float
	c_double	double
	c_long_double	long double
complex	c_float_complex	float _Complex
	c_double_complex	double _Complex
	c_long_double_complex	long double _Complex
logical	c_bool	_Bool
character	c_char	char

The following named constants (with the obvious meanings) are provided:

```
c_null_char  c_alert                      c_backspace       c_form_feed
c_new_line   c_carriage_return c_horizontal_tab c_vertical_tab
```

They are all of type character with length one and kind c_char (or default kind if c_char has the value -1).

20.3   Interoperability with C pointer types

For interoperating with C pointers (which are just addresses), the module contains the derived types c_ptr and c_funptr that are interoperable with C object and function pointer types, respectively. Their components are private. There are named constants c_null_ptr and c_null_funptr for the corresponding null values of C.

The module also contains the following procedures, which, apart from c_f_pointer and c_f_procpointer, are all pure:

c_associated (c_ptr_1 [, c_ptr_2]) is a function for scalars of type c_ptr or c_funptr. It returns a default logical scalar. It has the value false if c_ptr_1 is a C null pointer or if c_ptr_2 is present with a different value; otherwise, it has the value true.

c_f_pointer (cptr, fptr [, shape]) is a subroutine where

cptr is a scalar of type c_ptr with intent in. Its value is either
  i) the C address of an interoperable data entity;
  ii) the result of a reference to c_loc with a non-interoperable argument; or
  iii) the C address of a storage sequence that is not in use by any other Fortran entity.

  It must not be the C address of a Fortran variable that does not have the target attribute.

fptr is a pointer with intent out. It must not have a deferred type parameter. It must not be coindexed.
  i) If cptr is the C address of an interoperable entity, fptr must be a data pointer of the type and type parameters of the entity and it becomes pointer associated with the target of cptr.
  ii) If cptr was returned by a call of c_loc with a non-interoperable argument x, fptr must be a non-polymorphic pointer of the type and type parameters of x. x or its target if it is a pointer shall not have since been deallocated or have become undefined due to execution of a return or an end statement. fptr becomes pointer associated with x or its target.
  iii) If cptr is the C address of a storage sequence that is not in use by any other Fortran entity, the fptr argument becomes pointer associated with that storage sequence. The purpose of this is to allow for interaction with memory allocators written in C.

shape (optional) is a rank-one array of type integer with intent in. If present, its size is equal to the rank of fptr. It must be present if fptr is an array; in this case, the shape of fptr is specified by shape and each lower bound is 1.[1]

c_f_procpointer (cptr, fptr) is a subroutine where

cptr is a scalar of type c_funptr with intent in. Its value is the C address of a procedure that is interoperable, or the result of a call to c_funloc with a non-interoperable argument.

fptr is a procedure pointer with intent out. Its interface must be that of the target of cptr and it becomes pointer-associated with that target.

c_loc (x) is a function that returns a scalar of type c_ptr that holds the C address of its argument. The argument must be a pointer or have the target attribute. It must be a variable with interoperable type and kind type parameters, be an assumed-type variable (Section 20.15), or be a non-polymorphic variable with no length type parameters. If it is an array, it must be contiguous and have nonzero size. It shall not be a zero-length string. If it is allocatable, it must be allocated. If it is a pointer it must be associated. It must not be coindexed.

If the argument is non-interoperable, no use of the data can be made from C, but a Fortran pointer to it can be recovered by c_f_pointer, which allows the fptr argument to be a pointer of non-interoperable type. The purpose here is to facilitate exchange of information between two Fortran procedures when the intervening procedure is C. An illustration for reconstructing a Fortran pointer to a non-interoperable object is provided in Figure 20.8, Section 20.12.

c_funloc (x) is a function that returns the C address of a procedure. The argument x is permitted to be a procedure that is interoperable (see Section 20.9) or a pointer associated with such a procedure. In a pure subprogram c_funloc must not be invoked with an impure argument.

The argument may also be a non-interoperable procedure. As in the previous item, the purpose here is to facilitate information passing between two Fortran procedures via C.

A Fortran pointer or allocatable variable, and most Fortran arrays, do not interoperate directly with any C entity because C does not have quite the same concepts; for example, unlike a Fortran array pointer, a C array pointer cannot describe a discontiguous array section.[2] However, this does not prevent such entities being passed to C via argument association since Fortran compilers already perform copy-in copy-out when this is necessary. Also, the function c_loc may be used to obtain the C address of an allocated allocatable array, which is useful if the C part of the program wishes to maintain a pointer to this array. Similarly, the address of an array allocated in C may be passed to Fortran and c_f_pointer used to construct a Fortran pointer whose target is the C array. There is an illustration of this in Figure 20.7, Section 20.12.

---

[1]Fortran 2023 permits explicit lower bounds specification, see Section 23.4.1.
[2]But this deficiency is remedied by use of the C descriptor, see Section 21.2.

Care must be exercised in using these procedures because of their potential for subverting the type system. It is the programmer's responsibility to make sure this does not happen by supplying the required type or interface information.

20.4 Interoperability of derived types

For a derived type to be interoperable, it must have the `bind` attribute:

```
type, bind(c) :: mytype
   ⋮
end type mytype
```

It must not be a sequence type (Appendix A.8), have type parameters, have the `extends` attribute (Section 15.2), or have any type-bound procedures (Section 15.8). Each component must have interoperable type and type parameters, must not be a zero-sized array, must not be a pointer, and must not be allocatable.

These restrictions allow the type to interoperate with a C structure type that has the same number of components. The components correspond by position in their definitions. Each Fortran component must be interoperable with the corresponding C component. Here is a simple example:

```
typedef struct {
    int m, n;
    float r;
} myctype;
```

is interoperable with

```
use, intrinsic :: iso_c_binding
type, bind(c) :: myftype
    integer(c_int) :: i, j
    real(c_float) :: s
end type myftype
```

The name of the type and the names of the components are not significant for interoperability. If two equivalent definitions of an interoperable derived type are made in separate scoping units, they interoperate with the same C type (but it is usually preferable to define one type in a module and access it by `use` statements).

No Fortran type is interoperable with a C union type, a C structure type that contains a bit field, or a C structure type that contains a flexible array member.

20.5 Shape and character length disagreement

Before discussing interoperability of variables, we need to introduce the concept of associating an actual argument that is an array or an array element with a dummy array whose rank and bounds are declared afresh. This is known as **sequence association** because it depends on the sequence of array elements in array element order. An important case concerns the **assumed-size** array, which is a dummy array whose size is assumed from the size of the corresponding actual argument.

The concept dates from Fortran 77, and is usually implemented by passing just the address of the first element of the array. It provides a good match for C semantics, but we do not recommend its use in pure Fortran programs because it is error-prone compared with the assumed-shape array, allocatable array, and array section, where the whole shape is passed and compilers can check for valid indices when in debugging mode. Because the standard allows only the address of the actual argument to be passed, and requires that only the address be passed for an interoperable procedure, it might be better to call the dummy argument an unknown-size array.

For an assumed-size array, the final upper bound is declared with an asterisk. Any other bounds must be given in a declaration that is as for an explicit-shape array. The bounds of the corresponding actual argument are ignored. The elements of the assumed-size array, in array-element order, are associated with the elements of the corresponding actual argument, in array-element order. We illustrate this in Figure 20.1, where the array elements aa(1,1), aa(2,1), aa(1,2), ... are associated with da(1), da(2), da(3), ..., and the array elements ab(1), ab(2), ab(3), ... are associated with db(1,1), db(2,1), db(1,2), ...

Figure 20.1 Passing arrays to assumed-size arrays.

```
real :: aa(2,3), ab(6)
   ⋮
call sub (aa, ab)
   ⋮
subroutine sub(da, db)
real :: da(*), db(2,*)
   ⋮
```

Because an assumed-size array has no upper bound in its last dimension, it does not have a shape and therefore must not be used as a whole array, except as an argument to a procedure that does not require its shape. However, if an array section of it is formed with an explicit upper bound in the last dimension, this has a shape and may be used as a whole array.

Because it would require an action for every element of an array of unknown size, an assumed-size array is not permitted to have intent out if it is polymorphic, of a finalizable type, or of a derived type with default initialization or an ultimate allocatable component.

It is also permitted to associate an element of an array that is neither a pointer, of assumed size, nor polymorphic (such an array is always contiguous) with an explicit-shape or assumed-size array. This is interpreted as associating the subarray consisting of all the elements of the array from the given element onwards in array-element order with the dummy array. Figure 20.2 illustrates this. Here, only the last 49 elements of a are available to b, as the first array element of a which is passed to sub is a(52). Within sub, this element is referenced as b(1) and it is illegal to address b(50) in any way, as that would be beyond the declared length of a in the calling procedure. Similarly, associating aa(10) with b(50) is illegal. In

Fortran 77 this mechanism provided a limited form of array section. We do not recommend its use in modern Fortran, which has the far better feature of the array section.

Figure 20.2 Passing an array element to an assumed-size or explicit-shape array.

```
real :: a(100), aa(50)
   ⋮
call sub (a(52), aa(10))
   ⋮
subroutine sub(b, c)
real :: b(*), c(50)
   ⋮
```

In the case of default character type, agreement of character length is not required. For a scalar dummy argument of character length *len*, the actual argument may have a greater character length and its leftmost *len* characters are associated with the dummy argument. For example, if chasub has a single dummy argument of character length 1,

```
call chasub(word(3:4))
```

is a valid call statement. For a dummy argument that is an array, the restriction is on the total number of characters in the array. An array element or array element substring is regarded as a sequence of characters from its first character to the last character of the array. The size is the number of characters in the sequence divided by the character length of the dummy argument.

Shape or character length disagreement cannot occur when a dummy argument is assumed shape (by definition, the shape is assumed from the actual argument). It can occur for explicit-shape and assumed-size arrays. Implementations usually receive explicit-shape and assumed-size arrays in contiguous storage, but permit any uniform spacing of the elements of an assumed-shape array. They will need to make a copy of any array argument that is not stored contiguously (for example, the section a(1:10:2)), unless the dummy argument is assumed shape.

The rules on character length disagreement include character(kind=c_char) (which will often be the same as default character) and treat any other scalar actual argument of type default character or character(kind=c_char) as if it were an array of size one. This includes the case where the argument is an element of an assumed-shape array or an array pointer, or a subobject thereof; note that just that element or subobject is passed, not the rest of the array.

When a procedure is invoked through a generic name, as a defined operation, or as a defined assignment, rank agreement between the actual and the dummy arguments is required. In this case, only a scalar dummy argument may be associated with a scalar actual argument.

20.6   Interoperability of variables

A scalar Fortran variable is interoperable if it is of interoperable type and type parameters, and is neither a pointer nor allocatable. It is interoperable with a C scalar if the Fortran type and type parameters are interoperable with the C type.

An array Fortran variable is interoperable if its size is nonzero, it is of interoperable type and type parameters, and it is of explicit shape or assumed size.

For a Fortran array of rank one to interoperate with a C array, the Fortran array elements must be interoperable with the C array elements. If the Fortran array is of explicit size, the C array must have the same size. If the Fortran array is of assumed size, the C array must not have a specified size.

A Fortran array a of rank greater than one and of shape $(e_1, e_2, \ldots, e_r)$ is interoperable with a C array of size e_r with elements that are interoperable with a Fortran array of the same type as a and of shape $(e_1, e_2, \ldots, e_{r-1})$. For ranks greater than two, this rule is applied recursively. For example, the Fortran arrays declared as

```
integer(c_int) :: fa(18, 3:7), fb(18, 3:7, 4)
```

are interoperable with C arrays declared as

```
int ca[5][18], cb[4][5][18];
```

and the elements correspond. Note that the subscript order is reversed.

An assumed-size Fortran array of rank greater than one is interoperable with a C array of unspecified size if its elements are related to the Fortran array in the same way as in the explicit-size case. For example, the Fortran arrays declared as

```
integer(c_int) :: fa(18, *), fb(18, 3:7, *)
```

are interoperable with C arrays declared as

```
int ca[ ][18], cb[ ][5][18];
```

20.7   Function c_sizeof

The intrinsic module iso_c_binding also contains the inquiry function c_sizeof. It provides similar functionality to that of the sizeof operator in C.

c_sizeof (x) If x is scalar, this returns the value that the companion processor returns for the C sizeof operator applied to an object of a type that interoperates with the type and type parameters of x. If x is an array, the result is the value returned for an element of the array multiplied by its number of elements. x must be interoperable (see Section 20.6), and is not permitted to be an assumed-size array (see Section 20.5).

For example,

```
use iso_c_binding
integer(c_int64_t) x
print *, c_sizeof(x)
```

will print the value 8, and if n has the value 10,

```
subroutine s(y, n)
   use iso_c_binding
   integer(c_int64_t) y(n)
   print *, c_sizeof(y)
end subroutine
```

will print the value 80.

Caution is required when doing mixed-language programming in both C and Fortran, as this is not quite what the C `sizeof` operator does; in many contexts (such as being a dummy argument) a C array 'decays' to a pointer and then `sizeof` will return the size of the pointer (not the whole array) in bytes.

20.8 The `value` attribute

For the sake of interoperability, there exists an attribute, `value`, for dummy arguments. It may be specified in a type declaration statement for the argument or separately in a `value` statement:

```
function func(a, i, j) bind(c)
   real(c_float) func, a
   integer(c_int), value :: i, j
   value :: a
```

When the procedure is invoked, a copy of the actual argument is made. The dummy argument is a variable that may be altered during execution of the procedure, but on return no copy back takes place. By implication, it acquires the intent `in` attribute. If the type has any length type parameter (character type or a parameterized derived type, see Section 13.2), its value need not be known at compile time. The argument can be an explicit-shape or assumed-shape array, but cannot be a pointer, be allocatable, have intent `out` or `inout`, be a procedure, have the `volatile` attribute (Section A.15), be a coarray, or have a coarray ultimate component. If the procedure has the `bind` attribute, an argument with the `value` attribute is not permitted to be `optional`.

The `value` attribute is not limited to procedures with the `bind` attribute; it may be used in any procedure. This is useful for a particular programming style; for example, in Figure 20.3, the argument n is locally decreased until it reaches zero, without affecting the actual argument or requiring an extra temporary variable. Because the attribute alters the argument-passing mechanism, a procedure with a `value` dummy argument is required to have an explicit interface.

In the context of a call from C, the absence of the `value` attribute indicates that it expects the actual argument to be an object pointer to an object of the specified type or a function pointer whose target has a prototype that is interoperable with the specified interface (see next section).

Figure 20.3 Find the *n*th word in a string.

```
integer function nth_word_position(string, n) result(pos)
    character(*), intent(in) :: string
    integer, value           :: n
    logical                  :: in_word
    in_word = .false.
    do pos = 1, len(string)
        if (string(pos:pos)==' ')then
            in_word = .false.
        else if (.not.in_word) then
            in_word = .true.    ! At first character of a word.
            n = n - 1
            if (n==0) return    ! Found nth one, return position.
        end if
    end do
    pos = 0                     ! n words not found, return zero.
end function
```

20.9   Interoperability of procedures

A Fortran procedure is interoperable if it has an explicit interface and is declared with the bind attribute:

```
function func(i, j, k, l, m) bind(c)
subroutine subr () bind(c)
procedure (proc_interface), bind(c) :: proc
```

Note that for a subroutine with no arguments the parentheses are required. Each dummy argument must be interoperable and neither optional nor an array with the value attribute. For a function, the result must be scalar and interoperable. In a procedure statement with bind(c), the procedure interface referenced must be interoperable and the procedure named must not be a dummy argument and must not be given the pointer attribute.

The procedure usually has a *binding label*, which has global scope and is the name by which it is known to the C processor.[3] By default, it is the lower-case version of the Fortran name. For example, the function in the previous paragraph has the binding label func. An alternative binding label may be specified for an external or module procedure:

```
function func(i, j, k, l, m) bind(c, name='c_func')
```

but this is not permitted for an abstract interface, dummy procedure, or internal procedure. The value following the name= must be a scalar default character constant expression. The binding label is the result of discarding all its leading and trailing blanks and the case is significant, but if the character expression has zero length or is all blanks, there is no binding label.

---

[3] If an external procedure has a binding label, the procedure name has local scope.

A binding label is not an alias for the procedure name for an ordinary Fortran invocation. It is for use only from C. Two different entities must not have the same binding label.

A procedure with no binding label may still be invoked from C through a procedure pointer and, if this is the only way it will be invoked, it is not appropriate to give it a binding label. Similarly, it is not appropriate to give a binding label to a module procedure with the `private` attribute because this means that it can be called from C despite being private to the module.

An interoperable Fortran procedure interface is interoperable with a C function prototype that has the same number of arguments and does not have variable arguments denoted by the ellipsis (...). For a function, the result must be interoperable with the prototype result. For a subroutine, the prototype must have a void result. A dummy argument with the `value` attribute must be interoperable with the corresponding formal parameter. A dummy argument without the `value` attribute must correspond to a formal parameter of a pointer type and be interoperable with an entity of the referenced type of the formal parameter. Note that a Fortran explicit-shape or assumed-shape array without the `value` attribute can interoperate with a C array since this is automatically of a pointer type.

Here is an example of procedure interface interoperability. The Fortran interface in Figure 20.4 is interoperable with the C function prototype

```
short int func(int i, double *j, int *k, int l[10], void *m);
```

If a C function with this prototype is to be called from Fortran, the Fortran code must access an interface such as this. The call itself is handled in just the same way as if an external Fortran procedure with an explicit interface were being called. This means, for example, that the array section `larray(1:20:2)` might be the actual argument corresponding to the dummy array `l`; in this case, copy-in copy-out takes place.

Figure 20.4 A Fortran interface for a C function.

```
interface
    function func(i, j, k, l, m) bind(c)
    use, intrinsic        :: iso_c_binding
        integer(c_short)       :: func
        integer(c_int), value :: i
        real(c_double)        :: j
        integer(c_int)        :: k, l(10)
        type(c_ptr), value    :: m
    end function func
end interface
```

Similarly, if a Fortran function with the interface of the previous paragraph is to be called from C, the C code must have a prototype such as that of the previous paragraph.

If a C function is called from Fortran, it must not change the handling of any exception that is being handled by the Fortran processor, and it must not alter the floating-point status (Section 19.6.4) other than by setting an exception flag to signaling. The values of the floating-point exception flags on entry to a C function are processor dependent.

20.10   Interoperability of global data

An interoperable module variable[4] may be given the `bind` attribute in a type declaration statement or in a `bind` statement:

```
module bound_data
   use iso_c_binding
   integer(c_int), bind(c) :: c_extern
   integer(c_long) :: c2
   bind(c, name='myvariable') :: c2
end module
```

It has a binding label defined by the same rules as for procedures, and interoperates with a C variable with external linkage that is of a corresponding type. If a binding label is specified in a statement, the statement must define a single variable.

A variable with the `bind` attribute also has the `save` attribute (which may be confirmed explicitly). A change to the variable in either language affects the value of the corresponding variable in the other language. This is known as **linkage association**. A C variable is not permitted to interoperate with more than one Fortran variable.

The `bind` statement is available only for this purpose; it is not available, for instance, to specify the `bind` attribute for a module procedure. Also, the `bind` attribute must not be specified for a variable that is not a module variable (that is, it is not available to confirm that a variable is interoperable), and it must not be specified for a module variable that is in a common block.

The double colon in a `bind` statement is optional.

20.11   Invoking a C function from Fortran

If a C function is to be invoked from Fortran, it must have external linkage and be describable by a C prototype that is interoperable with an accessible Fortran interface that has the same binding label.

If it is required to pass a Fortran array to C, the interface may specify the array to be of explicit or assumed size and the usual Fortran mechanisms, perhaps involving copy-in copy-out, ensure that a contiguous array is received by the C code. Here is an example involving both an assumed-size array and an allocatable array. The C prototype is

```
int c_library_function(int expl[100], float alloc[], int len_alloc);
```

and the Fortran code is shown in Figure 20.5.

The rules on shape and character length disagreement (Section 20.5) allow entities specified as `character(kind=c_char)` of any length to be associated with an assumed-size or explicit-shape array, and thus to be passed to and from C. For example, the C function with prototype

```
void Copy(char in[], char out[]);
```

may be invoked by the Fortran code in Figure 20.6.

---

[4]Or a common block, Appendix B.1.7, with interoperable members

Figure 20.5 Passing Fortran arrays to a C function.

```
use iso_c_binding
interface
   integer (c_int) function c_library_function        &
                  (expl, alloc, len_alloc) bind(c)
      use iso_c_binding
      integer(c_int)        :: expl(100)
      real(c_float)         :: alloc(*)
      integer(c_int), value :: len_alloc
   end function c_library_function
end interface
integer(c_int)                :: expl(100), len_alloc, x1
real(c_float), allocatable :: alloc(:)
   :
len_alloc = 200
allocate (alloc(len_alloc))
   :
x1 = c_library_function(expl, alloc, len_alloc)
   :
```

Figure 20.6 Passing Fortran character strings to a C function.

```
use, intrinsic :: iso_c_binding, only: c_char, c_null_char
interface
   subroutine copy(in, out) bind(c, name='Copy')
      use, intrinsic :: iso_c_binding, only: c_char
      character(kind=c_char), dimension(*) :: in, out
   end subroutine copy
end interface
character(len=10, kind=c_char) :: &
            digit_string = c_char_'123456789' // c_null_char
character(kind=c_char) :: digit_arr(10)
call copy(digit_string, digit_arr)
print '(1x, a1)', digit_arr(1:9)
end
```

This code works because Fortran allows the character variable `digit_string` to be associated with the assumed-size dummy array `in`. We have also taken the opportunity here to illustrate the use of a binding label to call a C procedure whose name includes an upper-case letter.

20.12   Invoking Fortran from C

A reference in C to a procedure that has the `bind` attribute, has the same binding label, and is defined by means of Fortran, causes the Fortran procedure to be invoked.

Figure 20.7 shows an example of a Fortran procedure that is called from C and uses a structure to enable arrays allocated in C to be accessed in Fortran. The corresponding C structure declaration is:

```
structure pass { int lenc, lenf;  float *c, *f; };
```

the C function prototype is:

```
void simulation(struct pass *arrays);
```

and the C calling statement might be:

```
simulation(&arrays);
```

Figure 20.7 Accessing in Fortran an array that was allocated in C.

```
subroutine simulation(arrays) bind(c)
   use iso_c_binding
   type, bind(c) :: pass
      integer (c_int) :: lenc, lenf
      type (c_ptr)    :: c, f
   end type pass
   type (pass), intent(in) :: arrays
   real (c_float), pointer :: c_array(:)
   :
   ! associate c_array with an array allocated in C
   call c_f_pointer(arrays%c, c_array, (/arrays%lenc/) )
   :
end subroutine simulation
```

It is not uncommon for a Fortran library module to have an initialization procedure that establishes a data structure to hold all the data for a particular problem that is to be solved. Subsequent calls to other procedures in the module provide data about the problem or receive data about its solution. The data structure is likely to be of a type that is not interoperable, for example because it has components that are allocatable arrays.

The procedures `c_loc` and `c_f_pointer` have been designed to support this situation. The Fortran code in Figure 20.8 illustrates this. The type `problem_struct` holds an allocatable array of the size of the problem, and lots more. When the C code calls `new_problem`, it passes the size. The Fortran code allocates a structure and an array component within it of the relevant size; it then returns a pointer to the structure. The C code later calls add and passes additional data together with the pointer that it received from new_problem. The Fortran procedure add uses c_f_pointer to establish a Fortran pointer for the relevant structure and performs calculations using it. Note that the C code may call new_problem several times if it wishes to work simultaneously with several problems; each will have a separate structure of type problem_struct and be accessible through its own 'handle' of type (c_ptr). When a problem is complete, the C code calls goodbye to deallocate its structure.

Figure 20.8 Providing access in C to a Fortran structure that is not interoperable.

```fortran
module lib_code
   use iso_c_binding
   type :: problem_struct
      real, allocatable :: a(:)
      : ! More stuff
   end type
contains
   type(c_ptr) function new_problem(problem_size) bind(c)
      integer(c_size_t), value      :: problem_size
      type(problem_struct), pointer :: problem_ptr
      allocate(problem_ptr)
      allocate(problem_ptr%a(problem_size))
      new_problem = c_loc(problem_ptr)
   end function new_problem
   subroutine add(problem,...) bind(c)
      type(c_ptr), intent(in), value :: problem
      type(problem_struct), pointer  :: problem_ptr
      :
      call c_f_pointer(problem, problem_ptr)
      :
   end subroutine add
   subroutine goodbye(problem) bind(c)
      type(c_ptr), intent(in), value :: problem
      type(problem_struct), pointer  :: problem_ptr
      call c_f_pointer(problem, problem_ptr)
      deallocate(problem_ptr)
   end subroutine goodbye
end module lib_code
```

20.13   Enumerations

An enumeration is a set of integer constants (enumerators) that is appropriate for interoperating with C. The kind of the enumerators corresponds to the integer type that C would choose for the same set of constants. Here is an example:

```
enum, bind(c)
   enumerator :: red = 4, blue = 9
   enumerator yellow
end enum
```

This declares the named constants `red`, `blue`, and `yellow` with values 4, 9, and 10, respectively.

If a value is not specified for an enumerator, it is taken as one greater than the previous enumerator or zero if it is the first.

To declare a variable of the enumeration type, use the `kind` intrinsic function on one of the constants. An example using the above `enum` definition is:

```
integer(kind(red)) :: background_colour
```

20.14   Optional arguments

The C programming language does not have any direct equivalent of Fortran optional arguments. However, a widespread programming idiom in C with a similar effect to optional arguments is to pass the argument by reference, with a null pointer to indicate that the argument is not present. For example, the second argument of the standard C library function `strtod` is permitted to be a null pointer; if it is not a null pointer, it is assigned a value through the pointer.

Fortran 2018 adopted this idiom for permitting Fortran interoperable procedures to have optional arguments. As the idiom needs the argument to be passed by reference, an optional argument of an interoperable procedure is not permitted to have the `value` attribute.

For example, the interface

```
function strtod(string, final) bind(C)
   use iso_c_binding
   character(kind=c_char) string(*)
   type(c_ptr),optional :: final
   real(c_double) strtod
end function strtod
```

allows calling the standard C library function `strtod` directly, without needing to explicitly code the passing of a null pointer, or the address of the second argument. For example,

```
real(c_double) x, y
type(c_ptr) yfinal
   :
x = strtod(xstring//c_null_char)
y = strtod(ystring//c_null_char, yfinal)
```

In a call from Fortran, when the actual argument is absent, a null pointer will be passed to the C procedure. Similarly, if the interoperable procedure is written in Fortran, when called from C with a null pointer for the optional argument it will be treated as absent.

For example, for the Fortran procedure

```
subroutine fortran_note(main, minor) bind(c)
   use iso_c_binding
   integer(c_int), value :: main
   integer(c_int), optional :: minor
   if (present(minor)) then
      print '(1X,"Note ",I0,".",I0)', main, minor
   else
      print '(1X,"Note ",I0)', main
   end if
end subroutine fortran_note
```

the invocation from C

```
fortran_note(10, (int *)0);
```

would print 'Note 10', whereas the invocation

```
int subnote = 3;
fortran_stop(11,&subnote);
```

would print 'Note 10.3'.

20.15 Assumed-type dummy arguments

The frequent need to pass C pointers for low-level operations (using `c_loc`) can lead to ugly code that is difficult to understand. For greater convenience in such situations, a dummy argument that is either scalar or assumed size (Section 20.5) can be declared to have **assumed type** with the syntax `type(*)`.

For example, a C function that produces a 64-bit checksum of a block of memory of arbitrary size might have the signature

```
extern uint64_t checksum(void *block,size_t nbytes);
```

and this could be used in Fortran as shown in Figure 20.9.

Like `class(*)`, a `type(*)` variable is **unlimited polymorphic** and has no declared type. But unlike `class(*)`, a `type(*)` dummy argument is not allowed to be associated with a type that has type parameters or a type-bound procedure part. Also, only dummy arguments can be declared `type(*)`.[5]

An assumed-type dummy argument is not allowed to be allocatable, a coarray, a pointer, or an explicit-shape array; the first three of these require structure beyond a simple address, and the last one is not needed because assumed size is available. Furthermore, it is not permitted to have intent `out` or the `value` attribute. However, it may be assumed shape or assumed rank, in which case it will interoperate with a C descriptor instead of a `void *` pointer; this is described in Section 21.3.2.

---

[5]This may also be done with an `implicit` statement (Section A.9).

Figure 20.9 A C function that produces a 64-bit checksum.

```
interface
   function checksum(block, nbytes) bind(c)
      use iso_c_binding
      type(*),intent(in)        :: block(*)
      integer(c_size_t),value :: nbytes
      integer(c_int64_t)        :: checksum
   end function checksum
end interface
   ⋮
integer(c_int64_t) blocksum
   ⋮
blocksum = checksum(x, c_sizeof(x))
```

Because an assumed-type dummy argument essentially has no type information associated with it, its usage within Fortran is extremely limited. It is only allowed to appear as an actual argument corresponding to a dummy argument that is also assumed type, or as the first argument of the intrinsic functions present and rank, and the argument of the function c_loc from the intrinsic module iso_c_binding. If it is an array, it is also permitted as the first argument of one of the intrinsic functions is_contiguous, lbound, shape, size, and ubound.

An assumed-type variable that is not assumed shape or assumed rank must not be further passed on to an assumed-rank dummy argument. This is because in the scalar and assumed-size cases there is no type information available, but both assumed shape and assumed rank require type information.

The interface of a procedure with an assumed-type dummy argument is required to be explicit, even if it is not interoperable (that is, does not have the bind(c) attribute).

20.16   Assumed character length

One of the annoyances when calling C functions directly from Fortran used to be the need to remember to append a null character to the end when passing character strings, as the character length is not passed.   Fortran 2018 addressed this by permitting an interoperable procedure to have a dummy argument with assumed character length, which is passed by reference to as a structure of the type CFI_desc_type defined in the source file ISO_Fortran_binding.h.

For example, a C function to write a Fortran character string to C's standard error file, with a newline character to end the line, could have the interface

```
interface
   subroutine err_msg(string) bind(c)
      use iso_c_binding
      character(*,c_char),intent(in) :: string
   end subroutine err_msg
end interface
```

and could be implemented by the C function

```
#include <stdio.h>
#include "ISO_Fortran_binding.h"
void err_msg(CFI_cdesc_t *string)
{
   int len;
   char *p = string->base_addr;
   for (len=string->elem_len; len>0; len--) putc(*p++,stderr);
   putc('\n',stderr);
}
```

When invoked by

```
call err_msg('oops')
```

the message 'oops' would be written (with a newline).

 C descriptors are far more powerful than this, and enable many other advanced techniques, but for simple string handling this is all we need. Chapter 21 describes C descriptors in full detail.

Exercises

1. Write a generic Fortran interface block for the standard C libm error functions `erf` and `erff`.

2. Write Fortran functions to compute the dot product of two vectors, suitable for being called from C.

3. On a POSIX-compliant system, the API call `getpwnam()` allows the extraction of data associated with a specific user name from the authentication database. Write a Fortran module that contains the necessary type definition, `bind(c)` interfaces, and module procedures to produce the path name of the user's home directory. The manual page for `getpwnam()` provides needed information.

21. Interoperating with C using descriptors

21.1 Introduction

The design of the basic C interoperability features, described in the previous chapter, follows the principle that only very similar features can be interoperable. Although this leads to semantics that are relatively easy to understand, and to write interoperable procedures for, there are some very useful Fortran features that do not follow this principle; in particular,

- assumed-length character dummy arguments,
- assumed-shape arrays,
- allocatable dummy arguments, and
- pointer dummy arguments.

These Fortran features require additional information to be passed to/from C functions. This is done with a **C descriptor**, and the Fortran processor provides mechanisms for the C functions to use such descriptors.

21.2 C descriptors

21.2.1 Introduction

In order to be able to handle Fortran objects that have no equivalent concept in C, Fortran has the **C descriptor**. This is a C structure containing all the information needed to describe the supported Fortran objects. The exact contents of the structure are processor dependent, though some members are fully standardized. The Fortran processor provides a source file ISO_Fortran_binding.h that contains the definition of this structure, together with macro definitions and function prototypes for a C function to use when interacting with the C descriptor.

The Fortran objects that C descriptors are used for are all dummy arguments,[1] and are assumed character length, assumed shape, assumed rank (Section 7.18), allocatable, or pointers. They must have interoperable type and kind type parameter (if any), except that an assumed-shape or assumed-rank dummy argument is also permitted to be assumed type

---

[1] Both for Fortran calling C (the dummy arguments in the interface), and for C calling Fortran.

Modern Fortran Explained, 3rd Edition. M. Metcalf, J. Reid, M. Cohen, and R. Bader. Oxford University Press (2024). © M. Metcalf, J. Reid, M. Cohen, and R. Bader (2024). DOI 10.1093/oso/9780198876571.001.0021

(Section 20.15). An allocatable or pointer dummy argument must not be of a type with default initialization, and if of type character must have deferred character length.

The information about the object contained in the descriptor includes its address, rank, type, size (if scalar) or element size (if an array), and whether it is allocatable, a pointer, or neither. For an array the descriptor additionally contains information describing the shape, bounds (only relevant for allocatable and pointer), and memory layout; the latter is particularly important for arrays that are assumed shape, assumed rank, or pointers, as these can have discontiguous memory storage.

In order to preserve Fortran semantics, it is forbidden to use the C descriptor to access memory that is not part of the object the descriptor describes; for example, memory that is before the object, after the object, or, when the object is discontiguous, in the gaps in the object.

Also, a C descriptor must not be modified, or even copied, except by using the functions provided by ISO_Fortran_binding.h; this is to preserve the integrity of the descriptor and any associated run-time data structures.

21.2.2 Standard members

A C descriptor is a C structure with type CFI_cdesc_t. Its first three members are:

void * base_addr; the address of the object; for allocatables and pointers, it is the address of the allocated memory or associated target. For an unallocated allocatable or a disassociated pointer, base_addr is a null pointer; otherwise, base_addr is not null, though for an object of zero size, it is not the address of any usable memory.

size_t elem_len; storage size in bytes of the object (if scalar) or an array element (if an array).

int version; version number of the descriptor. It is the value of CFI_VERSION in the file ISO_Fortran_binding.h that defined the format and meaning of this C descriptor.

A C descriptor also contains the following members, in any order, possibly interspersed with non-standard processor-dependent members.

CFI_attribute_t attribute; argument classification, indicating whether the object is allocatable, a pointer, or neither, according to Table 21.1.

CFI_rank_t rank; rank of the object (zero if scalar). The maximum rank supported is provided by the value of CFI_MAX_RANK.

CFI_type_t type; data type of the object; this is an integer type code from Table 21.2.

The last member in a C descriptor is

CFI_dim_t dim[]; member describing the shape, bounds, and memory layout of an array object. This is described in Section 21.2.5.

The typedef names CFI_attribute_t, CFI_rank_t, and CFI_type_t are defined in ISO_Fortran_binding.h, and specify standard signed integer types, each with a size suitable for its purpose.

21.2.3   Argument classification (attribute codes)

The macros in Table 21.1 provide integer values for the `attribute` member of a C descriptor; they are non-negative and distinct.

The value passed in the `attribute` member when invoked from Fortran describes the dummy argument in the Fortran interface; thus, the only real use for this member is for error checking. For example, if the C function is expecting to be passed an allocatable array, it can check that the interface in the Fortran invocation really did specify that it was allocatable.

Table 21.1: Macros for attribute codes.

Macro name	Attribute
`CFI_attribute_allocatable`	Allocatable
`CFI_attribute_pointer`	Data pointer
`CFI_attribute_other`	Non-allocatable non-pointer

21.2.4   Argument data type

The `type` member of a C descriptor specifies the data type of the actual argument in C terms. This will be the same as that of the dummy argument in the interface unless the dummy argument is of assumed type or of an interoperable C derived type.

The first part of Table 21.2 lists the C type specifiers that have type codes of the form `CFI_type_`*type*, where *type* is the C type specifier (replacing internal spaces with underscore, and removing underscores at the beginnings of words); for example, `CFI_type_int`, `CFI_type_Bool`, and `CFI_type_long_double_Complex`. There are three additional type codes that do not fit this pattern, shown in the second part of the table.

All the `CFI_type_` type code macros have constant integer values. If a C type has no Fortran equivalent, its type code will be negative. A Fortran type with no C equivalent has the type code `CFI_type_other`, which is also negative and distinct from the other type codes. All other type codes are positive, but in general need not be distinct; for example, `CFI_type_long_long`, `CFI_type_int64_t`, `CFI_type_int_least64_t`, `CFI_type_int_fast64_t`, and `CFI_type_intmax_t` are likely to all be the same. The type code `CFI_type_struct` specifies an interoperable C structure (or Fortran `bind(c)` type); its value is distinct. Also, an implementation may support Fortran intrinsic types that are interoperable with C types not listed in the table by adding additional type codes to `ISO_Fortran_binding.h`. For example, if a Fortran integer kind exists that interoperates with an `int128_t`, the type code `CFI_type_int128_t` might be defined, with a distinct positive value.

21.2.5   Array layout information

Array information is provided by the `dim` member of a C descriptor; this is an array of structures with type `CFI_dim_t`, with one element for each rank of the array, the elements

Table 21.2: `CFI_type_` macros for type codes.

C type specifiers with `CFI_type_`*type* macros		
`signed char`	`intptr_t`	`int_least16_t`
`short`	`ptrdiff_t`	`int_least32_t`
`int`	`int8_t`	`int_least64_t`
`long`	`int16_t`	`int_fast8_t`
`long long`	`int32_t`	`int_fast16_t`
`size_t`	`int64_t`	`int_fast32_t`
`intmax_t`	`int_least8_t`	`int_fast64_t`
`float`	`double`	`long double`
`float _Complex`	`double _Complex`	`long double _Complex`
`_Bool`	`char`	

Additional `CFI_type_` macros	
Macro name	Meaning
`CFI_type_cptr`	void * or any C data pointer
`CFI_type_struct`	Interoperable C structure
`CFI_type_other`	Fortran non-interoperable type

being in Fortran dimension order (first element is for the first Fortran dimension). For each dimension, the structure members are, in any order:

`CFI_index_t lower_bound;` the lower bound. This only has meaning for an array pointer or allocatable array; for all others, it has the value zero.

`CFI_index_t extent;` size of the dimension, or -1 for the final dimension of an assumed-size array.

`CFI_index_t sm;` spacing between elements, or 'memory stride'; that is, the difference in bytes between the addresses of successive elements in that dimension.

where `CFI_index_t` is a typedef name for a standard signed integer type capable of representing a memory address difference in bytes. The structure could have additional, non-standard, members.

21.3 Accessing Fortran objects

21.3.1 Traversing contiguous Fortran arrays

It is very straightforward to access every element of a contiguous Fortran array of any rank; all that is needed is to calculate the number of elements from the C descriptor, and then step through the elements. For example,

```
int64_t xor64(CFI_cdesc_t *cdesc) {
    int i;
    size_t nelts = 1;
    int64_t result = 0;
    int64_t *a = (int64_t *)cdesc->base_addr;
    for (i=0; i<cdesc->rank; i++) nelts *= cdesc->dim[i].extent;
    while (nelts-->0) result ^= *a++;
    return result;
}
```

computes the elementwise bitwise exclusive or of a contiguous Fortran array with element storage size of 64 bits, first calculating the number of elements from the array extents, then initializing the result, and using bitwise exclusive-or assignment (^=) with every element. This C function could be used from Fortran on three-dimensional arrays by using the interface

```
interface
    pure function xor64(array) bind(c) result(r)
        use iso_c_binding
        integer(c_int64_t), intent(in), contiguous :: array(:,:,:)
        integer(c_int64_t) r
    end function xor64
end interface
```

Note that because the function assumes the array is contiguous, the contiguous attribute is necessary in the interface, so that a reference to the function from Fortran with a **discontiguous** actual argument will pass a contiguous copy of it to the function.

There is a utility function declared in ISO_Fortran_binding.h that can be used to check this assumption:

```
int CFI_is_contiguous(const CFI_cdesc_t *dv);
```

The argument dv must be the address of a descriptor that describes an array, and must not be an unallocated allocatable or disassociated pointer (so the base_addr member of the C descriptor must not be a null pointer). The result has the value one if the array described by dv is contiguous, and zero otherwise. In our previous example, we could use this to produce an error message if called with an inappropriate array:

```
if (!CFI_is_contiguous(cdesc)) {
    fputs("xor64 called with discontiguous array\n",stderr);
    exit(2);
}
```

Other than error checking, this function can be used to select a more efficient code path when a function that can handle discontiguous is called with a contiguous array argument; as we will see in Section 21.3.3, the code for traversing a discontiguous array is significantly more complicated with higher overheads.

Notice that the way the code inside the function is written, it would work without change on arrays of any rank, and even on scalars (though the result is not very interesting in the scalar case, being merely the argument value unchanged); see Section 21.3.6 for details of how this can be enabled.

21.3.2 Generic programming with assumed type

We have already seen, in Section 20.15, how **assumed-type** dummy arguments can be used for low-level C interoperability. When combined with assumed shape or assumed rank, however, assumed type works very differently; a C descriptor is passed, and this includes data type information that can be used by the invoked C procedure.

For example, the C function in Figure 21.1 could be invoked via the interface

```
interface
   subroutine negate_real(array) bind(c)
      type(*), intent(inout), contiguous :: array(:,:,:)
   end subroutine negate_real
end interface
```

to negate any three-dimensional real array.

Figure 21.1 A C function for negating a floating-point array.

```
void negate_real(CFI_cdesc_t *cdesc) {
    size_t i;
    size_t nelts = 1;
    for (i=0; i<cdesc->rank; i++) nelts *= cdesc->dim[i].extent;
    if (cdesc->type==CFI_type_float) {
        float *f = cdesc->base_addr;
        for (i=0; i<nelts; i++) f[i] = -f[i];
    }
    else if (cdesc->type==CFI_type_double) {
        double *d = cdesc->base_addr;
        for (i=0; i<nelts; i++) d[i] = -d[i];
    }
    else if (cdesc->type==CFI_type_long_double) {
        long double *ld = cdesc->base_addr;
        for (i=0; i<nelts; i++) ld[i] = -ld[i];
    }
    else {
        fputs("negate_real: not a supported real type\n",stderr);
        exit(2);
    }
}
```

As with the previous example, the `contiguous` attribute is used to avoid complicating the example with discontiguous array traversal code. Similarly, the code inside the function would work on scalars or arrays of any rank.

21.3.3   Traversing discontiguous Fortran arrays

Although the `contiguous` attribute is very useful for simplifying the objects that a C function needs to be able to handle, and the `CFI_is_contiguous` function can be used to check at run time that no mistake has been made, it is nearly always far more efficient to process a discontiguous array section directly than it is to make a copy of it (and potentially also copy it back afterwards). Fortunately, it is simple, though tedious, to traverse a discontiguous array using the information in the C descriptor.

The simplest way to do this is to have one pointer per array dimension. For example, with a three-dimensional array we could have a 'plane pointer' starting at the base address, being stepped by the memory stride of a whole plane, and inside that loop a 'column pointer' starting at the beginning of the current plane, being stepped by the memory stride of a whole column, and inside that loop an 'item pointer' starting at the beginning of the current column and being stepped by the memory stride of the innermost dimension.

This does have a lot of overhead, though, and accessing the extents and memory strides through a pointer is not conducive to optimization (for example, keeping these values and pointers in registers).

A better way is to use a single pointer to traverse the whole array, with precomputed step values, and copying the extents into local variables.

Figure 21.2 Assumed-shape array traversal.

```
int64_t xor64(CFI_cdesc_t *cdesc) {
    CFI_index_t i,j,k,step0,step1,step2,extent0,extent1,extent2;
    int64_t result = 0;
    char *ptr = cdesc->base_addr;
    step0 = cdesc->dim[0].sm;
    extent0 = cdesc->dim[0].extent;
    step1 = cdesc->dim[1].sm - cdesc->dim[0].sm*extent0;
    extent1 = cdesc->dim[1].extent;
    step2 = cdesc->dim[2].sm - cdesc->dim[1].sm*extent1;
    extent2 = cdesc->dim[2].extent;
    for (k=0; k<extent2; k++) {
        for (j=0; j<extent1; j++) {
            for (i=0; i<extent0; i++) {
                result ^= *(int64_t *)ptr;
                ptr += step0;
            }
            ptr += step1;
        }
        ptr += step2;
    }
    return result;
}
```

Figure 21.2 shows this method, extending the `int64_t` exclusive-or reduction we saw in Section 21.3.1 to handle discontiguous three-dimensional arrays. The interface for invoking this function from Fortran is the same as before, except that the `contiguous` attribute is omitted from the declaration of the dummy argument `array`.

The precomputed step values are derived straightforwardly from the memory strides (`sm` values) for each dimension. For the first (innermost) dimension, the step value $step_1$ is just the same as the memory stride sm_1, while for every subsequent dimension $i + 1$, the step value is the memory stride of that dimension minus the accumulated increments of all the inner dimensions, which is $sm_i \times extent_i$.

Optimization should keep the pointer and the `step` and `extent` values in registers, so despite the increased overhead compared with contiguous array traversal, with a simple operation like exclusive or the performance is likely to be dominated by the time taken to access the array from main memory.

We can also see that both the simplistic and this faster formulation of discontiguous traversal will only work for three-dimensional arrays, whereas the contiguous case worked for any rank of array.

21.3.4 Fortran pointer operations

Accessing a Fortran pointer is relatively easy, using the `base_addr` member of the C descriptor in conjunction with the array information if it is an array. For example, testing whether a Fortran pointer argument is disassociated is simple:

```
if (cdesc->base_addr)
    printf("Associated\n");
else
    printf("Disassociated\n");
```

For nullifying or pointer assigning, a function is provided by `ISO_Fortran_binding.h`:

```
int CFI_setpointer(CFI_cdesc_t *result, CFI_cdesc_t *source,
                   const CFI_index_t lower_bounds[]);
```

where `result` must be a C descriptor for a Fortran pointer, and `source` must be a null pointer or a C descriptor for either a similar pointer to `result` or an acceptable target for `result`. That is, if `source` is not a null pointer, its `type` and `rank` members must be the same as that of `result`; if the pointer is not of type character, this also applies for its `elem_len` member. If `source` does not describe an array, `lower_bounds` is ignored; otherwise, it may be a null pointer or the address of an array of size `source->rank`.

The result of the function is an **error indicator**; this is similar to a `stat=` value, being zero if the function executed successfully, and nonzero otherwise (Section 21.3.8).

If `source` is a null pointer, the effect is the same as the Fortran statement `nullify(result)`, and `result` becomes a disassociated pointer.

If `source` is not a null pointer, the effect is similar to the Fortran pointer assignment `result => source`. If `source` is a C descriptor for a Fortran pointer, `result` will have the same pointer association as `source` (either disassociated, or associated with the same target). Otherwise, `result` will become associated with the object described by `source`; if `source` describes an object of type character, the value of `source->elem_len` will be copied to `result->elem_len`. Note that if `source` is a Fortran entity, it must have the `target` attribute.

If `source` describes an array and `lower_bounds` is not a null pointer, the lower bounds of `result` are set to the values specified in `lower_bounds`, similar to the Fortran pointer assignment `result(lower₁:,lower₂:,...) => source` (as described in Section 7.14).

The simple example in Figure 21.3 chooses between two scalar integer pointers, choosing the non-null pointer that references the highest value; if an error occurs the result will be nullified.

Functions are also available for allocating and deallocating Fortran pointers via their C descriptors. These functions also operate on allocatable variables, so are described in the next section.

Figure 21.3 Choose between two scalar integer pointers.

```
interface
    subroutine set_ptr_to_max(result, inptr1, inptr2) bind(c)
        use iso_c_binding
        integer(c_int), pointer, intent(out) :: result
        integer(c_int), pointer, intent(in) :: inptr1, inptr2
    end subroutine set_ptr_to_max
end interface

#include "ISO_Fortran_binding.h"
void set_ptr_to_max(CFI_cdesc_t *result,CFI_cdesc_t *inptr1,
                    CFI_cdesc_t *inptr2) {
    int res;
    if (!inptr1->base_addr)
        res = CFI_setpointer(result,inptr2,(CFI_index_t*)0);
    else if (!inptr2->base_addr ||
             *(int *)inptr1->base_addr>=*(int *)inptr2->base_addr)
        res = CFI_setpointer(result,inptr1,(CFI_index_t*)0);
    else
        res = CFI_setpointer(result,inptr2,(CFI_index_t*)0);
    if (res)
        (void)CFI_setpointer(result,(CFI_cdesc_t*)0,(CFI_index_t*)0);
}
```

21.3.5   Allocatable objects

Like Fortran pointers, accessing Fortran allocatable objects using the `base_addr` member of the C descriptor is straightforward. For example, testing whether a Fortran allocatable argument is allocated is simple:

```
if (cdesc->base_addr)
    printf("Allocated\n");
else
    printf("Unallocated\n");
```

Deallocation is also very simple, using the function

```
int CFI_deallocate(CFI_cdesc_t *dv);
```

where dv is a C descriptor for a Fortran allocatable or pointer argument. The return value is an error indicator.

The memory must have been allocated using the Fortran memory allocator, that is, by the function CFI_allocate or a Fortran allocate statement. If deallocation is successful, the base_addr member of dv will become a null pointer.

Note that in the pointer case the conditions for successful deallocation by a Fortran deallocate statement must be satisfied. For example, the pointer must be associated with the entirety of an allocated object, with the array elements in the same order.

For allocation, array bounds and character length may have to be supplied. This is done by the function CFI_allocate:

```
int CFI_allocate(CFI_cdesc_t *dv, const CFI_index_t lower_bounds[],
                 const CFI_index_t upper_bounds[], size_t elem_len);
```

where dv is a C descriptor for an unallocated allocatable or disassociated pointer. If the object is an array, lower_bounds and upper_bounds must be arrays of size dv->rank; these arguments are ignored for a scalar. If the object is a Fortran character type, elem_len is the size of each element in bytes, and dv->elem_len is overwritten with its value; otherwise, elem_len is ignored. The return value is an error indicator.

For example, the C function in Figure 21.4, if invoked with the interface

```
interface
    subroutine get_home(home) bind(c)
        use iso_c_binding
        character(:,c_char), allocatable, intent(inout) :: home
    end subroutine get_home
end interface
```

will allocate a deferred-length character scalar to hold the value of the environment variable HOME (usually the user's home directory path), deallocating it first if necessary. On any error, or if the environment variable does not exist, the argument will become unallocated.

If the interface for a C function has an allocatable dummy argument with intent out, calling that function from Fortran with an allocated actual argument will automatically deallocate the argument before invocation. Similarly, if a Fortran procedure with an intent out allocatable dummy argument is called with an allocated actual argument from C, the argument will automatically be deallocated on entry to the procedure. This preserves Fortran semantics for allocatable intent out dummy arguments whenever either procedure in a call is Fortran. Keeping the conditionally executed CFI_deallocate statement in the code of Figure 21.4 is advisable, since it may get executed in calls to the function from C, which have no awareness of Fortran's intent out semantics.

Figure 21.4 A C function that allocates a Fortran character variable.

```
#include <string.h>
#include "ISO_Fortran_binding.h"
void get_home(CFI_cdesc_t *home) {
    int r;
    char *homename = getenv("HOME");
    if (home->base_addr) {
        r = CFI_deallocate(home);
        if (r) return; /* failed. */
    }
    if (homename) {
        size_t len = strlen(homename);
        r = CFI_allocate(home,(CFI_index_t*)0,(CFI_index_t*)0,len);
        if (r) return; /* failed. */
        memcpy(home->base_addr,homename,len);
    }
}
```

Finally, note that if the CFI_allocate function is applied to a descriptor for an allocatable object that was created by an invocation of CFI_establish (Section 21.4.2), that object will not be automatically deallocated once the descriptor's scope ceases to exist; a call to CFI_deallocate (or a Fortran deallocate statement) will be needed to avoid orphaning the memory associated with that object.

21.3.6   Handling arrays of any rank

We saw in Section 21.3.1 that the C code for elementwise processing of a contiguous array can be the same for any rank, but that the Fortran interface is different. The **assumed rank** feature allows us to write a single interface for a C function that accepts any rank of array. This feature is described fully in Section 7.18, but all we need here is the declaration for the interface. For example, an interface that would let us call the earlier xor64 example with an array of any rank is

```
interface
    pure function xor64(array) bind(c) result(r)
        use iso_c_binding
        integer(c_int64_t), intent(in), contiguous :: array(..)
        integer(c_int64_t) r
    end function xor64
end interface
```

The only change to the interface is the use of '..' to denote assumed rank.

This may be combined with the assumed type feature for generic programming, in the same way. So the interface

```
interface
   subroutine negate_real(array) bind(c)
      type(*), intent(inout), contiguous :: array(..)
   end subroutine negate_real
end interface
```

would let us negate a real array of any rank.

There is one fly in this ointment: if the dummy argument is assumed rank and not assumed type, the actual argument is allowed to be an assumed-size array. This is a problem because an assumed-size array has no upper bound in the last dimension, rendering the calculation of the number of elements impossible. To avoid incorrect results or attempting to process all memory (and likely crashing), this should be checked for and an appropriate action taken.

This problem does not arise for `negate_real`, because its array argument is assumed type, but is a problem for `xor64`. To detect an assumed-size argument at run time, it suffices to check for the final extent of the C descriptor being less than zero (and this should be done before attempting to calculate the number of elements). For example,

```
if (cdesc->rank>0 && cdesc->dim[cdesc->rank-1].extent<0) {
   fputs("Actual argument to xor64 is assumed-size\n",stderr);
   exit(2);
}
```

When it comes to discontiguous arrays of any rank, the interface remains simple – just leave off the `contiguous` attribute – but the traversal is more complicated. If the intention is only to handle, say, arrays up to rank five, a reasonable solution is to use the method shown in Figure 21.2 with a deep loop nest, pretending that any missing dimension has extent one.

A better way is to use a single loop together with an array of indices to keep track of how much of each dimension has been traversed. Figure 21.5 shows how this can be done. The dimension-collapsing phases are not actually necessary, but improve performance by reducing the bookkeeping needed during traversal. This is especially important for the innermost dimensions, since that is the only loop that could be vectorized. If the array is contiguous, dimension collapse will reduce the internal array description to rank one, and only the innermost loop will execute more than once.

21.3.7 Accessing individual array elements via a C descriptor

It is not difficult to manually calculate the address of a single element of an array that is passed as a C descriptor, using subscripts; simply subtract the lower bounds from the subscripts, multiply each by its 'memory stride' (`sm` member), and add to the base address cast to a `char *` pointer. For a non-allocatable non-pointer dummy argument, the lower bounds in a C function are always zero, so the subtraction can be omitted in that case.

For example, to access an element `a(i,j)` for a non-allocatable non-pointer dummy argument `a` passed as the C descriptor `a`, the address is simply

```
(i*a->dim[0].sm + j*a->dim[1].sm) + (char *)(a->base_addr)
```

For accessing an element of an array that is allocatable or a pointer, with subscripts based on the actual lower bounds, the complete formulation

Figure 21.5 Traversing a discontiguous array of any supported rank.

```c
int64_t xor64(CFI_cdesc_t *cdesc) {
    int i;
    CFI_index_t extent[CFI_MAX_RANK],ii[CFI_MAX_RANK],ix,step[CFI_MAX_RANK];
    int64_t result = 0;
    int rank = cdesc->rank;
    char *ptr = cdesc->base_addr;
    if (rank==0) return *(int64_t *)ptr; /* Scalar result is trivial. */
    if (cdesc->dim[rank-1].extent<0) {
        fputs("Actual argument to xor64 is assumed-size\n",stderr);
        exit(2);
    }
    /* Establish step values for the array, and save the extents locally. */
    step[0] = cdesc->dim[0].sm;
    extent[0] = cdesc->dim[0].extent;
    for (i=1; i<rank; i++) {
        ii[i] = 0;
        step[i] = cdesc->dim[i].sm - cdesc->dim[i-1].sm*extent[i-1];
        extent[i] = cdesc->dim[i].extent;
        if (extent[i]==0) return 0; /* Zero-sized array has zero result. */
    }
    /* Collapse outermost and innermost contiguous dimensions. */
    while (step[rank-1]==0) {
        extent[rank-2] *= extent[rank-1];
        rank--;
    }
    while (rank>2 && step[1]==0) {
        extent[0] *= extent[1];
        rank--;
        for (i=1; i<rank; i++) {
            extent[i] = extent[i+1];
            step[i] = step[i+1];
        }
    }
    /* Now traverse the array. */
    for (;;) {
        for (ix=0; ix<extent[0]; ix++,ptr+=step[0]) {
            result ^= *(int64_t *)ptr;
        }
        /* Step through the outer dimensions. */
        for (i=1; i<rank; i++) {
            ptr+=step[i];
            ii[i]++;
            if (ii[i]<extent[i]) break;
            ii[i] = 0;
        }
        if (i==rank) return result;
    }
}
```

```
((i-a->dim[0].lower)*a->dim[0].sm +
 (j-a->dim[1].lower)*a->dim[1].sm) + (char *)(a->base_addr)
```
will be necessary.

A convenience function is available in ISO_Fortran_binding.h to simplify this:

```
void * CFI_address(const CFI_cdesc_t *dv,const CFI_index_t subs[]);
```
where dv is a C descriptor for an object other than an unallocated allocatable or disassociated pointer, and subs is an array of subscripts of size dv->rank; the address of the array element is returned. If dv is scalar, subs is ignored and may be a null pointer. As in the manual case, the subscripts must be within the bounds of the array and, if the array is assumed size, must specify an element within the size of the array. However, as there is no error handling provided by CFI_address, it has little advantage over the manual method, unless the rank is high or the subscripts are already in an array of type CFI_index_t.

Once the address has been calculated, whether manually or using CFI_address, referencing or defining the element will also need to cast the address to the data type of the array. For example, the C function

```
void zero_element(CFI_cdesc_t *a,int i,int j) {
    char *elt = a->base_addr;
    elt += (i*a->dim[0].sm + j*a->dim[1].sm);
    switch (a->type) {
    case CFI_float:
        *(float *)elt = 0.0f;
        break;
    case CFI_double:
        *(double *)elt = 0.0;
        break;
    default:
        fputs("zero_element: unsupported data type\n",stderr);
        exit(2);
    }
}
```

could be invoked through the interface

```
interface
    subroutine zero_element(a,i,j) bind(c)
        use iso_c_binding
        type(*),intent(inout) :: a(:,:)
        integer(c_int),value :: i,j
    end subroutine zero_element
end interface
```

to set a selected element of a floating-point array to zero, noting that the indices are zero-based; for example

```
        call zero_element(r,i-1,j-1)
```

would zero the Fortran array element r(i,j).

21.3.8   Handling errors from CFI functions

Many of the functions provided by ISO_Fortran_binding.h return success or failure as an **error indicator**. This is an integer value akin to the stat= specifier of a Fortran statement or stat argument of some intrinsic procedures, and is zero to indicate that the function was successful (no error), and nonzero to indicate which error was detected. If an error was detected, the function will not have modified any of its arguments.

The macro CFI_SUCCESS is defined to be zero for use in testing for success. Standard error conditions, with corresponding macro names, are shown in Table 21.3. All values are nonzero and distinct. If a processor can detect additional error conditions, they will return values that differ from the standard ones.

Table 21.3: Macros for error codes.

Macro name	Error condition
CFI_ERROR_BASE_ADDR_NULL	The base_addr member of a C descriptor is a null pointer, but a non-null pointer is required.
CFI_ERROR_BASE_ADDR_NOT_NULL	The base_addr member of a C descriptor is not a null pointer, but a null pointer is required.
CFI_INVALID_ELEM_LEN	Invalid value for elem_len member of a C descriptor.
CFI_INVALID_RANK	Invalid value for rank member of a C descriptor.
CFI_INVALID_TYPE	Invalid value for type member of a C descriptor.
CFI_INVALID_ATTRIBUTE	Invalid value for attribute member of a C descriptor.
CFI_INVALID_EXTENT	Invalid value for extent member of a CFI_dim_t structure.
CFI_INVALID_DESCRIPTOR	A C descriptor is invalid in some way.
CFI_ERROR_MEM_ALLOCATION	Memory allocation failed.
CFI_ERROR_OUT_OF_BOUNDS	A reference is out of bounds.

21.4   Calling Fortran with C descriptors

21.4.1   Allocating storage for a C descriptor

When a C function is called from Fortran, the Fortran processor takes care of setting up the C descriptor. But when calling Fortran from a C function, unless simply passing on an argument received from Fortran, storage will need to be allocated for the descriptor itself.

To allocate the storage for a C descriptor, the macro CFI_CDESC_T is provided. It is a function-like macro that takes one argument and evaluates to an unqualified type of suitable size and alignment for holding a C descriptor. The argument must be an integer constant expression that specifies the rank of the object to be described. The requirement for the rank to be constant means that, in order to be able to handle an array of any rank, the maximum possible rank would need to be specified here, for example

```
CFI_CDESC_T(CFI_MAX_RANK) pointer_to_any_array;
```

To use the allocated storage, it is usually necessary to cast the address of the storage to the descriptor type CFI_cdesc_t *, for example

```
CFI_CDESC_T(5) object;
int stat;
CFI_cdesc_t * dv = (CFI_cdesc_t *)&object;
stat = CFI_establish(dv, ...); /* See Section 21.4.2. */
```

The C descriptor may be allocated statically, on the stack, or it may be a component of a structure which might then be allocated on the heap, for example:

```
struct fortran_string_list_item {
    CFI_CDESC_T(0) string;
    struct fortran_string_list_item *next;
};
```

Before any C descriptor is used, whether its memory was allocated statically or on the stack by CFI_CDESC_T, or on the heap as part of a structure allocated by malloc, it must be established.

21.4.2 Establishing a C descriptor

Before a C descriptor can be used, it must be **established**; this sets up its internal data, both standard members and any processor-dependent members. When a C function is invoked from Fortran, the Fortran processor establishes the C descriptors for the relevant arguments, but when a Fortran processor is invoked from C, the C function needs to do this. This is done by the function CFI_establish, which has the prototype

```
int CFI_establish(CFI_cdesc_t *dv, void *base_addr,
                  CFI_attribute_t attribute, CFI_type_t type,
                  size_t elem_len, CFI_rank_t rank,
                  const CFI_index_t extents[]);
```

where dv is the C descriptor to be established; it must not be associated with a Fortran dummy or actual argument, and it is highly recommended that the CFI_CDESC_T macro be used to allocate its storage. The remaining arguments provide the characteristics of the object that the C descriptor will describe, and the return value is an error indicator.

base_addr the address of the object, or a null pointer.

attribute whether the object is allocatable, a pointer, or an ordinary data object.

type a type code in Table 21.2 or the type code of an interoperable C type.

elem_len provides the size in bytes of the object, or an element of the object when it is an array, if type identifies a structure or character type, and is ignored otherwise.

rank specifies the desired rank.

extents is an array of size rank of non-negative values specifying the array shape, when rank is nonzero and base_addr is not a null pointer; otherwise, it is ignored.

When attribute is CFI_attribute_allocatable, base_addr must be a null pointer signifying an unallocated allocatable. When attribute is CFI_attribute_pointer, base_addr may be a null pointer to signify a disassociated Fortran pointer, or the address of a suitable target. When attribute is CFI_attribute_other, base_addr may be a null pointer signifying that the descriptor is unfinished (it must be completed by CFI_section or CFI_select_part before use), or the address of a suitable object.

When the object is an array and base_addr is not a null pointer, the lower bounds are zero.

This looks a bit complicated, but it is not too hard to use it to construct C descriptors for calling Fortran procedures that have assumed-shape or assumed character length arguments. For example, given the C array and string

```
float x[10][20];
char *name;
```

and the Fortran procedure with the interface

```
interface
   subroutine process(name,a) bind(c)
      use iso_c_binding
      character(*,c_char),intent(in) :: name
      real(c_float) a(:,:)
   end subroutine
end subroutine
```

we can pass x and name to the assumed-shape array and assumed-length character arguments of the Fortran procedure with the code shown in Figure 21.6.

Figure 21.6 Pass arguments to a Fortran procedure.

```
extern void process(CFI_cdesc_t *,CFI_cdesc_t *);
CFI_CDESC_T(2) xdesc;
CFI_CDESC_T(0) namedesc;
CFI_index_t xshape[2] = { 20,10 };
if (CFI_establish((CFI_cdesc_t *)&xdesc,&x,CFI_attribute_other,
                  CFI_type_float,0,2,xshape) ||
    CFI_establish((CFI_cdesc_t *)&namedesc,name,CFI_attribute_other,
                  CFI_type_char,strlen(name),0,0)) {
   fputs("Unexpected CFI_establish failure\n",stderr);
   exit(2);
}
process((CFI_cdesc_t *)&namedesc,(CFI_cdesc_t *)&xdesc);
```

21.4.3 Constructing an array section

In Fortran a section of an array, possibly discontiguous, can be passed as an actual argument using array section notation. This facility is also available from C when the called procedure expects a C descriptor, using the CFI_section function. This function needs to have an established C descriptor that will be modified to describe the section, a C descriptor that describes the base array, and all the information to describe the array section.

```
int CFI_section(CFI_cdesc_t *result, const CFI_cdesc_t *source,
                const CFI_index_t lower[], const CFI_index_t upper[],
                const CFI_index_t stride[]);
```

where result points to an established C descriptor that will be modified to describe the section, source points to a C descriptor that describes the base array, and lower, upper, and stride are arrays of size source->rank that describe the section. The return value is an error indicator.

The lower, upper, and stride arrays may be null pointers to take the default for all dimensions: the lower bounds and upper bounds from source, and the value one for the stride. Note that upper must not be null when the base array is associated with an assumed-size array. The type and elem_size members of result must be the same as those of source.

Note that CFI_attribute_allocatable is not permitted for the attribute member of result, because allocatable arrays are always contiguous and array sections are usually discontiguous. If result has CFI_attribute_pointer, the effect is as if it were pointer associated with the section; otherwise the effect is as if it were argument associated. If result does not have CFI_attribute_pointer, it must not be currently associated with a Fortran actual or dummy argument.[2]

The section described can be characterized in Fortran as

```
source(lower₁:upper₁:stride₁, ... )
```

where if $stride_i$ is zero, $lower_i$ must be equal to $upper_i$ and that dimension acts as if specified with the simple subscript $lower_i$. The rank of this array section is the number of dimensions with nonzero stride, and result must have this rank.

A simple example without error checking is shown in Figure 21.7 This example passes the first 23 elements of the 100th Fortran row (C column) of x to process_vector via a C descriptor for a discontiguous vector. It is notable that even this simple example has a lot of tedious bookkeeping that is easy to get wrong.

The process_vector procedure would have a Fortran interface like

```
interface
    subroutine process_vector(array) bind(c)
        use iso_c_binding
        integer(c_int64_t), intent(inout) :: array(:)
    end subroutine
end interface
```

---

[2]Using this function to do pointer assignment instead of CFI_setpointer is opaque and not recommended.

Figure 21.7 Use of a C descriptor for a discontiguous vector.

```
extern void process_vector(CFI_cdesc_t *);
static int64_t x[200][100];
   ⋮
CFI_CDESC_T(2) xdesc;
CFI_CDESC_T(1) rowdesc;
CFI_index_t lower[2],upper[2],stride[2],xshape[2];
xshape[0] = 100;
xshape[1] = 200;
(void)CFI_establish((CFI_cdesc_t *)&xdesc,&x,CFI_attribute_other,
                    CFI_type_int64_t,0,2,xshape);
(void)CFI_establish((CFI_cdesc_t *)&rowdesc,0,CFI_attribute_other,
                    CFI_type_int64_t,0,1,0);
lower[0] = upper[0] = 99; stride[0] = 0;
lower[1] = 0; upper[1] = 22; stride[1] = 1;
(void)CFI_section((CFI_cdesc_t *)&rowdesc,(CFI_cdesc_t *)&xdesc,
                  lower,upper,stride);
process_vector((CFI_cdesc_t *)&rowdesc);
```

A more interesting example is shown in Figure 21.8. This is a C function that accepts an assumed-shape 64-bit integer array of rank three, and invokes the same Fortran interoperable procedure process_vector on every row of the third plane of that array.

21.4.4   Accessing components

In Fortran, an array section can also be constructed by taking a component, substring, or complex part of a parent array, for example

```
array%re
```

This is also possible from C when producing a C descriptor with the CFI_select_part function, which has the prototype:

```
int CFI_select_part(CFI_cdesc_t *result, const CFI_cdesc_t *source,
                    size_t offset, size_t elem_len);
```

It returns an error indicator, and its arguments are as follows:

result is an established C descriptor with the same rank as source, and will be updated to describe the array section. The type information is unchanged, so this must already describe the type of the component or part being selected. As with CFI_section, it must not be allocatable, and if it is currently associated with a Fortran argument, must be a pointer (and as with CFI_section, this is not recommended).

source is the C descriptor that describes the base array; it must not be unallocated, disassociated, or assumed size.

Figure 21.8 Constructing and passing array sections with C descriptors.

```
extern void process_vector(CFI_cdesc_t *);
void process_rows_of_plane_3(CFI_cdesc_t *src) {
  CFI_index_t i,lower[3],upper[3],stride[3];
  CFI_CDESC_T(1) row;
  if (src->attribute!=CFI_attribute_other || src->rank!=3 ||
      src->dim[2].extent<3 || src->type!=CFI_type_int64_t ) {
    fputs("Wrong class/rank/last extent/type of src\n",stderr);
    exit(2);
  }
  if (CFI_establish((CFI_cdesc_t *)&row,0,CFI_attribute_other,
    CFI_type_int64_t,0,1,0)) {
    fputs("Unexpected CFI_establish error\n",stderr);
    exit(2);
  }
  stride[0] = 0;
  lower[1] = src->dim[1].lower_bound;  /* Size of row */
  upper[1] = src->dim[1].extent - 1;   /* is no of cols */
  stride[1] = 1;
  lower[2] = upper[2] = 2; stride[2] = 0;
  for (i=0; i<src->dim[0].extent; i++) {
    lower[0] = upper[0] = i;                /* column index */
    if (CFI_section((CFI_cdesc_t *)&row,src,lower,upper,stride)) {
      fputs("Unexpected CFI_section error\n",stderr);
      exit(2);
    }
    process_vector((CFI_cdesc_t *)&row);
  }
}
```

offset is the distance in bytes between the beginning of each array element and the component or part being described. If the base array is of an interoperable C structure, the offsetof macro may be used to obtain the correct value.

elem_len is the storage size in bytes of the part being selected if it is a Fortran character type, and this will override the information in source; otherwise, elem_len is ignored.

The part being selected must be a component, substring, or the real or imaginary part of a complex variable, and all sizes and the offset must be valid; for example, the selected part must lie entirely within each array element and must be properly aligned for its type.

This is most useful for interoperable structures; for example,

```
typedef struct {
    double r,i,j,k;
} Quaternion;
void process_j(CFI_cdesc_t *q) {
    CFI_CDESC_T(CFI_MAX_RANK) qj;
    (void)CFI_establish((CFI_cdesc_t *)&qj,0,CFI_attribute_other,
                        CFI_type_double,0,q->rank,0);
    (void)CFI_select_part((CFI_cdesc_t *)&qj,q,
                          offsetof(Quaternion,j),0);
    process_doubles((CFI_cdesc_t *)&qj);
}
```

constructs a C descriptor (without error checking) for the array section that is the `j` member of a `Quaternion` array, and passes it to another function for processing.

21.5   Restrictions

21.5.1   Other limitations on C descriptors

If the address of a C descriptor is a formal parameter that corresponds to a Fortran actual argument or a C actual argument that corresponds to a Fortran dummy argument,

- the C descriptor must not be modified if either the corresponding dummy argument in the Fortran interface has intent `in` or the C descriptor is for a non-allocatable non-pointer object, and
- the `base_addr` member of the C descriptor must not be accessed before it is given a value if the corresponding dummy argument in the Fortran interface is a pointer and has intent `out`.

If a C descriptor is passed to an assumed-shape Fortran dummy argument, the C descriptor must have the correct rank and must not describe an assumed-size array.

21.5.2   Lifetimes of C descriptors

A C descriptor or C pointer that is associated with any part of a Fortran object becomes undefined in the same circumstances that would cause a Fortran pointer to it to become undefined (for example, if it is deallocated, or if it is an unsaved local variable and its containing procedure returns).

A C descriptor whose address is a formal parameter that corresponds to a Fortran dummy argument becomes undefined on return from a call to the function from Fortran. Furthermore, if the dummy argument does not have either the `target` or `asynchronous` attribute, all C pointers to any part of the object become undefined on return from the call.

No further use may be made of a C descriptor or pointer that becomes undefined in any of these ways.

22. Generic programming

22.1 Introduction

Generic programming is a rather nebulous term. The basic idea is that the same source code can be used, unchanged, to perform the equivalent operations for things that are different in some way, usually different data types.

Fortran 90's introduction of user-defined generic procedures and operators was a simple example of this kind of genericity. Here, the same executable statements can be used for different types (and kinds), with only the declarations needing to be changed.

Fortran 95's user-defined elemental procedures were an example of a different kind of genericity: genericity in rank. The same executable statements can be used for any rank of array, or scalar, as long as the operations are to be performed elementwise.

Fortran 2003's object-oriented features provided genericity within a family of types. Unlike the other generic features, execution time genericity is the main focus of object orientation. Thus heterogeneous data structures are easy to write, but they may lack the efficiency or convenience of homogeneous data structures.

Generic code can be used and/or written using various technologies, from the most primitive to the most sophisticated:

- manual inclusion/modification with a text editor;
- simple code inclusion with `include` files, described in Section A.2;
- macro processing;
- automatic code generation (either as a separate processing phase using an external program, or a built-in feature like C++ templates).

The last of those is also sometimes called meta-programming, though that term could also be applied to macro processing.

Fortran 2023 contains several new features that are oriented towards generic programming, and these can work whether the technology being used is primitive (e.g., `include` files) or more advanced (e.g., macro processing). As Fortran does not yet contain templates (a form of templates is being proposed for a future revision), we will use the macro processor fpp (or cpp) in some examples.[1]

---

[1] The usability of cpp and other non-Fortran-aware macro processors has been improved by the free-form source changes in Fortran 2023, see Section 23.2.1. Either fpp or cpp is frequently built in to a Fortran compiler. If not, the source code of fpp is available on www.netlib.org, and cpp can be accessed using a C compiler (typically, the -E option).

Modern Fortran Explained, 3rd Edition. M. Metcalf, J. Reid, M. Cohen, and R. Bader. Oxford University Press (2024). © M. Metcalf, J. Reid, M. Cohen, and R. Bader (2024). DOI 10.1093/oso/9780198876571.001.0022

The basic idea of macro processing is that the user's source file is input to the macro processor, and it is the macro processor output that is actually compiled.

With fpp or cpp, macros are defined on a line beginning with #define, and specify replacement text that is applied to all lines from then on. The general form of a #define line is

```
#define name text
```

or

```
#define name(arguments) text
```

where arguments is a comma-separated list of names. The argumentless form replaces name with text on all following lines, while the argumented form replaces name (*actual-arguments*) with the text modified by replacing each argument within text by the corresponding *actual-argument*.

Although a macro is conceptually defined on a single line, that line may be continued by ending it with a backslash character (\).

Unlike Fortran names, the names of macros are case sensitive. Upper case is widely recommended for macro names, so that their use is clearly indicated.

22.2 Genericity in type

22.2.1 Copying the type and type parameters of another object

To facilitate generic inline code, there are two new type specifiers: typeof and classof. The typeof type specifier specifies the same type and type parameters as a given data object. The classof type specifier is used to declare polymorphic entities whose declared type and type parameters are the same as the declared type and type parameters of a given data object.

The typeof type specifier has the general form

```
typeof ( data-ref )
```

where *data-ref* is a reference to a data object whose type and type parameters have already been specified.[2] The syntax of classof is similar, with only the keyword being changed.

If a length type parameter is deferred, that type parameter is also deferred in the object being declared by typeof or classof. Otherwise, it is the value of the type parameter that is used. Thus, in

```
subroutine sub(a, b, c)
  character(*), intent(in) :: a
  typeof(a), intent(inout) :: b, c
```

the character lengths of b and c are not assumed, but declared to be equal to a%len. Note that if the *data-ref* is an optional dummy argument, it must not have any deferred or assumed type parameter.

In the case of typeof, the *data-ref* may be polymorphic, but cannot be unlimited polymorphic (class (*) or type (*)), or of abstract type. In the case of classof, the *data-ref* cannot be type (*) or of intrinsic type, but may be class (*) or of abstract type.

---

[2]Or established by the implicit typing rules, see Section 8.2.1.

The `typeof` and `classof` type specifiers can only be used in a type declaration statement or a component definition statement. Any object declared with `classof` must have the `allocatable` or `pointer` attribute, or be a dummy argument.

The `typeof` specifier is particularly useful for declaring temporary variables that need to have the same type and type parameters as another variable. For example, the following macro declares such a temporary and uses it to swap the values of two variables.

```
! Works for ordinary scalar a and b of any type,
! as long as neither "a" nor "b" are called "inline_swap_tmp".
#define INLINE_SWAP(a,b) \
  block; \
    typeof(a) :: inline_swap_tmp; \
    inline_swap_tmp = a; \
    a = b; \
    b = inline_swap_tmp; \
  end block
  ⋮
complex :: x, y
  ⋮
INLINE_SWAP(x,y)
```

Using `classof`, one can similarly write a macro to swap the allocations of two polymorphic allocatable variables with inline code.

```
! Swap the allocations of two scalar allocatable variables.
#define SWAP_ALLOC(a,b) \
block; \
  classof(a), allocatable :: swap_alloc_tmp; \
  call move_alloc (a, swap_alloc_tmp); \
  call move_alloc (b, a); \
  call move_alloc (swap_alloc_tmp, b); \
end block
```

22.2.2 Type-independent programming

Type-independent programming can often be performed with simple use of `include` files, or with more of the boilerplate handled by macro processing.

For example, a homogeneous list type (where each item on the list has the same type and is not polymorphic) can be programmed with an `include` file, as shown in Figure 22.1. Conceptually, the `include` file has an 'argument' named `item_type`, and so the type the user wants to have a list of needs to be renamed (on the `use` statement) to `item_type`. The file defines a type named `list`, so in practical use, one would also rename that as well, in a `use` statement.

Figure 22.1 Creating a homogeneous list type for any item type.

The include file:

```
type, public :: list
  type(item_type) :: value
  type(list), pointer :: next => null()
end type
interface next
  module procedure listnext
end interface
private listnext, null
public next
contains
  function listnext(item)
    type(list), intent(in) :: item
    type(list), pointer :: listnext
    listnext => item%next
  end function
```

This can be instantiated as follows:

```
module mytype_list_module
  use mytype_module, only: item_type=>mytype
  include 'define-list.inc'
end module
```

This module can then be used:

```
use mytype_list_module, mytype_list=>list
```

If macro processing with cpp is available, it could be programmed and used as shown below. This uses a feature of cpp (sadly unavailable in fpp), namely the ## token to concatenate tokens; thus, typename##_list expands typename, then appends _list, resulting (in this example) in the name mytype_list.

```
#define DEFINE_LIST_MODULE(typemodule, typename) \
  module typename##list_module; \
    use typemodule; \
    type typename##_list; \
      type(typename) value; \
      type(typename##_list), pointer :: next; \
    end type; \
    : ! (similarly to the include file)
  :
DEFINE_LIST_MODULE(mytype_module, mytype)
```

The macro version has the slight advantage that it works for intrinsic types as well as derived types, as type renaming in the use statement is not available for intrinsic types, and the slight disadvantage when cpp is being used in that the macro itself should not expand to more than 10 000 characters, as the expansion is always a single line.

22.3 Genericity in rank

Fortran 2023 provides syntax for declaring and accessing arrays where the syntax does not depend on the rank, but instead on the size of one or more vectors. The size is required to be constant, so this is compile-time, not run-time, genericity.

This makes it possible to write an entire procedure that processes arrays, and instantiate it (whether with an include file, macro processing, or any form of automatic code generation) for any particular desired rank.

We call this *rank-independent* syntax.

For complete genericity in rank, we need rank-independent syntax both for declarations and for usage. Whole-array operations are already rank independent, and Fortran 2023 provides such syntax for array elements and array sections.

22.3.1 Rank-independent array declarations

There are two new forms of array declarations that do not vary by the rank. They can be used to declare explicit-shape, deferred-shape, and assumed-shape arrays. There is no rank-independent version of assumed size. Assumed rank already accepts any rank, and is thus inherently rank independent.

When the operations to be performed on the arrays are already rank independent, that is, whole-array operations or elemental operations, providing rank-independent declarations is all that is required. Otherwise, the declarations are not very useful in themselves, but they are an essential first step.

Rank-independent deferred-shape (and assumed-shape)

The first form is the rank clause. This can be used to declare allocatable and pointer entities of any rank (including zero), and can also be used to declare dummy arguments that are scalar, or assumed-shape with the default value (1) for all lower bounds. It specifies the dimension attribute, and can be used in a type declaration statement or a component definition statement. It has the syntax

```
rank (scalar-int-constant-expr)
```

and the scalar integer constant expression must have a value that is a valid rank, from zero (for scalar) up to the maximum value supported by the processor.

For example,

```
real, rank (n), pointer :: r
```

declares r to be a pointer of type real and rank n; if n is equal to 3, this has the same effect as

```
real, dimension (:, :, :), pointer :: r
```

Rank-independent explicit-shape (and assumed-shape)

The second rank-independent declaration form is simply to allow the lower and/or upper bounds for all dimensions to be supplied by a supplementary array. Such an array is required to have constant size, but may have non-constant value (as with, for instance, automatic arrays). This syntax can be used in an *array-spec* or a dimension attribute; the syntax for the attribute is thus

```
dimension ([ lower-bounds-expr : ] upper-bounds-expr)
```

for explicit shape, or

```
dimension (lower-bounds-expr :)
```

for assumed shape, where at least one of the *lower-bounds-expr* or *upper-bounds-expr* is a vector. If one is a vector and the other is scalar, the scalar is broadcast to the shape of the vector.

For example,

```
integer, parameter, dimension(3) :: zeroes_3 = 0
real, dimension( [1,2,3] : [4,5,6] ) :: x
real :: y([-10, 10]:n), z(zeroes_3:)
```

has the same effect as

```
integer, parameter, dimension(3) :: zeroes_3 = 0
real, dimension( 1:4, 2:5, 3:6 ) :: x
real :: y( -10:n, 10:n ), z(0:, 0:, 0:)
```

This is actually useful outside of generic programming, as it makes it easy to declare a multi-dimensional array with the same shape, or even the same bounds, as another array, for example

```
subroutine sub(a,b)
  real :: a(:,:,:),b
  real(shape(a)) temp
  ⋮
```

Note that all of the lower or upper bounds must be supplied as a single array expression, that is, either the bounds are supplied individually by scalars, or all together as an array. This is not a functional limitation, however, as one may simply use an array constructor to extend the bounds, for example,

```
real :: y( [ shape(x),n ] )
```

declares an array y which has rank one greater than that of x, with the upper bound in the last dimension being n.

22.3.2   Rank-independent allocation

An allocate statement could already be rank independent when the array bounds were being taken from another array, using the source= or mold= specifiers. However, that is only available when the type and type parameters of the other array are the same as the array being allocated.

Thus, Fortran 2023 permits the bounds in an *allocation* to be specified by vectors, or by a vector and a scalar, exactly the same as when declaring an explicit-shape array. At least one of the vectors must have constant size, equal to the rank of the object being allocated. For example,

```
real, allocatable, dimension(:,:,:) :: x, y, z
integer :: lower(3), upper(3)
: ! Code that assigns values to lower and upper
allocate(x(:upper), y(lower:upper), z(0:upper))
```

Similarly, the lower bounds in a bounds-remapping pointer assignment, or the lower and/or upper bounds in a rank-remapping pointer assignment, can be specified by a vector instead of a comma-separated list of scalars. Just as in allocation, at least one of the vectors must have constant size, equal to the rank of the pointer being associated. For example, creating a pointer, with lower bounds zero, to an existing pointer whose bounds may be different, with rank given by the named constant n:

```
real, rank(n), intent(in), pointer :: arrayptr
real, rank(n), pointer :: zb_array
integer, dimension(rank(n)) :: zb_lower
zb_lower = 0
zb_array(zb_lower:) => arrayptr
```

An example of using this feature for rank-independent rank remapping is in Figure 22.2, in which the rank of the data is determined by the named constant n. This example establishes, in a rank-independent way, the multi-dimensional `array` as a pointer to a block of memory, and `diag` as a pointer to its diagonal elements.

Figure 22.2 Rank-independent code for rank-remapping pointer assignment.

```
real, target :: memoryblock(:)
real, rank(n), pointer :: array
real, rank(1), pointer :: hdiag
integer diagstep, datashape(n)
: ! Assign the desired shape of the data to the vector datashape.
array(1:datashape) => memoryblock
diagstep = 1 + sum([(product(datashape(1:i)),i=1,n-1)])
diag(1:minval(datashape)) => mydata(::diagstep)
```

22.3.3 Rank-independent array accessing

For whole-array operations, including elemental and conditional operations on the whole array, Fortran 90 was already rank independent.

For array element accesses, the same simple approach as for declarations (using a vector instead of a list of scalars) does not work, as that would be ambiguous with a vector subscript. Therefore, Fortran 2023 introduces new syntax which serves not only to disambiguate it for the compiler, but is also a valuable signal to the reader that something unusual is occurring. This new syntax uses the @ sign as a marker token.

Array elements

The basic idea is simple: to use an integer vector (array of rank one), preceded by the @ marker, as a list of subscripts. For example, in

```
a( @ idx )
```

each element of idx, which must be an integer vector of constant size, contributes a subscript. This is called a **multiple subscript**. The supplied vector may be an expression, but even so, must still have constant size.

In that example, the size of idx must be equal to the rank of a, but there is no requirement for the multiple subscript to be the only subscript in the array element. For example, in

```
a( @idx, j )
```

the vector idx provides the subscripts for an element of the jth hypercube of a; in this case the size of idx will need to be one less than the rank of a.

Here is a simple example to find the maximum value along the diagonal of a multidimensional array; when the lower bounds are all equal, the diagonal elements are the elements with all subscripts equal. The rank is provided by the named constant n, but the function itself is identical for every possible rank.

```
real function maxdiag(array)
  real, rank(n) :: array
  integer :: i, subs(n)
  maxdiag = -huge(array)
  do i=1, minval(shape(array))
    subs = i
    maxdiag = max(maxdiag, array(@subs))
  end do
end function maxdiag
```

Even though where constructs, whole-array operations, and elemental procedures (including impure elemental procedures) enable a lot of rank-independent functionality, it is sometimes necessary to traverse an entire array, doing something to each element, in a way that is not captured by the existing features.

Figure 22.3 shows an example of how to traverse a whole array, of any rank, in array element order. The traverse loop processes the whole array, using the process_vector loop to process each vector in the first dimension, and the advance loop to step through the subscripts for the higher dimensions.

Figure 22.3 Rank-independent traversal of a whole array.

```
integer :: i, j, subs(rank(array))
subs = lbound(array)
if (size(array)>0) then
  traverse: do
    process_vector: do i=lbound(array, 1), ubound(array, 1)
      subs(1) = i
      print *,'Processing element (',subs,') =',array(@subs)
    end do process_vector
    advance: do j=2, rank(array)
      subs(j) = subs(j) + 1
      if (subs(j)<=ubound(array,j)) exit advance
      subs(j) = lbound(array, j)
    end do advance
    ! If we did not exit the advance loop early,
    ! we have processed the whole array.
    if (j>rank(array)) exit traverse
  end do traverse
end if
```

Array sections

A multiple subscript can be used in an array section designator. It acts exactly the same as in an array element designator: it provides a list of scalar subscripts. For example,

```
array(@v1, :, @v2)
```

is a rank-one array section, along the dimension $size(v1)+1$. The rank of `array` must be equal to $size(v1)+size(v2)+1$.

More powerfully, there is also a **multiple subscript triplet,** which specifies a sequence of subscript triplets. The syntax of a multiple subscript triplet is

@ *[int-expr]* : *[int-expr] [* : *int-expr]*

where each *int-expr* is an integer expression, at least one *int-expr* appears, and at least one of them must be a vector. An *int-expr* that is scalar is broadcast to the size of the vector one. If more than one is a vector, it must have the same size as the other vectors. The first *int-expr* specifies the lower bounds, the second the upper bounds, and the third the stride, in the same way (and with the same defaults) as a normal subscript triplet.

As in a multiple subscript, the vector(s) of a multiple subscript triplet must have constant size.

For example,

```
array( @ : [ 2, 3, 5 ] )
```

specifies the same array section as

```
array( :2, :3, :5 )
```

More than one multiple subscript triplet may appear, so long as the sum of the number of subscripts, number of subscript triplets, size of multiple subscripts, and size of multiple subscript triplets, is equal to the rank of the base array. The rank of the section is equal to the number of subscript triplets and vector subscripts plus the sizes of the multiple subscript triplets.

A simple example is remote access to a whole-array coarray on another image; the syntax for whole-array operations requires the array section to be explicit, that is, a bare array name followed by an image selector, for example, `array [`*cosubscripts*`]`, is not permitted. This is, however, easy to do in a rank-indendent manner by using a multiple subscript triplet:

```
array ( @ lbound(a) : ) [ cosubscripts ]
```

A more elaborate example would be processing an array as a set of independent hypercubes:

```
associate(lb=>[ (lbound(a,i),i=1,rank(a)-1) ])
  do concurrent (i=lbound(a,rank(a)):ubound(a:rank(a)))
    associate(sec=>a(@lb:,i))
      : ! process the hypercube described by sec
    end associate
  end do
end associate
```

22.4   Future directions

Generic programming remains a hot topic in Fortran standardization, with proposals for templates, macro processing, and additional rank-independent features all being considered for the next revision of the standard.

23. Other Fortran 2023 enhancements

23.1 Introduction

This chapter describes the Fortran 2023 extensions to Fortran 2018, other than those tailored for generic programming.

23.2 Language elements

23.2.1 Longer lines and overall statement length

The maximum length of a free-form source line has been increased from 132 characters of default kind to 10 000 characters of any kind. There is no longer any limit to the number of consecutive continuation lines; however, the total length of a single statement is limited to one million characters. The increase in the length of a line is long overdue; the new limits are very high to support simplistic methods of mechanical program generation.

In the past, some compilers silently truncated source lines to the standard length. Therefore the Fortran 2023 standard explicitly requires diagnostic messages, which may be warnings or errors, if these limits are exceeded.

23.2.2 Automatic allocation of lengths of character variables

When a deferred-length allocatable variable is defined by an intrinsic assignment, as in the example

```
character(:), allocatable :: quotation
   :
quotation = 'Now is the winter of our discontent.'
```

it is allocated by the processor to the correct length. This behaviour is now extended to scalar allocatable deferred-length character variables in `iomsg=` and `errmsg=` specifiers, as internal files in `write` statements, and as actual arguments to intrinsic procedures such as `get_command`. Note that writing to a deferred-length character array as an internal file is unchanged; the array must be allocated to the desired length before the `write` statement.

It should be noted that this changes the semantics of existing programs that used preallocated deferred-length allocatable variables in such contexts, as they may now be

Modern Fortran Explained, 3rd Edition. M. Metcalf, J. Reid, M. Cohen, and R. Bader. Oxford University Press (2024). © M. Metcalf, J. Reid, M. Cohen, and R. Bader (2024). DOI 10.1093/oso/9780198876571.001.0023

reallocated to be a different length. The automatic length reallocation was viewed as so desirable that this minor incompatibility was considered to be acceptable.

If automatic reallocation of a deferred-length allocatable variable in these contexts is not desired, substring notation selecting the whole of the variable can be used. For example,

```
character(:), allocatable :: msg
allocate(character(256) :: msg)
   :
   :
write (msg, '(a)') 'Now is the winter of our discontent'
```

will reallocate msg to length 35, whereas

```
write (msg(:), '(a)') 'Now is the summer of our happiness'
```

will leave msg at its allocated length (256), with blank padding.

23.2.3   Conditional expressions and arguments

Conditional expressions, expressions whose value is one of several alternatives, are added. A simple example is

```
value = ( a>0.0 ? a : 0.0 )
```

which is a short way of writing

```
if (a>0.0) then
   value = a
else
   value = 0.0
end if
```

The general form of a conditional expression is

(*condition* ? *expression* [: *condition* ? *expression*] ... : *expression*)

where each *condition* is a scalar logical expression, and *expression* has the same declared type, kind type parameters, and rank. During execution, each *condition* in succession is evaluated until either

- one with the value true is found, in which case no further *condition* is evaluated and the conditional expression takes the value of the following *expression*, or
- all are found to be false, in which case the conditional expression takes the value of the final *expression*.

A conditional expression is a primary, just as is an expression in parentheses, and can be used in the same way as any other primary. In particular, it may be used as an *expression* in a conditional expression, that is, nesting is permitted. It is polymorphic if and only if one or more *expressions* are polymorphic. Its dynamic type is the dynamic type of the chosen expression. Its declared type, kind type parameters, and rank do not depend on which expression is chosen, because they are required to be the same for each expression.

A simple use is the case where an array element is to be tested for being non-negative, but not if the subscript is out of range:

```
if (( n<1 .or. n>size(a) ? .false. : a(n)<0)) stop 'Negative element'
```

Note that the parentheses are part of the syntax of the conditional expression. Requiring these avoids ambiguity, making it clear that the question mark and colon are not operators, and thus do not need to have any operator priority.

For actual arguments, there is a need to allow for variables and absent arguments, so a conditional expression is not itself permitted to be an actual argument, and instead there is a conditional argument. This has the syntax

(*condition* ? *consequent* [: *condition* ? *consequent*] ... : *consequent*)

where each *consequent* is an expression, a variable, or the token .nil. to specify absence. On execution, as for a conditional expression, the *condition*s are evaluated in turn. If one that is true is found, its *consequent* is used and no remaining *condition* is evaluated; otherwise, the final *consequent* is used. Nesting of conditional arguments is not allowed, that is, a *consequent* is not permitted to be a conditional argument, but it may be a conditional expression.

We will refer to a *consequent* that is an expression or a variable as a **consequent argument**. Each consequent argument in a conditional argument must have the same declared type and kind type parameters. Each must have the same rank or each must be of assumed rank. Each must satisfy all the requirements for being an actual argument corresponding to its dummy argument, apart from those that can vary at run time. For example, each must be a variable if the dummy argument has intent out or inout.

To ensure that the resolution of a generic call does not depend on which consequent argument is chosen, if any consequent argument in a conditional argument has the allocatable or pointer attribute, they must all have that attribute.

Here is an example using conditional arguments in a procedure reference:

```
call sub((x>0? x : y>0? y : z), (edge>0? edge : mode==3? 1.0 : .nil.))
```

where the interface of sub is

```
subroutine sub(x, bnd)
  real, intent(inout) :: x
  real, intent(in), optional :: bnd
```

23.2.4 More use of binary, octal, and hexadecimal constants

The use of binary, octal, and hexadecimal ('boz') constants is very limited in Fortran 2018. They are allowed only in data statements, and as actual arguments to type conversion and bit manipulation intrinsic functions. A 'boz' constant is now allowed

- in the initialization of a named constant or variable of type integer or real,
- as the right-hand side of an intrinsic assignment where the variable is integer or real,
- as a value in an array constructor with a *type-spec* that specifies integer or real, or
- as an integer value in an enum constructor (Section 23.8.3).

Where the destination (constant, variable, etc.) is of type real, the 'boz' constant is converted as if by the intrinsic real, and otherwise as if by the intrinsic int. In the real case, the 'boz' constant must be a valid representation for the specified kind of real, the same as when the intrinsic `real` is explicitly used. For example,

```
integer :: byte_2_mask = z'ff00'
```

has the same effect as

```
integer :: byte_2_mask = int(z'ff00')
```

and

```
real :: badx = z'ff00ff00ff00ff00ff00'
```

is invalid if the default real kind has less than ten bytes, as the bits truncated from the left are not all zero.

23.3   Intrinsic procedures and intrinsic modules

23.3.1   Extracting tokens from a string

Two intrinsic subroutines, both simple (see Section 23.7.1), have been added to facilitate the extraction of tokens from a string. For example, if the separator characters are spaces and semicolons, the string

```
"one two three;"
```

is taken to contain the three tokens one, two, three, and a fourth token of zero length between the semicolon and the end of the string. The tokens can be extracted one at a time by the positions of the separators or all at once by an array of tokens or by arrays of starting and ending positions of tokens. No allocatable argument is permitted to be a coarray or a coindexed object. Note that there are two overloaded forms of `tokenize`.

call split (string, set, pos [,back])   updates the integer pos to the position of the next (or previous) separator in string.

> **string**   is a scalar character object with intent in.
>
> **set**   is a scalar character object with intent in. It has the same kind value as string and holds a set of separator characters.
>
> **pos**   is a scalar integer variable with intent inout. If back is absent or is present with the value false, pos must have a value in the range $0 \leq \text{pos} \leq \text{len(string)}$ on entry and is given the position of the first separator in string after position pos or len(string)+1 if there is no such separator. If back is present with the value true, pos must have a value in the range $1 \leq \text{pos} \leq \text{len(string)}+1$ on entry and is given the position of the last separator in string before position pos or 0 if there is no such separator.
>
> **back**   is an optional scalar object of type logical and intent in.

Figure 23.1 shows how to use `split` to break a string into tokens.

Figure 23.1 Using `split` to tokenize a string.

The variable containing the text to be tokenized is called `line`.

```
integer :: pos1, pos2
pos1 = 0
pos2 = 0
do
   call split(line, ',', pos2)
   print *,'Token "',line(pos1+1:pos2-1),'"'
   if (pos2>len(line)) exit
   pos1 = pos2
end do
```

If `line` is equal to `'one,two,,three'`, this will print

```
Token "one"
Token "two"
Token ""
Token "three"
```

---

call tokenize (string, set, tokens [, separator])   finds all the tokens in a string.

> **string** is a scalar character object with intent `in`.
>
> **set** is a scalar character object with intent `in`. It has the same kind value as `string` and holds a set of separator characters.
>
> **tokens** is an allocatable deferred-length character array of rank one, with the same kind as `string`. It has intent `out`, and is allocated by `tokenize` to have character length equal to the length of the longest token, lower bound one, and upper bound the number of tokens in `string`. It is assigned the values of the tokens, in order.
>
> **separator** is an allocatable deferred-length character array of rank one, with the same kind as `string`. It has intent `out`, and if present, is allocated by `tokenize` to have character length one, lower bound one, and upper bound the number of separators in `string`. The elements are assigned the values of the separators, in order.

This form tokenizes the whole string in one call, but when blank is not a separator character, it cannot distinguish between tokens with varying numbers of trailing blanks. An example showing the use of this form of tokenize is in Figure 23.2.

call tokenize (string, set, first, last)   finds the starts and ends of all the tokens in a string.

> **string** is a scalar character object with intent `in`.
>
> **set** is a scalar character object with intent `in`. It has the same kind value as `string` and holds a set of separator characters.

Figure 23.2 Using the first form of `tokenize` to tokenize a string.

The variable containing the text to be tokenized is called `line`.

```
character(:), allocatable :: tokens(:)
integer :: i
call tokenize(line, ',', tokens)
print '(a)', (' Token "', trim(tokens(i)), '"', i=1, size(tokens))
```

If `line` is equal to `'one,two,,three'`, this will print

```
Token "one"
Token "two"
Token ""
Token "three"
```

---

first is an allocatable integer array of rank one and intent `out`. It is allocated by `tokenize` to have lower bound one and upper bound the number of tokens in `string`. The elements are assigned the starting positions of the tokens, in order. The position of a zero-length token is that of the separator that terminated it, or `len(string)+1` if it is terminated by the end of the string.

last is an allocatable integer array of rank one and intent `out`. It is allocated by `tokenize` to have lower bound one and upper bound the number of tokens in `string`. The elements are assigned the ending positions of the tokens, in order. For a token of zero length, the ending position is one less than its starting position.

For each token except the last, the separator character that terminated token number i is equal to `string(last(i)+1:last(i+1))`.

Figure 23.3 shows the use of this form of `tokenize` to tokenize a string.

Figure 23.3 Using the second form of `tokenize` to tokenize a string.

The variable containing the text to be tokenized is called `line`.

```
integer, allocatable :: from(:), to(:)
integer :: i
call tokenize(line, ',', from, to)
print '(a)', (' Token "', line(from(i):to(i)), '"', i=1, size(from))
```

If `line` is equal to `'one,two,,three'`, this will print

```
Token "one"
Token "two"
Token ""
Token "three"
```

23.3.2 Trigonometric functions that work in degrees

The following are new elemental functions that evaluate elementary mathematical functions for real arguments, working in degrees rather than radians. Each result is real with the kind type parameter of the first argument. They have been added because they have been widely implemented as extensions to previous standards, and are widely used.

acosd (x) returns the arc cosine (inverse cosine) function value for a real value x such that $|x| \leq 1$, expressed in degrees in the range $0 \leq \text{acosd}(x) \leq 180$.

asind (x) returns the arc sine (inverse sine) function value for a real value x such that $|x| \leq 1$, expressed in degrees such that $-90 \leq \text{asind}(x) \leq 90$.

atand (x) returns the arc tangent (inverse tangent) function value for a real value x, expressed in degrees in the range $-90 \leq \text{atand}(x) \leq 90$.

atand (y, x) returns the arc tangent (inverse tangent) function value in degrees for a pair of real values, x and y, with the same kind type parameter. They must not both be zero. The result is expressed in degrees in the range $-180 \leq \text{atand}(y,x) \leq 180$. It has a value equal to a processor-dependent approximation to $\text{atan}(y,x) \times 180/\pi$.

atan2d (y, x) is the same as $\text{atand}(y, x)$.

cosd (x) returns the cosine function value for real values x in degrees.

sind (x) returns the sine function value for real values x in degrees.

tand (x) returns the tangent function value for real values x in degrees.

23.3.3 Trigonometric functions that work with half revolutions

The following are elemental functions that evaluate elementary mathematical functions for real arguments, working in half revolutions (180 degrees, or π radians). Each result is real with the kind type parameter of the first argument. They have been added because they are mentioned in the IEEE standard for floating-point arithmetic and have been implemented as extensions to previous standards.

acospi (x) returns the arc cosine (inverse cosine) function value for a real value x such that $|x| \leq 1$, expressed in half revolutions in the range $0 \leq \text{acospi}(x) \leq 1$.

asinpi (x) returns the arc sine (inverse sine) function value for a real value x such that $|x| \leq 1$, expressed in half revolutions in the range $-0.5 \leq \text{asinpi}(x) \leq 0.5$.

atanpi (x) returns the arc tangent (inverse tangent) function value for a real value x, expressed in half revolutions in the range $-0.5 \leq \text{atanpi}(x) \leq 0.5$.

atanpi (y, x) returns the arc tangent (inverse tangent) function value in half revolutions for a pair of real values, x and y, with the same kind type parameter. They must not both be zero. The result is expressed in half revolutions in the range $-1 \leq \text{atanpi}(y,x) \leq 1$. It has a value equal to a processor-dependent approximation to $\text{atan}(y,x)/\pi$.

atan2pi (y, x) is the same as atanpi(y, x).

cospi (x) returns the cosine function value for real values x in half revolutions.

sinpi (x) returns the sine function value for real values x in half revolutions.

tanpi (x) returns the tangent function value for real values x in half revolutions.

23.3.4   The function selected_logical_kind

The function selected_logical_kind has been added to match the existing functions for choosing kind type parameter values.

selected_logical_kind (bits) is a transformational function that returns as a default integer scalar the value of a kind type parameter of a logical data type whose storage size in bits is at least bits, or -1 if none is available. If there is more than one such kind, the one with the smallest storage size is returned; if there are several such kinds, the smallest of their kind type parameter value is returned.

23.3.5   Changes to system_clock

In Fortran 2018, users are free to use integers of any kind as actual arguments for system_clock. It was intended that this would allow the use of long integers for accurate timing on modern hardware with fast clocks. However, this left vendors unclear about how to accommodate short integers – some provided an imprecise clock, while others produced compilation errors. There was also uncertainty over the effect of disagreement in kinds within a single call.

These problems are now remedied by requiring that all integer arguments in a single call have the same kind. Furthermore, their decimal exponent range is required to be at least that of default integer, and implementations are recommended to support long integers (decimal exponent range at least 18). The processor is permitted to provide more than one clock. Which clock is referenced depends on the kind of the integer arguments. Whether an image has no clock, has one or more clocks of its own, or shares a clock with another image is processor dependent.

The standard recommends that all calls to system_clock use integer arguments with a decimal exponent range of at least 18; that is, 64-bit integers. This maximizes the precision of the clock whilst minimizing the risk of wraparound.

23.3.6   Changes for conformance with new IEEE standard

Changes have been made to the intrinsic module ieee_arithmetic for conformance with the new IEEE standard, ISO/IEC 60559:2020.

Four elemental functions have been added for the IEEE operations of maximum, maximumMagnitude, minimum, and minimumMagnitude:

```
ieee_max (x, y)        ieee_min (x, y)
ieee_max_mag (x, y)    ieee_min_mag (x, y)
```

where x and y are real with the same kind type parameter. If ieee_support_datatype is false for that kind, the functions are defined but they must not be invoked.

If either x or y is a NaN, the result is a quiet NaN; otherwise, the result is the value of the argument that has the indicated maximum, maximum magnitude, minimum, or minimum magnitude, respectively, where positive zero is treated as being greater than negative zero. If x or y is a signaling NaN, ieee_invalid signals; otherwise, no exception is signaled.

The four elemental functions ieee_max_num, ieee_max_num_mag, ieee_min_num, and ieee_min_num_mag now conform to the operations maximumNumber, maximumMagnitudeNumber, minimumNumber, and minimumMagnitudeNumber in ISO/IEC 60559:2020; the changes affect the treatment of zeros, which are as for ieee_max, ieee_max_mag, ieee_min, and ieee_min_mag (see previous paragraph), and also the treatment of NaNs: if either of x or y is a signaling NaN, ieee_invalid signals (whether the result is NaN or not).

23.3.7 Additional named constants to specify kinds

Additional named constants are available in the module iso_fortran_env for kind type parameter values of types of given storage sizes in bits:

logical8	8-bit logical
logical16	16-bit logical
logical32	32-bit logical
logical64	64-bit logical
real16	16-bit real

They are default integer scalars. If the compiler supports more than one kind with a particular size, which one is chosen is processor dependent. If the compiler does not support a kind with a particular size, that constant has a value of -2 if it supports a kind with a larger size, and -1 if it does not support any larger size.

23.4 Interoperability with C

23.4.1 Extension to the intrinsic procedure `c_f_pointer`

An extra optional argument lower has been added at the end of the argument list of the subroutine c_f_pointer from the intrinsic module iso_c_binding. It has intent in and is an integer array of rank one with the same size as the argument shape; it can be present only if shape is present. If present, it specifies the lower bounds of the argument fptr, which otherwise all have the value one.

This makes c_f_pointer as capable as pointer assignment, which allows the lower bounds of an array pointer to be specified. For example, in

```
real, dimension(10, 10), target :: x
real, dimension(:, :), pointer :: xp, yp
type(c_ptr) :: y
y = c_loc(x)
xp(-10:,20:) => x
call c_f_pointer(y, yp, shape(x), [ -10, 20 ])
```

the pointers xp and yp have not only the same targets, but the same bounds.

23.4.2   Procedures for converting between Fortran and C strings

The procedures described here have been added to the intrinsic module iso_c_binding to facilitate passing strings between C functions and Fortran procedures. If the processor does not support the C character kind (that is, c_char is equal to -1), these generic names are still in the module, but have no specific procedure. Note that there are two overloaded forms of c_f_strpointer.

f_c_string (string [, asis]) is a simple transformational function; string is an intent in character scalar of kind c_char, and asis is an optional intent in logical scalar. It returns a character scalar of the same type and kind as string with a value of string//c_null_char if asis is present with the value true, and trim(string)//c_null_char otherwise.

For example,

```
character(256) buf
  ⋮
n = strlen(f_c_string(buf))
```

where strlen is the string length function from the standard C library, will assign the same value to n as

```
n = len_trim(buf)
```

call c_f_strpointer (cstrarray, fstrptr [, nchars]) is an impure subroutine with arguments:

cstrarray is an intent in rank-one character array of kind c_char and character length one. Its actual argument must be simply contiguous and have the target attribute.

fstrptr is an intent out scalar deferred-length character pointer of kind c_ptr. It becomes associated with the leading elements of cstrarray, its character length *len* being set to the largest value which results in no C null character appearing in the sequence, and which is also less than nchars if nchars is present, and less than the size of cstrarray otherwise.

nchars is an optional integer scalar with intent in. If cstrarray is assumed size, nchars must be present. If nchars is present, its value must be non-negative and not greater than the size of cstrarray.

For example, if cfun is a C function with the interface

```
interface
  subroutine cfun(buffer, bufferlen) bind(c)
    use iso_c_binding
    character, intent(out) :: buffer(*)
    integer(c_size_t), value :: bufferlen
  end subroutine
end interface
```

and it writes a null-terminated string to buffer (with a maximum of bufferlen characters), then the invocation

```
character, target :: buffer(1000)
character(:), pointer :: result
call cfun(buffer, int(size(buffer), c_size_t))
call c_f_strpointer(buffer, result)
```

will associate result with the data written by cfun.

This is not as useful as it appears, as one could simply have passed a large scalar character variable instead of an array, and searched within that for a C null character, obviating the need for the target attribute and a pointer. For example,

```
character(1000) buffer
integer resultlen
call cfun(buffer, int(len(buffer), c_size_t))
resultlen = index(buffer, c_null_char) - 1
if (resultlen<0) resultlen = len(buffer)
! The data returned by cfun is now result(:resultlen)
```

call c_f_strpointer (cstrptr, fstrptr, nchars) is an impure subroutine with arguments:

cstrptr is a scalar of type c_ptr. It is an intent in argument holding the C address of a contiguous array *S* of nchars characters. Its value must not be the C address of a Fortran variable that does not have the target attribute.

fstrptr is an intent out deferred-length character pointer of kind c_char. It becomes pointer associated with the leftmost characters (in array element order) of the array *S*, its length being set to the largest value for which no C null character appears in fstrptr and which is also less than or equal to nchars.

nchars is an intent in integer scalar with a non-negative value.

This is particularly useful for dealing with those C functions that return a pointer to a null-terminated character string, such as getenv, as the Fortran interface for such a function has no choice but to declare the return type as c_ptr. For example:

```
interface
  function getenv(name) bind(c)
    use iso_c_binding
    character, intent(in) :: name(*)
    type(c_ptr) :: getenv
  end function
end interface
character(:), pointer :: envval
type(c_ptr) :: envres
  ⋮
envres = getenv(f_c_string('HOME'))
if (c_associated(envres)) then
  call c_f_pointer(envres, envval, 32768)
  print *, 'HOME is ', envval
else
  print *,'No HOME'
end if
```

23.5   Input/output

23.5.1   The `at` edit descriptor

The `at` edit descriptor has been added to provide a convenient way to output a character string omitting trailing blanks, as if the intrinsic function `trim` had been applied to the output item. It is not available for input.

For example, in

```
character(132) :: line
  ⋮
print '("Line is: ",at)', line
```

the effect of the `print` statement is identical to that of

```
print '("Line is: ",a)', trim(line)
```

23.5.2   Control over leading zeros in output of real values

Whether a leading zero appears before the decimal symbol in output with the f, e, and g edit descriptors is now controlled by the **leading-zero mode**. If the mode is `print`, leading zeros are printed. If the mode is `suppress`, leading zeros are suppressed. If the mode is `processor_defined` (the default), it is processor dependent whether leading zeros are printed or suppressed. For example, printing the value -0.5 with format f6.2 will print b-0.50 with mode `print`, bb-.50 with mode `suppress`, and either one will be printed with mode `processor_defined`.

The leading-zero mode may be specified by the `leading_zero=` specifier on an `open` statement, which can take the value `print`, `suppress`, or `processor_defined`. If no `leading_zero=` specifier appears, or for output to an internal file, the leading-zero mode is `processor_defined`.

The leading-zero mode for a connection may be overridden by a `leading_zero=` specifier in a `write` statement with one ofthese values.

There is a corresponding specifier in the `inquire` statement.If the connection is for formatted input/output, the specifier is assigned the value `PRINT`, `SUPPRESS`, or `PROCESSOR_DEFINED`, as appropriate; otherwise, it is assigned the value `UNDEFINED`.

The leading-zero mode may be temporarily changed within an output statement to `print`, `suppress`, or `processor_defined` by the `lzp`, `lzs`, or `lz` edit descriptor, respectively. An `lzp`, `lzs`, or `lz` will remain in force for the remainder of the format specification, unless another `lzp`, `lzs`, or `lz` edit descriptor is met. These edit descriptors are permitted, but have no effect, during the execution of an input statement.

23.5.3   Namelist with private variables

In Fortran 90, a public namelist group was not allowed to have a private variable as a group object. This was for safety – preventing the user from inadvertently permitting access to the contents of a private variable through namelist input/output.

This was occasionally inconvenient, for example, a module could not simply export a namelist to permit save/restore of a group of settings if any of the settings was private to the module. Thus this restriction is lifted in Fortran 2023, as it was considered the reasonable uses outweighed any minor affect on safety.

For example, with the module

```
module settings_m
   integer, public :: a = 1, b = 2
   integer, private, protected :: c = 3
   namelist /settings_n/ a, b, c
   ⋮
end module
```

a program unit that uses the module can use the `settings_n` namelist in a `write` statement to save or display the settings, but because c has the `protected` attribute, cannot use `settings_n` in a `read` statement to directly modify them.

23.6   Coarrays

23.6.1   An array or allocatable object may have a coarray component

For simplicity, Fortran 2018 requires that a variable or component that is of a type with a coarray ultimate component be a non-pointer non-allocatable non-coarray scalar. Fortran 2023 permits such a variable or component to be an array and/or allocatable, and thus arrays of coarrays, or dynamically created structures with coarray components, are possible.

This can happen at any depth of component selection, so an array or allocatable object can have a potential subobject component that is a coarray. Most of the restrictions on types or objects with a coarray ultimate component are generalized to apply to ones with a coarray potential subobject component, as there may now be an intervening non-coarray allocatable component between the base object and the coarray component.

For example, given the type

```
type t
    real, allocatable, codimension[:], dimension(:) :: data
end type
```

the variable

```
type(t) x(100)
```

is an array with 100 allocatable coarray components. A list of coarrays can also be created using this new facility, as shown in Figure 23.4.

Figure 23.4 Allocatable list of coarrays.

A type, and object, for a list of allocatable coarrays:

```
type covalue
    real, allocatable, codimension[:], dimension(:) :: data
    type(covalue), allocatable :: next
end type
type(covalue), allocatable :: co_list
```

A procedure that displays the list:

```
recursive subroutine show_covalues(covlist)
    type(covalue), allocatable, intent(in) :: covlist
    if (allocated(covlist)) then
      if (allocated(covlist%data)) then
        print *, 'data:',covlist%data
      else
        print *, 'no data:'
      end if
      if (allocated(covlist%next)) then
        print *, 'next...'
        call show_covalues(covlist%next)
      end if
    end if
end subroutine
```

An `allocate` statement for an object with a coarray potential subobject component is not an image control statement, that is, it involves only the executing image, because the coarray is unallocated and no image is able to reference it. Fortran 2018 has the constraint that an `allocate` statement must not have a `source=` specifier that is an object with a coarray ultimate component, because the need to copy the coarray if it is allocated would require the statement to be an image control statement. This restriction is retained, generalized to an object with a coarray potential subobject component. However, the corresponding constraint for a *type-spec* or mold= specifier is not needed and is removed, except for the case where the object is unlimited polymorphic (see the paragraph after next).

In an intrinsic assignment for an object with a coarray potential subobject component, the component is subject to the rules of intrinsic assignment for a coarray so the statement is not an image control statement.

On the other hand, a deallocate statement for an object of a type that has a coarray potential subobject component involves the deallocation of all its coarray subobjects. It therefore needs to be an image control statement and involve synchronization of all the images in the current team. To make sure that an unlimited polymorphic object does not get a dynamic type with a coarray potential subobject component, thus causing its deallocate statement to become an image control statement, an allocate statement that allocates an unlimited polymorphic object is not permitted to have a *type-spec* that specifies a type with a coarray potential subobject component, or a mold= specifier with an object of such a type.

An allocate statement for a coarray component remains an image control statement and involves image synchronization. For an allocate or deallocate statement of a coarray component, additional conditions must be satisfied to ensure that the coarrays on each image properly correspond. On each active image of the current team, if the coarray is an ultimate component of:

- a dummy argument, the ultimate arguments must be declared with the same name in the same scoping unit, and, if the ultimate argument is an unsaved local variable of a recursive procedure, the depth of recursion of that procedure must be the same;
- an array element, the element must have the same position in array element order;
- an unsaved local variable of a recursive procedure, the depth of recursion of that procedure must be the same.[1]

Copy-in copy-out needs to be avoided for a dummy argument that is a non-allocatable array of a type with a coarray potential subobject component. Here, the actual argument is required to be simply contiguous or an element of a simply contiguous array.

23.6.2 Put with notify

Put with notify is an efficient synchronization technique for data transfers between images. The basic idea is to combine a put (definition of a variable on a different image) with a notification mechanism that allows the receiving image to know that the data has arrived. It is especially efficient if network hardware can ensure that the data arrive before the notification occurs.

The derived type notify_type is defined by the intrinsic module iso_fortran_env. It is an extensible type with private components, and all non-allocatable components are default-initialized. A scalar variable of this type is a **notify variable**.

Here is a simple example of the new feature:

---

[1] This also clarifies the behaviour for the existing Fortran 2018 functionality.

```
use iso_fortran_env
type(notify_type) nx[*]
:
me = this_image()
if (me == 1) then
   x[10, notify=nx] = y
else if (me == 10) then
   notify wait (nx, until_count=1)
   z = x
end if
```

Here, nx is a notify variable. On every image it contains a count that has the initial value zero. After an image has assigned its value of y to x on image 10, the count value of nx[10] is incremented by one. Image 10 waits until the count of its nx is at least one, then decrements the count by one and continues execution.

The value of a notify variable includes its notify count, which is default-initialized to zero, and can be altered only by

- execution of an intrinsic assignment statement whose left-hand-side variable has an image selector that selects the image of the notify variable and has a notify= specifier that specifies the notify variable, or
- execution of a notify wait statement on the image of the notify variable.

The intrinsic assignment statement increases the count by one. The notify wait statement has a threshold value that is its until_count specifier if it appears, or one otherwise, and waits until the count is at least the threshold value and then decreases the count by the threshold value. Here is an example:

```
use iso_fortran_env
type(notify_type) nx[*]
:
me = this_image()
if (me <= 4) then
   x(me)[10, notify=nx] = y
else if (me == 10) then
   notify wait (nx, until_count=4)
   z(1:4) = x(1:4)
end if
```

The effect of each update of a notify variable is as if the intrinsic subroutine atomic_add were executed for a variable that stores the notify count and has type integer with kind atomic_int_kind. The rule on referencing variables in unordered segments does not apply to the coarray that is assigned a value in the assignment statement being referenced on its own image after the notify statement, provided its value cannot meanwhile have been altered in another way.

To ensure that the counts of notify variables are initialized to zero and not altered except as explained in the previous paragraphs, the only additional ways a notify variable is permitted to appear in a variable definition context are in an `allocate` statement without a `source=` specifier, in a `deallocate` statement, or as an actual argument corresponding to a dummy argument with intent `inout` in a reference to a procedure with an explicit interface. Furthermore, a variable with a non-pointer subobject of type `notify_type` is permitted to appear in a variable definition context only in an `allocate` statement without a `source=` specifier, in a `deallocate` statement, or as an actual argument corresponding to a dummy argument with intent `inout` in a reference to a procedure with an explicit interface.

An entity with declared type `notify_type`, or which has a non-coarray potential subobject component with declared type `notify_type`, must be a coarray or a data component.

The `notify wait` statement is not an image control statement. The general form is

notify wait (*notify-variable [, event-wait-spec-list]*)

where an *event-wait-spec* is an `until_count=` specifier, `stat=` specifier, or `errmsg=` specifier.

23.6.3 Error conditions in collectives

The Fortran standard defines a collective subroutine as an 'intrinsic subroutine that performs a calculation on a team of images without requiring synchronization'. For performance reasons, it was intended that once an image has completed its part of the calculation it should be free to execute other statements without waiting for the other images to do so. Fortran 2018 did not recognize that a collective might be successful on one image, while encountering an error condition on another image. Fortran 2023 acknowledges this and allows different images to have different error conditions, or for only a subset of images to receive an error condition.

For example, the source image on a `co_broadcast` might simply send a message containing the value to all other images, and continue execution without waiting for any response.

23.7 Procedures

23.7.1 Simple procedures

A pure procedure may change variables outside its scope only through its arguments. This allows it to be used in parallel constructs, where concurrency issues would otherwise prevent its use. A simple procedure is a pure procedure with the additional restriction that it may refer to variables outside its scope only through its arguments. It represents an entirely local calculation. This allows it to be executed in parallel in an array assignment statement that assigns to an array that might be visible from a pure procedure.

All the intrinsic functions are simple. All the module functions in all of the intrinsic modules are simple. The intrinsic subroutines `mvbits`, `split` (Section 23.3.1), and `tokenize` (Section 23.3.1) are simple. The intrinsic subroutine `move_alloc` is simple when the argument `from` is not a coarray. The module subroutines

```
ieee_get_flag                ieee_set_flag
ieee_get_halting_mode        ieee_set_halting_mode
ieee_get_modes               ieee_set_modes
ieee_get_rounding_mode       ieee_set_rounding_mode
ieee_get_status              ieee_set_status
ieee_get_underflow_mode      ieee_set_underflow_mode
```

are simple. The elemental operators `==` and `/=` for values of type `ieee_class_type` or `ieee_round_type` are simple.

A procedure is specified as simple by using the keyword `simple` in the prefix of its `function` or `subroutine` statement, for example:

```
real simple elemental function f(a, b)
   real, intent(in) :: a, b
   f = min(cosh(a), sinh(b))
end function
```

Of course, if `simple` is specified, neither `pure` nor `impure` may be specified. A simple procedure automatically has the property of being pure. A simple procedure may also be elemental. A dummy procedure or a procedure pointer may be specified to be simple. A type-bound procedure is simple if it is bound to a simple procedure or, if it is deferred, its interface specifies it to be simple; a simple type-bound procedure can only be overridden by another simple type-bound procedure.

A simple procedure must satisfy all the requirements of a pure procedure. In addition,

- it must not reference a variable by use or host association,[2]
- all its dummy procedures must be simple,
- all its internal procedures must be simple,
- all procedures it references must be simple, and
- when used in a context that requires it to be simple, its interface must be explicit and specify that it is simple.

23.7.2   Reduction specifier for do concurrent

A named variable may be declared in a `do concurrent` construct to have reduce locality within it, for example,

```
real :: a, b, x(n)
   ⋮
a = 0.
b = -huge(b)
do concurrent (i = 1, n) reduce(+:a) reduce(max:b)
   a = a + x(i)**2
   b = max(b,x(i))
end do
```

---

[2]Also, it must not contain an `entry` statement (Appendix B.1.12) or reference a variable in a common block (Appendix B.1.7).

allows the computations $\sum_{i=1}^{n} x_i^2$ and $\max_{i=1}^{n} x_i$ to be parallelized. We will refer to a variable that computes such a result as a **reduction variable**. It must be a named variable. Its name and the operator or function name must be specified in a `reduce` clause on the `do concurrent` statement.

A named variable with `reduce` locality must be of an intrinsic type that is suitable for its operator or function, and permitted to appear in a variable definition context. It must not be a coarray or an assumed-size array. It must not be asynchronous or volatile. It must not be an optional argument.

A reduction variable is allowed to appear within its `do concurrent` construct only in an intrinsic assignment of one of the forms *var = var op expr* and *var = expr op var* where *op* is one of the intrinsic operators +, *, .and., .or., .eqv., or .neqv., or as the left-hand side of an intrinsic assignment whose right-hand side is one of the intrinsic functions `max`, `min`, `iand`, `ieor`, or `ior` with the reduction variable as an argument. All occurrences in the construct must have the same form.

The effect is as if each iteration has a separate reduction variable with exactly the same properties as its original, the 'outside' variable. If the outside variable is allocatable it must be allocated. If it is a pointer, it must have a target. The inside variable does not have the `allocatable`, `bind`, `intent`, `pointer`, `protected`, `save`, `target`, or `value` attribute, even if the outside variable does. Each inside variable is initialized at the start of execution of its iteration to

- 0 for +, `ieor`, or `ior`;
- 1 for *;
- all 1s for `iand`;
- .true. for .and. or .eqv.;
- .false. for .or. or .neqv.;
- the least representable value of the type and kind for `max`; and
- the largest representable value of the type and kind for `min`.

On termination of the `do concurrent` construct, the outside variable, or its target if it is a pointer, is updated by using the reduction operation to combine it with the values of all the inner variables. The updates may be performed in any order.

A `reduce` clause may be given a list of variables for a single reduction, and there may be any number of `reduce` clauses, as shown in

```
do concurrent (i = 1, n) reduce(+:a, b, c) reduce(max:d, e, f)
```

23.8 Enumerations

23.8.1 Introduction

Fortran 2023 supports two variants of enumeration types. One variant extends the pre-existing C-interoperability feature of `enum` and the other one provides Fortran-specific enumeration types with additional semantics that are incompatible with those of C-style enumerations. We start with the description of the Fortran-specific variant.

23.8.2   Enumeration types

An **enumeration type** is a user-defined type that does not define a structure (unlike a derived type), but has a finite number of values that form an ordered set.

A scalar variable of the type has one of those values (or is undefined). A simple example of using an enumeration type variable to control an action is:

```
enumeration type :: colour
    enumerator :: red, orange, green
end enumeration type
type(colour) light
   ⋮
if (light==red) ...
```

The possible values of an enumeration type are defined in the enumeration type definition, which has the general form

```
enumeration type [ [, access-spec ] :: enumeration-type-name
    enumerator [ :: ] enumerator-name-list
    [ enumerator [ :: ] enumerator-name-list ]...
end enumeration [ enumeration-type-name ]
```

The name of the type is *enumeration-type-name*, and each *enumerator-name* is a scalar named constant of that type. They have the same scoping rules as other type names and named constants. The order in which the enumerator names appear defines the ordering of the values; the **ordinal value** of an enumerator is the position of its name in the type definition, which is in the range 1 (for the first enumerator) up to the number of enumerators in the definition.

The *access-spec*, which is public or private, is only permitted in the specification part of a module, where it specifies the accessibility of the enumeration type name, and the default accessibility of its enumerators. The accessibility of an enumerator can be confirmed or overridden by a separate public or private statement. If the accessibility of an enumerator is not declared in these ways, its accessibility is the default for entities of the module.

An enumeration type is not an extensible type, and thus the class type specifier cannot be used – there are no polymorphic enumerations. An enumeration type is also not an interoperable type, as there is no corresponding concept in C. Other than that, anything that can have a type can have enumeration type; for example, variables (including coarrays), components, and functions can all be of an enumeration type.

Intrinsic assignment of an enumeration type is available only when the variable (left-hand side) and expression (right-hand side) are of the same enumeration type.

The intrinsic operations ==, /=, <, <=, >, and >= are provided for both operands being of the same enumeration type; the result is of type default logical. The inequalities use the ordering of the values, with each enumerator listed in the type definition comparing less than all subsequent enumerators. This is equivalent to the ordering of their ordinal values. No other intrinsic operations are provided for enumeration types.

A value of enumeration type can be created using an **enumeration constructor**; this takes an integer, and returns the enumeration value with that ordinal value. For example, colour(2) has the value orange. The integer must be a valid ordinal value (from 1 to the number of enumerators in the type). This allows the enumerators to be accessed in a do loop such as:

```
do i = 1,3
   light = colour(i)
      ⋮
end do
```

The general form of an enumeration constructor is

   type (*expression*)

where *expression* is a scalar integer expression. If *expression* is constant, the enumeration constructor is also a constant expression, and may appear in a data statement.

An array constructor with the usual syntax is available to construct an array of an enumeration type, for example, [red, (colour(i),i=2,3)]. The values must all be of the same enumeration type.

An enumeration type may be used in a select case construct, for example,

```
select case (light)
   case (red)
      ⋮
   case (orange:green)
      ⋮
end select
```

As with other types in select case, the test is equality (==) when there is no colon in the case range, and <= and/or >= when there is a colon in the case range.

The intrinsic function int is extended to be the inverse of the enumeration constructor; that is, if the a argument is of enumeration type, it returns its ordinal value.

The intrinsic function huge is extended to return the greatest value (last enumerator in the type definition) when its argument is of an enumeration type. There is no corresponding function to return the first enumerator, as that is straightforwardly available by using the integer value 1 in the enumeration constructor.

There are two new elemental intrinsic functions that are useful for traversing the sequence of values of an enumeration type.

next (a *[,* **stat** *])* returns the next enumerator after a, or a itself if it is equal to the last enumerator.

   a is of enumeration type and has intent in.

   stat is an optional intent out integer scalar with a decimal exponent range of at least four. If present, it is assigned a processor-dependent positive value if a is equal to the last enumerator of its type, and the value zero otherwise.

previous (a *[,* **stat** *])* returns the previous enumerator before a, or a itself if it is equal to the first enumerator.

   a is of enumeration type and has intent in.

stat is an optional intent out integer scalar with a decimal exponent range of at least four. If present, it is assigned a processor-dependent positive value if a is equal to the first enumerator of its type, and the value zero otherwise.

For either function, if stat is not present but would have been assigned a nonzero value if present, error termination is initiated. This is to minimize the chance of the programmer accidentally writing an infinite loop.

Using these functions, we can rewrite the earlier do loop example without needing to use an extra integer variable as a counter:

```
light = colour(1)
do
    ⋮ ! Process light
    if (light==huge(light)) exit
    light = next(light)
end do
```

An object of enumeration type is not permitted in list-directed or namelist input/output. In formatted input/output, it must correspond to an i, b, o, or z edit descriptor. For output of a value of enumeration type, its ordinal value is transferred. For input, the value must be positive and less than or equal to the number of enumerators in its type definition, and the enumerator with that ordinal value is transferred.[3]

23.8.3   Enum types

An interoperable enumeration (in Fortran 2018, just called an 'enumeration', Section 20.13) is a collection of enumerators, which are named integer constants with the same kind. That kind is the one that interoperates with the companion C processor's matching enumerated type.

This has the disadvantage that there is no type safety: in particular, a value of one interoperable enumeration can be passed to a dummy argument of a different enumeration, should the integer kinds be the same (which they usually are). It is also inconvenient to have to write integer(kind(*an-enumerator-name*)) to declare suitable variables.

Fortran 2023 allows the definition of an interoperable enumeration to create an **enum type**, that also interoperates with a C enumerated type. This is done by adding the new type name to the enum statement, for example,

```
enum, bind(c) :: season
    enumerator :: spring=5, summer=7, autumn, winter
end enum
```

defines the enum type season. This name can be used to declare variables of this type, for example:

```
type(season) my_season
```

---

[3]The reason there are strict requirements to use integer edit descriptors in Fortran 2023 is to make it easier to extend to use something else, e.g., enumerator names, in a future standard.

Nothing else about the interoperable enumeration is changed: the enumerators are still named integer constants. The name of the enum type follows the same scoping rules as other user-defined type names. The values of the enum type include equivalents to all the representable integer values of the kind of the enumeration, they are not limited to the listed enumerators.

Even though the enum type is not an integer type, we want to permit free usage of its enumerators in contexts that expect the enum type. This is done by the concept of **type conformance** of expressions: any expression that has one of the enumerators as a primary is permitted to be used instead of a value of the enum type. Thus, using the example above,

```
my_season = winter
my_season = winter + 137
```

are allowed, but

```
my_season = 137
```

is not. This principle applies to assignment, argument correspondance, initialization, and so on. In particular, different enum types are not in conformance, to preserve a modicum of type safety.

The **integer value** of an entity of enum type is the value it would have when passed to C. For example, `winter` has the value 9, and so after the second valid assignment above, the integer value is 146. This value is returned by the intrinsic function `int` when its argument a is of enum type. The reverse conversion, from integer to the enum type, is provided by the **enum constructor**, which has the form *enum-type-name* (*expr*), where *expr* is a scalar integer expression or a 'boz' constant; the value of *expr* must be representable in the integer kind of the enumeration. Thus, `season(summer)` is the enum type value corresponding to the enumerator `summer` (which is an integer named constant). An enum constructor whose *expr* is constant is a constant expression, and may appear in a `data` statement.

The intrinsic operations ==, /=, <, <=, >, and >= are available when both operands are objects of the same enum type, or one is of enum type and the other is an integer expression that is in type conformance with it. They compare the integer values. No other intrinsic operations are available for objects of an enum type.

An array constructor with the usual syntax is available to construct an array of an enum type, for example, `[spring, (season(i),i=2,4)]`. All the array elements specified must be of a single enum type or be integer expressions that are in type conformance with it.

Enum types can be used in a `select case` construct; if the expression in a `select case` statement is of enum type, the expressions in the `case` statements must be in type conformance with it. For example:

```
select case (my_season)
   case (:spring)
      ⋮
   case (summer)
      ⋮
   case (winter+1:)
      ⋮
end select
```

As with other types in `select case`, the test is equality (==) when there is no colon in the case range, and <= and/or >= when there is a colon in the case range.

For input/output, an effective item of enum type is treated as if it were of type integer, with the interoperating kind type parameter. Thus it can be used in namelist or list-directed input/output, and an explicit format can use the i, b, o, z, or g edit descriptor.

An enum type that is declared in a module has the `private` or `public` attribute that is the default for its module unless it is included in a `private` or `public` statement.

A. Deprecated features

A.1 Introduction

In this appendix we describe features that are still fully part of the language but are now redundant. The authors have chosen to place them here and not in the main part of the book, Chapters 2 to 23, because we deprecate their use. They might become obsolescent in a future revision, but this is a decision that can be made only within the standardization process. We note that this decision to group certain features into an appendix and to deprecate their use is ours alone, and does not have the actual or implied approval of either WG5 or J3. We include descriptions of how they may be replaced by better features.

A.2 The `include` line

It is sometimes useful to be able to include source text from somewhere else into the source stream presented to the compiler. This facility is possible using an `include` line:

    `include` *char-literal-constant*

where *char-literal-constant* must not have a kind parameter that is a named constant. This line must appear as a single source line where a statement may occur. However, it is not in itself a statement, and thus cannot have a statement label, nor may another statement appear on the same line separated by a semicolon, but it may have a trailing comment. It will be replaced by material in a processor-dependent way determined by the character string *char-literal-constant*, which usually refers to a file. The included text may itself contain `include` lines, which are similarly replaced. An `include` line must not reference itself, directly or indirectly. When an `include` line is resolved, the first included line must not be a continuation line and the last line must not be continued. The included text may not contain incomplete statements.

The `include` line was available as an extension to many Fortran 77 systems and was often used to ensure that every occurrence of global data in a `common` block was identical. In modern Fortran the same effect is better achieved by placing global data in a module (Section 5.5). This cannot lead to accidental declarations of local variables in each procedure.

This feature is useful when identical executable statements are needed for more than one type, for example in a set of procedures for sorting data values of various types. The executable statements can be maintained in an include file that is referenced inside each instance of the sort procedure.

A.3   Alternative form of complex constant

A complex constant may be written with a named constant of type real or integer for its real part, imaginary part, or both. For example,

```
real, parameter    :: zero = 0, one = 1
complex, parameter :: i = (zero, one)
```

However, no sign is allowed with a name, so although $(0,-1)$ is a perfectly good complex constant, `(zero,-one)` is invalid.

Since the intrinsic function `cmplx` is permitted to appear in a constant expression there is very little use for this feature.

A.4   Double precision real

Another *type* that may be used in a type declaration, `function`, `implicit`, or component declaration statement is `double precision`, which specifies double precision real. The precision is greater than that of default real.

Literal constants written with the exponent letter d (or D) are of type double precision real by default; no kind parameter may be specified if this exponent letter is used. Thus, `1d0` is of type double precision real. If `dp` is an integer named constant with the value `kind(1d0)`, `double precision` is synonymous with `real(kind=dp)`.

There is a d (or D) edit descriptor that was originally intended for double precision quantities, but now is identical to the e edit descriptor except that the output form may have a D instead of an E as its exponent letter. A double precision real literal constant, with exponent letter d, is acceptable on input whenever any other real literal constant is acceptable.

There are two elemental intrinsic functions which were not described in Chapter 9 because they have a result of type double precision real:

dble (a)   for a of type integer, real, complex, or a 'boz' constant returns the double precision real value `real(a, kind(0d0))`.

dprod (x, y)   returns the product x*y for x and y of type default real as a double precision real result.

The type specifier `double precision` is permitted in a `select type` construct (Section 15.7).

The double precision real data type has been replaced by the real type of kind `kind(1d0)`.

A.5   Type statement for declaring an entity of intrinsic type

The `type` keyword can be used with an intrinsic type specification instead of a derived type specification, and this declares the entities to be of that intrinsic type. For example,

```
type(complex(kind(0d0))) :: a, b, c
```

declares a, b, and c to be of intrinsic type `complex` with kind type parameter equal to `kind(0d0)`, that is, double precision complex. This syntax is completely redundant and the example is equivalent to

```
complex(kind(0d0)) :: a, b, c
```

This feature was added for consistency with the `type is` statement in the `select type` construct: in that statement an intrinsic type is specified by its keyword, but a derived type is specified simply by its type name without the `type` keyword (or the concomitant parentheses).

We consider that this feature adds nothing to the language; furthermore, it might confuse a reader into thinking that an intrinsic type is really a derived type, so we do not recommend its use.

A.6   The `dimension, codimension,` and `parameter` statements

To declare entities we normally use type specifications. However, if all the entities involved are arrays, they may be declared *without* type specifications in a `dimension` statement:

```
dimension i(10), b(50,50), c(n,m) ! n and m are integer dummy
                                  ! arguments or named constants
```

The general form is

> `dimension` *[::] array-name (array-spec) [, array-name (array-spec)]* ...

Here, the type may either be specified in a type declaration statement such as

```
integer i
```

that does not specify the dimension information, or may be declared implicitly. Our view is that neither of these is sound practice; the type declaration statement looks like a declaration of a scalar, and we explained in Section 8.2.1 that we regard implicit typing as dangerous. Therefore, the use of the `dimension` statement is not recommended.

The same remark applies to the `codimension` statement for declaring coarrays with the syntax

> `codimension` *[::] coarray-name [coarray-spec] [, array-name [coarray-spec]]* ...

An alternative way to specify a named constant is by the `parameter` statement. It has the general form

> `parameter` (*named-constant-definition-list*)

where each *named-constant-definition* is

> *constant-name = constant-expr*

Each constant named must either have been typed in a previous type declaration statement in the scoping unit, or take its type from the first letter of its name according to the implicit typing rule of the scoping unit. In the case of implicit typing, an appearance of the named constant in a subsequent type declaration statement in the scoping unit must confirm the type and type parameters, and there must not be an `implicit` statement for the letter subsequently in the scoping unit. Similarly, the shape must have been specified previously or be scalar. Each named constant in the list is defined with the value of the corresponding expression according to the rules of intrinsic assignment.

An example using implicit typing and a constant expression including a named constant that is defined in the same statement is

```
implicit integer (a, p)
parameter (apple = 3, pear = apple**2)
```

For the same reasons as for `dimension`, we recommend avoiding the `parameter` statement.

A.7   Sequence types

A derived type may be declared as a **sequence type** by including a `sequence` statement in its type declaration before any of its components, for example

```
type storage
   sequence
   integer i
   real a(0:999)
end type storage
```

The main purpose of sequence types is in association with `equivalence` statements (Section B.1.6) and `common` blocks (Section B.1.7) but they have not been classed as obsolescent so are included in this appendix.

Should any other derived types appear in the definition of a sequence type, they too must be sequence types. A sequence type is not allowed to have type parameters (Section 13.2) or type-bound procedures (Section 15.8).

A `private` statement may be added to a sequence derived-type definition, making its components private. The `private` and `sequence` statements may be interchanged but must be the second and third statements of the derived-type definition.

Two derived-type definitions in different scoping units define the same data type if they have the same name,[1] both have the `sequence` attribute, and have components that are not `private` and agree in order, name, and attributes. Relying on this is prone to error and offers no advantage over having a single definition in a module that is accessed by use association.

A.8   Storage association

Storage units are the fixed units of physical storage allocated to certain data. Again, their main purpose is in association with `equivalence` statements and `common` blocks but they have not been classed as obsolescent.

There is a storage unit called **numeric** for any non-pointer scalar of the types default real, default integer, and default logical, and a storage unit called **character** for any non-pointer scalar of type default character and character length one. Non-pointer scalars of type default complex or double precision real (Appendix A.4) occupy two contiguous numeric storage units. Non-pointer scalars of type default character and length *len* occupy *len* contiguous character storage units.

As well as numeric and character storage units, there are a large number of **unspecified** storage units. A non-pointer scalar object of type non-default integer, real other than default or double precision, non-default logical, non-default complex, or non-default character of any particular length occupies a single unspecified storage unit that is different for each case. A data object with the `pointer` attribute has an unspecified storage unit, different from that of any non-pointer object and different for each combination of type, type parameters, and rank. The standard makes no statement about the relative sizes of all these storage units and permits storage association to take place only between objects with the same category of storage unit.

---

[1] If one or both types have been accessed by use association and renamed, it is the original names that must agree.

A non-pointer array occupies a sequence of contiguous storage sequences, one for each element, in array element order.

Objects of derived type have no storage association, each occupying an unspecified storage unit that is different in each case, except where the type is a sequence type. A sequence type is a **numeric sequence type** if no component is a pointer or allocatable, and each component is of type default integer, default real, double precision real, default complex, or default logical. A component may also be of a previously defined numeric sequence type. This implies that the ultimate components occupy numeric storage units and the type itself has **numeric storage association**. Similarly, a sequence type is a **character sequence type** if no component is a pointer or allocatable, and each component is of type default character or a previously defined character sequence type. Such a type has **character storage association**.

A scalar of numeric or character sequence type occupies a storage sequence that consists of the concatenation of the storage sequences of its components. A scalar of any other sequence type occupies a single unspecified storage unit that is unique for each type.

A sequence type is permitted to have an allocatable component, which permits independent declarations of the same type in different scopes, but such a type, like a pointer, has an unspecified storage unit.

The intrinsic module `iso_fortran_env` provides information about the Fortran storage environment, in the form of named constants as follows:

`character_storage_size` The size in bits of a character storage unit.

`numeric_storage_size` The size in bits of a numeric storage unit.

Use of storage association is error prone and we do not recommend it, except where its use is necessary in the context of interoperating with C.

A.9 Non-default mapping for implicit typing

The default for implicit typing (Section 8.2.1) is that entities whose names begin with one of the letters i, j, ..., n are of type default integer, and variables beginning with the letters a, b, ..., h or o, p, ..., z are of type default real. If implicit typing with a different rule is desired in a given scoping unit, the `implicit` statement may be employed. This changes the mapping between the letters and the types with statements such as

```
implicit integer (a-h)
implicit real(selected_real_kind(10)) (r,s)
implicit type(entry) (u,x-z)
```

The letters are specified as a list in which a set of adjacent letters in the alphabet may be abbreviated, as in a-h. No letter may appear twice in the `implicit` statements of a scoping unit and, if the scoping unit contains the `implicit none` or equivalent `implicit none(type)` statement, there must be no other `implicit` statement in the scoping unit. For a letter not included in the `implicit` statements, the mapping between the letter and a type is the default mapping.

In the case of a scoping unit other than a program unit or an interface block, for example a module subprogram, the default mapping for each letter in an inner scoping unit is the

Figure A.1 Illustration of the rules of implicit typing.

```fortran
module example_mod
   implicit none
   ⋮
   interface
      function fun(i)     ! i is implicitly
          integer :: fun  ! declared integer.
      end function fun
   end interface
contains
   function jfun(j)        ! All data entities must
      integer :: jfun, j   ! be declared explicitly.
      ⋮
   end function jfun
end module example_mod
subroutine sub
   implicit complex (c)
   c = (3.0,2.0)           ! c is implicitly declared complex
   ⋮
contains
   subroutine sub1
      implicit integer (a,c)
      c = (0.0,0.0) ! c is host associated and of type complex
      z = 1.0       ! z is implicitly declared real.
      a = 2         ! a is implicitly declared integer.
      cc = 1.0      ! cc is implicitly declared integer.
      ⋮
   end subroutine sub1
   subroutine sub2
      z = 2.0          ! z is implicitly declared real and is
                       ! different from the variable z in sub1.
      ⋮
   end subroutine sub2
   subroutine sub3
      use example_mod    ! Access the integer function fun.
      q = fun(k)         ! q is implicitly declared real and
                         ! k is implicitly declared integer.
      ⋮
   end subroutine sub3
end subroutine sub
```

mapping for the letter in the immediate host. If the host contains an `implicit none` or `implicit none(type)` statement, the default mapping is null and the effect may be that implicit typing is available for some letters, because of an additional `implicit` statement in the inner scope, but not for all of them. The mapping may be to a derived type even when that type is not otherwise accessible in the inner scoping unit because of a declaration there of another entity with the same name.

Implicit typing does not apply to an entity accessed by use or host association because its type is the same as in the module or the host. Figure A.1 provides a comprehensive illustration of the rules of implicit typing.

The general form of the `implicit` typing statement is

```
implicit type (letter-spec-list) [ , type (letter-spec-list) ]...
```

where *type* specifies the type and type parameters (Section 8.16) and each *letter-spec* is

letter [– letter]

The `implicit` statement may be used for a derived type. For example, given access to the type

```
type posn
   real    :: x, y
   integer :: z
end type posn
```

and given the statement

```
implicit type(posn) (a,b), integer (c-z)
```

variables beginning with the letters a and b are implicitly typed posn and variables beginning with the letters c, d, ..., z are implicitly typed integer.

An `implicit` typing statement may be preceded within a scoping unit only by `use`, `parameter`, and `format` statements.

A.10   Alternative form of relational operator

The relational operators have alternative and redundant forms

```
.lt.   for   <    less than
.le.   for   <=   less than or equal
.eq.   for   =    equal
.ne.   for   /=   not equal
.gt.   for   >    greater than
.ge.   for   >=   greater than or equal
```

Corresponding operator forms are considered to be the same operator; for example, providing a generic interface for `.ne.` also provides it for `/=`.

A.11   The do while statement

In Section 4.4 a form of the do construct was described that may be written as

```
do
    if (scalar-logical-expr) exit
    ⋮
end do
```

An alternative, but redundant, form of this is its representation using a do while statement:

```
[ name: ] do [,] while (.not.scalar-logical-expr)
```

We prefer the form that uses the exit statement because this can be placed anywhere in the loop, whereas the do while statement always performs its test at the loop start. If *scalar-logical-expr* becomes false in the middle of the loop, the rest of the loop is still executed. Potential optimization penalties that the use of do while entails are fully described in Chapter 10 of *Optimizing Supercompilers for Supercomputers*, M. Wolfe, Pitman, 1989.

A.12   Control of execution flow by branching

A.12.1   The go to statement

Prior to Fortran 2008, especially when dealing with error conditions, the available control constructs could be inadequate for the programmer's needs. The remedy was to use the most disputed statement in programming languages – the go to statement – to **branch** to another statement. It is generally accepted that it is difficult to understand a program which is interrupted by many branches, especially if there is a large number of backward branches – those returning control to a statement preceding the branch itself. For this reason, we recommend against using this statement.

The form of the unconditional go to statement is

```
go to label
```

where *label* is a statement label. This statement label must be present on an **executable statement** (a statement that can be executed, as opposed to one of an informative nature, like a declaration). An example is

```
    x = y + 3.0
    go to 4
3   x = x + 2.0
4   z = x + y
```

in which we note that after execution of the first statement, a branch is taken to the last statement, labelled 4. This is a **branch target statement**. The statement labelled 3 is jumped over, and can be executed only if there is a branch to the label 3 somewhere else. If the statement following an unconditional go to is unlabelled, it can never be reached and executed, thus is **dead code**, normally a sign of incorrect coding.

The statements within a block of a construct may be labelled, but unless it is a block construct (Section 8.12) the labels must not be referenced in such a fashion as to pass control into a block from outside it, to an `else if` statement, or to an `else` statement. It is permitted to pass control from a statement in a construct to the terminal statement of the construct, or to a statement outside its construct.

A `go to` statement may be the *action-stmt* in an `if` statement, for example

```
if (flag) go to 6
```

A.12.2   The `continue` statement

The `continue` statement is an executable statement that does nothing. It purpose is solely to provide a placeholder for a label. It is useful only for terminating a labelled `do` construct (Section B.1.14).

A.12.3   Branching in input/output statements

The following optional specifiers can appear in input/output statements:

- `end=`*end-label* as an input specifier for sequential or stream-access `read`,
- `eor=`*eor-label* as an input specifier for non-advancing `read`, and
- `err=`*error-label* as a specifier for `read`, `write`, the file positioning statements of Section 12.2, and the file control statements `open`, `close`, and `inquire`.

The meaning and effect of these specifiers is as follows:

- if `end=` appears, *end-label* must be a statement label of a statement in the same inclusive scope, to which control will be transferred in the event of the end of the file being reached;
- if `eor=` appears, *eor-label* must be a statement label of a statement in the same inclusive scope, to which control will be transferred in the event of the end of the input record being reached; and
- if `err=` appears, *error-label* is a statement label in the same inclusive scope, to which control will be transferred in the event of any other exception occurring.

The *end-label* and the *eor-label* may have the same value, and this may be the same value as that of the *error-label*.

Because the `iostat=` facility is available for all input/output statements, we recommend avoiding these variants of branching.

A.13   Implicit interfaces

An external, dummy, or pointer procedure may be invoked without access to its interface. In this case, the interface is said to be **implicit**. All the compiler has is the information that is implicit in the statements in the environment of the invocation, that is, the number of arguments and their types, and the type of the function result in the case of a function. To specify that a name is that of a procedure with an implicit interface, the `external` statement is available. It has the form

```
external external-name-list
```

and appears with other specification statements, after any `use` or `implicit` statements (Section 8.2) and before any executable statements. It is not permitted for a procedure having an accessible interface. The type and type parameters of a function with an implicit interface are usually specified by a type declaration statement for the function name, optionally with the `external` attribute, for example,

```
real, external :: y
```

An alternative is by the rules of implicit typing (Section 8.2.1) applied to the name, but this is not available in a module unless the function has the `private` attribute (see Section 8.6.1).

Alternatively, the `procedure` statement can be used to declare procedures that have implicit interfaces; instead of putting the name of a procedure interface inside the parentheses, either nothing or a type specification is used. For example,

```
procedure() x
procedure(real) y
procedure(complex(kind(0.0d0))) z
```

declares x to be a procedure (which might be a subroutine or an implicitly typed function[2]), y to be a `real` function, and z to be a (double) `complex` function. This is exactly equivalent to

```
external :: x
real, external :: y
complex(kind(0.0d0)), external :: z
```

For these cases the `procedure` statement offers no useful functionality over the `external` or type declaration statement; it really only comes into its own for declaring procedure pointers (see Section 14.2).

Note that if a procedure is accessible in a scoping unit, its interface is either explicit or implicit there. An external procedure may have an explicit interface in some scoping units and an implicit interface in others.

An explicit interface is required to invoke a procedure with an allocatable, pointer, or target dummy argument or a function with an array, allocatable, or pointer function result. It is needed so that the processor can make the appropriate linkage. Even when not strictly required, it gives the compiler an opportunity to examine data dependencies and thereby improve optimization. Explicit interfaces are also desirable because of the additional security that they provide.

An external or dummy procedure that is used as a pure procedure must have an interface block that specifies it as pure. However, the procedure may be used in other contexts without the use of an interface block or with an interface block that does not specify it as pure. This allows library procedures to be specified as pure without limiting them to being used as such.

When calling a procedure, the interface must be explicit if any of the dummy arguments are optional or keyword arguments are in use. Otherwise, the compiler will not be able to make the appropriate associations. In all cases where an interface block is provided, it is the names of the dummy arguments in the block that are used to resolve the associations.

It is straightforward to ensure that all interfaces are explicit. Implicit interfaces are error-prone; explicit interfaces (Sections 5.12 and 14.1) should be used instead. If an external

---

[2]The latter is not a possibility in a module or if `implicit none` is in effect.

procedure has an implicit interface the processor is permitted to interpret the name as that of an intrinsic procedure. This is needed for portability since processors are permitted to provide additional intrinsic procedures. Naming a procedure in an `external` statement makes all versions of an intrinsic procedure having the same name unavailable. The same is true for giving it an interface body (but not when the interface is generic, Section 5.18).

A.14 Denoting an absent non-pointer non-allocatable argument

A null pointer or an unallocated allocatable can be used to denote an absent non-allocatable non-pointer optional argument. For example, in

```
interface
  subroutine s(x)
    real, optional :: x
  end subroutine
end interface
  ⋮
call s(null())
```

the `null()` reference is treated as if it were not present.

This is useful in the slightly contrived situation where one has a procedure with many optional arguments, together with pointers or allocatables to be passed as actual arguments only if associated or allocated. In the absence of this facility one needs a 2^n-way set of nested `if` constructs, where n is the number of local variables in question. Figure A.2 provides an outline of how this process works. In that example, the new feature allows the call to `process_work` to be a single statement; without the feature that call would need to be the unreadably complicated nested `if` constructs shown in Figure A.3.

Figure A.2 Absent optional denotation.

```
subroutine top(x, a, b)
   real :: x
   real, optional, target :: a(:), b(:)
   real, allocatable :: worka(:), workb1(:), workb2(:)
   real, pointer :: pivotptr
   ⋮ ! Code to conditionally allocate worka etc. elided.
   call process_work(x, worka, workb1, workb2, pivotptr)
end subroutine
subroutine process_work(x, wa, wb1, wb2, pivot)
   real :: x
   real, optional :: wa(:), wb1(:) , wb2(:), pivot
```

It is true that in this example making the dummy variables in `process_work` variously `allocatable` or `pointer` would achieve the same ends, but other callers of `process_work` might have different mixtures of `allocatable` and `pointer`, or indeed wish to pass plain variables.

Figure A.3 Huge unreadable nested `if` constructs.

```
if (allocated(worka)) then
  if (allocated(workb1)) then
    if (allocated(workb2)) then
      if (associated(pivotptr)) then
        call process_work(x, worka, workb1, workb2, pivotptr)
      else
        call process_work(x, worka, workb1, workb2)
      end if
    else if (associated(pivotptr)) then
      call process_work(x, worka, workb1, pivot=pivotptr)
    else
      call process_work(x, worka, workb1)
    end if
  : ! Remainder of huge nested if construct elided.
```

The intrinsic functions `count`, `lbound`, `lcobound`, `ubound`, and `ucobound` have an optional argument `dim` whose presence affects the rank of the result. Therefore, the corresponding actual argument is not permitted to be an optional dummy argument name. For the same reason, it is not permitted to be a null pointer or an unallocated allocatable (that is, the feature described in this section is not operative).

We see it as confusing to use a null pointer to denote an absent non-pointer argument – if you see `null()` in an argument list you would rather expect that a null pointer will be passed. We therefore recommend avoiding this feature unless the number of optional arguments is large. In Fortran 2023, conditional arguments (Section 23.2.3) provide a good solution to the problem discussed in this section and are also available for pointer and allocatable optional arguments. The code of Figure A.3 may be replaced by

```
call process_work( x, (allocated(worka)? worka : .nil.), &
    (allocated(workb1)? workb1 : .nil.), &
    (allocated(workb2)? workb2 : .nil.), &
    (associated(pivotptr)? pivotptr : .nil.) )
```

A.15   The `volatile` attribute

The `volatile` attribute was new in Fortran 2003, but is considered by us now to be redundant: the semantics of the `asynchronous` attribute have since been extended such that the `volatile` attribute need no longer be used. Furthermore, its impact on performance is significantly larger than that of `asynchronous`. It may be applied only to variables and is conferred either by the `volatile` attribute in a type declaration statement, or by the `volatile` statement, which has the form

```
volatile [::] variable-name-list
```

For example,

```
integer, volatile :: x
real              :: y
volatile          :: y
```

declares two volatile variables x and y.

Being volatile indicates to the compiler that, at any time, the variable might be changed and/or examined from outside the Fortran program. This means that each reference to the variable will need to load its value from main memory (so, for example, it cannot be kept in a register in an inner loop). Similarly, each assignment to the variable must write the data to memory. Essentially, this disables most optimizations that might have been applicable to the object, making the program run slower but, one hopes, making it work with some special hardware or multi-processing software.

However, it is the responsibility of the programmer to effect any necessary synchronization; this is particularly relevant to multi-processor systems. Even if only one process is writing to the variable and the Fortran program is reading from it, because the variable is not automatically protected by a critical section (see Section 17.14) it is possible to read a partially updated (and thus an inconsistent or impossible) value. For example, if the variable is an IEEE floating-point variable, reading a partially updated value could return a signaling NaN; or if the variable is a pointer, its descriptor might be invalid. In either of these cases the program could be abruptly terminated, so this facility must be used with care.

Similarly, if two processes attempt to update a single `volatile` variable, the effects are completely processor dependent. The variable might end up with its original value, one of the values from an updating process, a partial conglomeration of values from the updating processes, or the program could even crash.

A simple use of this feature might be to handle some external (interrupt-driven) event, such as the user typing Control-C, in a controlled fashion. In the example in Figure A.4, `register_event_flag` is a subroutine, possibly written in another language, which ensures that `event_has_occurred` becomes true when the specified event occurs.

Figure A.4 Handling an external event.

```
logical, target, volatile :: event_has_occurred
  ⋮
event_has_occurred = .false.
call register_event_flag(event_has_occurred, ...)
  ⋮
do
    ⋮                                  ! some calculations
  if (event_has_occurred) exit ! exit loop if event happened
    ⋮                                  ! some more calculations
  if (...) exit                 ! Finished our calculations yet?
end do
  ⋮
```

If the variable is a pointer, the `volatile` attribute applies both to the descriptor and to the target. Even if the target does not have the `volatile` attribute, it is treated as having it when accessed via a pointer that has it. If the variable is allocatable, it applies both to the allocation and to the value. In both cases, if the variable is polymorphic (Section 15.3), the dynamic type may change by non-Fortran means.

If a variable has the `volatile` attribute, so do all of its subobjects. For example, in

```
logical, target             :: signal_state(100)
logical, pointer, volatile :: signal_flags(:)
   ⋮

signal_flags => signal_state
   ⋮

signal_flags(10) = .true.    ! A volatile reference
   ⋮

write (20) signal_state       ! A nonvolatile reference
```

the pointer (descriptor) of `signal_flags` is volatile, and access to each element of `signal_flags` is volatile; however, `signal_state` itself is not volatile.

A.15.1　Volatile scoping

If a variable only needs to be treated as volatile for a short time, the programmer has two options: either to pass it to a procedure to be acted on in a volatile manner (see Section A.15.2), or to access it by use or host association, using a `volatile` statement to declare it to be volatile only in the accessing scope. For example, in the code of Figure A.5, the `data` array is not volatile in `data_processing`, but is in `data_transfer`. Similarly, declaring a variable that is accessed by host association to be volatile is allowed.

Figure A.5 Using a procedure to limit the scope of a variable's volatility.

```
module data_module
   real, allocatable :: data(:,:), newdata(:,:)
      ⋮
contains
   subroutine data_processing
         ⋮
   end subroutine data_processing
   subroutine data_transfer
      volatile :: data
         ⋮
   end subroutine data_transfer
end module data_module
```

A.15.2   Volatile arguments

The volatility of an actual argument and its associated dummy argument may differ. This is important since volatility may be needed in one but not in the other. In particular, a volatile variable may be used as an actual argument in a call to an intrinsic procedure. However, while a volatile variable is associated with a non-volatile dummy argument, the programmer must ensure that the value is not altered by non-Fortran means. Note that, if the volatility of an actual argument persists through a procedure reference, this means that the procedure referenced must have an explicit interface and the corresponding dummy argument must be declared to be volatile.

If the procedure is non-elemental and the dummy argument is a volatile array, the actual argument must not be an array section with a vector subscript; furthermore, if the actual argument is an array section or an assumed-shape array, the dummy argument must be assumed shape and if the actual argument is an array pointer, the dummy argument must be a pointer or assumed shape. These restrictions are designed to allow the argument to be passed by reference; in particular, to avoid the need for a local copy being made as this would interfere with the volatility.

A dummy argument with intent in or the value attribute (Section 20.8) is not permitted to be volatile. This is because the value of such an argument is expected to remain fixed during the execution of the procedure.

If a dummy argument of a procedure is volatile, the interface must be explicit whenever it is called and the dummy argument must be declared as volatile in any interface body for the procedure.

A.15.3   Volatile coarrays

If a dummy coarray is volatile, so too must the corresponding actual argument be, and vice versa. Without this restriction, the value of a non-volatile coarray might be altered via another image by means not specified by the program, that is, behave as volatile.

Similarly, agreement of the attribute is required when accessing a coarray or a variable with a coarray potential subobject component by use association, host association, or in a block construct (see Section 8.12) from the scope containing it. Here, the restriction is simple; since the attribute is the same by default, it must not be respecified for an accessed variable.

For the same reason, agreement of the volatile attribute is required for pointer association with any part of a coarray.

A volatile coindexed object is not permitted to be an actual argument that corresponds to an asynchronous or volatile dummy argument unless the dummy argument has the value attribute. This is because passing a coindexed object as an actual argument is likely to be done by copy-in copy-out.

A.16   The sync memory statement

The execution of a sync memory statement defines a boundary on an image between two segments, each of which can be ordered in some user-defined way with respect to segments on other images. It is an image control statement, but unlike the other image control statements it does not have any in-built synchronization effect. In case there is some user-defined ordering between images, the compiler will probably avoid optimizations involving moving statements across the sync memory statement and will ensure that any changed data that the image holds in temporary memory, such as cache or registers, or even packets in transit between images, are made visible to other images. Also, any data from other images that are held in temporary memory will be treated as undefined until reloaded from the host image.

We see the construction of reliable and portable code in this way as very difficult – it is all too easy to introduce subtle bugs that manifest themselves only occasionally.

One way to effect user-defined ordering between images is by employing the atomic subroutines atomic_define and atomic_ref (Section 17.23). As an example of user-defined ordering, consider the code in Figure A.6, which is executed on images p and q. The

---

Figure A.6 Spin-wait loop.

```
use, intrinsic :: iso_fortran_env
logical(atomic_logical_kind) :: locked[*] = .true.
logical :: val
integer :: iam, p, q
  :
  :
iam = this_image()
if (iam == p) then
   sync memory
   call atomic_define(locked[q], .false.)
   ! Has the effect of locked[q]=.false.
else if (iam == q) then
   val = .true.
   ! Spin until val is false
   do while (val)
       call atomic_ref(val, locked)
       ! Has the effect of val=locked
   end do
   sync memory
end if
```

---

do loop is known as a spin-wait loop. Once image q starts executing it, it will continue until it finds the value .false. for val. The atomic_ref call ensures that the value is refreshed on each loop execution. The effect is that the segment on image p ahead of the first sync memory statement precedes the segment on image q that follows the second sync memory statement. The normative text of the standard does not specify how resources should be distributed between images, but a note expects that the sharing should be equitable. It is therefore just possible that a conforming implementation might give all its resources to the spin loop while doing nothing on image p, causing the program to hang.

Note that the segment in which `locked[q]` is altered is unordered with respect to the segment in which it is referenced. This is permissible by the rules in the penultimate paragraph of Section 17.11.

Given the atomic subroutines and the `sync memory` statement, customized synchronizations can be programmed in Fortran as procedures, but it may be difficult for the programmer to ensure that they will work correctly on all implementations.

A.17   Coarray components of type `c_ptr` or `c_funptr`

A coarray is permitted to have a component of type `c_ptr` or `c_funptr`, but a coindexed object is not permitted to be of either of these types because it is almost certain to involve a remote reference. Furthermore, intrinsic assignment for either of these types causes the variable to become undefined unless the variable and expression are on the same image. It is very hard to see good uses for this feature.

B. Obsolescent and deleted features

B.1 Obsolescent features

B.1.1 Introduction

In this section we describe the obsolescent features of Fortran 2018. They are still part of the language but the standard marks them as outmoded and redundant. They are candidates for deletion in a future revision of the standard. We deprecate their use and include descriptions of suitable replacements.

B.1.2 Fixed source form

In the fixed source form, each statement consists of one or more **lines** exactly 72 characters long,[1] and each line is divided into three **fields**. The first field consists of positions 1 to 5 and may contain a **statement label**. A Fortran statement may be written in the third fields of up to 256 consecutive lines (only 20 in versions prior to Fortran 2003). The first line of a multi-line statement is known as the **initial line** and the succeeding lines as **continuation lines**.

A non-comment line is an initial line or a continuation line depending on whether there is a character other than zero or blank in position 6 of the line, which is the second field. The first field of a continuation line must be blank. The ampersand is not used for continuation.

The third field, from positions 7 to 72, is reserved for the Fortran statements themselves. Note that if a construct is named, the name must be placed here and not in the label field.

Except in a character context, blanks are insignificant.

The presence of an asterisk (*) or a character c in position 1 of a line indicates that the whole line is commentary. An exclamation mark indicates the start of commentary, except in position 6, where it indicates continuation.

Several statements separated by a semicolon (;) may appear on one line. The semicolon may not, in this case, be in column 6, where it would indicate continuation. Only the first of the statements on a line may be labelled. A semicolon that is the last non-blank character of a line, or the last non-blank character ahead of commentary, is ignored.

A program unit end statement must not be continued, and any other statement with an initial line that appears to be a program unit end statement must not be continued.

A processor may restrict the appearance of its defined control characters, if any, in this source form.

---

[1]This limit is processor dependent if the line contains characters other than those of the default type.

In applications where a high degree of compatibility between the old and the new source forms is required, observance of the following rules can be of great help:

- confine statement labels to positions 1 to 5 and statements to positions 7 to 72;
- treat blanks as being significant;
- use only ! to indicate a comment (but not in position 6); and
- for continued statements, place an ampersand in both position 73 of a continued line and position 6 of a continuing line.

The fixed source form has been replaced by the free source form (Section 2.4).

B.1.3   Character length specification with `character*`

Alternatives for default characters to

    character ([len=] len-value)

as a *type* in a type declaration, `function`, `implicit`, or component definition statement are

    character* (len-value) [,]

and

    character*len [,]

where *len* is an integer literal constant without a specified kind value, and the optional comma is permitted only in a type declaration statement and only when : : is absent:

    character*20, word, letter*1

B.1.4   `data` statements among executables

The `data` statement may be placed among the executable statements, but such placement is rarely used and not recommended, since data initialization properly belongs with the specification statements.

B.1.5   Computed `go to`

A form of branch statement is the computed `go to`, which enables one path among many to be selected, depending on the value of a scalar integer expression. The general form is

    go to (sl₁, sl₂, sl₃, ...) [,] intexpr

where sl_1, sl_2, sl_3, etc. are labels of statements in the same inclusive scope, and *intexpr* is any scalar integer expression. The same statement label may appear more than once. An example is the statement

    go to (6,10,20) i(k) **2+j

which references three statement labels. When the statement is executed, if the value of the integer expression is 1, the first branch will be taken, and control is transferred to the statement labelled 6. If the value is 2, the second branch will be taken, and so on. If the value is less than 1, or greater than 3, no branch will be taken, and the next statement following the `go to` will be executed.

   This statement is replaced by the `case` construct (Section 4.3).

B.1.6   The `equivalence` statement

The `equivalence` statement specifies that a given storage area may be shared by two or more objects. For instance,

```
real aa, angle, alpha, a(3)
equivalence (aa, angle), (alpha, a(1))
```

allows `aa` and `angle` to be used interchangeably in the program text, as both names now refer to the same storage location. Similarly, `alpha` and `a(1)` may be used interchangeably.

It is possible to equivalence arrays together. In

```
real a(3,3), b(3,3), col1(3), col2(3), col3(3)
equivalence (col1, a, b), (col2, a(1,2)), (col3, a(1,3))
```

the two arrays `a` and `b` are equivalenced, and the columns of `a` (and hence of `b`) are equivalenced to the arrays `col1`, etc. We note in this example that more than two entities may be equivalenced together, even in a single declaration.

It is possible to equivalence variables of the same intrinsic type and kind type parameter or of the same derived type having the `sequence` attribute. It is also possible to equivalence variables of different types if both have numeric storage association or both have character storage association (see Appendix A.8). Default character variables need not have the same length, as in

```
character(len=4) a
character(len=3) b(2)
equivalence (a, b(1)(3:))
```

where the character variable `a` is equivalenced to the last four characters of the six characters of the character array `b`. Zero character length is not permitted. An example for different types is

```
integer i(100)
real x(100)
equivalence (i, x)
```

where the arrays `i` and `x` are equivalenced. This might be used, for instance, to save storage space if `i` is used in one part of a program unit and `x` separately in another part. This is a highly dangerous practice, as considerable confusion can arise when one storage area contains variables of two or more data types, and program changes may be made very difficult if the two uses of the one area are to be kept distinct.

Types with default initialization are permitted, provided each initialized component has the same type, type parameters, and value in any pair of equivalenced objects. Coarrays are not permitted. A variable that has the `bind` attribute or is a member of a `common` block (Section B.1.7) that has the `bind` attribute is not permitted.

All the various combinations of types that may be equivalenced have been described. No other is allowed. Also, apart from double precision real and the default numeric types, equivalencing objects that have different kind type parameters is not allowed. The general form of the statement is

```
equivalence (object, object-list) [ , (object, object-list ) ]...
```

where each *object* is a variable name, array element, or substring. An object must be a variable and must not be a dummy argument, a function result, a pointer, an object with a pointer component at any level of component selection, an allocatable object, an automatic object, a function, a structure component, a structure with an ultimate allocatable component, or a subobject of such an object. Each array subscript and character substring range must be a constant expression. The interpretation of an array name is identical to that of its first element. An equivalence object must not have the `target` attribute.

The objects in an equivalence set are said to be **storage associated**. Those of nonzero length share the same first storage unit. Those of zero length are associated with each other and with the first storage unit of those of nonzero length. An `equivalence` statement may cause other parts of the objects to be associated, but not such that different subobjects of the same object share storage. For example,

```
real a(2), b
equivalence (a(1), b), (a(2), b)  ! Prohibited
```

is not permitted. Also, objects declared in different scoping units must not be equivalenced. For example,

```
use my_module, only : xx
real bb
equivalence(xx, bb)              ! Prohibited
```

is not permitted.

The various uses to which `equivalence` was put may be replaced by automatic arrays, allocatable arrays, and pointers.

B.1.7   The `common` block

We have seen in Chapter 5 how two program units are able to communicate by passing variables or values of expressions between them via argument lists or by using modules. It is also possible to define areas of storage known as `common` blocks. Each has a storage sequence and may be either named or unnamed, as shown by the simplified syntax of the `common` specification statement:

> `common` *[/ [cname] /] vlist*

in which *cname* is an optional name and *vlist* is a list of variable names, each optionally followed by an array bounds specification. An unnamed `common` block is known as a **blank** `common` block. Examples of each are

```
common /hands/ nshuff, nplay, nhand, cards(52)
```

and

```
common // buffer(10000)
```

in which the named `common` block hands defines a data area containing the quantities which might be required by the subroutines of a card-playing program, and the blank `common` defines a large data area which might be used by different subroutines as a buffer area.

The name of a common block has global scope and must differ from that of any other global entity (external procedure, program unit, or common block). It may, however, be the same as that of a local entity other than a named constant or intrinsic procedure.

No object in a common block may have the parameter attribute or be a dummy argument, an automatic object, an allocatable object, a structure with an ultimate allocatable component, a coarray, a polymorphic pointer, or a function. An array may have its bounds declared either in the common statement or in a type declaration or dimension statement. If it is a non-pointer array, the bounds must be declared explicitly and with constant expressions. If it is an array pointer, however, the bounds may not be declared in the common statement itself. If an object is of derived type, the type must have the sequence or bind attribute and must not have default initialization.

In order for a subroutine to access the variables in the data area, it is sufficient to insert the common definition in each scoping unit which requires access to one or more of the entities in the list. In this fashion, the variables nshuff, nplay, nhand, and cards are made available to those scoping units. No variable may appear more than once in all the common blocks in a scoping unit.

Usually a common block contains identical variable names in all its appearances, but this is not necessary. In fact, the shared data area may be partitioned in quite different ways in different subroutines, using different variable names. They are said to be **storage associated**.. It is thus possible for one subroutine to contain a declaration

```
common /coords/ x, y, z, i(10)
```

and another to contain a declaration

```
common /coords/ i, j, a(11)
```

This means that a reference to i(1) in the first subroutine is equivalent to a reference to a(2) in the second. Through multiple references via use or host association, this can even happen in a single subroutine. This manner of coding is both untidy and dangerous, and every effort should be made to ensure that all declarations of a given common block declaration are identical in every respect. In particular, the presence or absence of the target attribute is required to be consistent, since otherwise a compiler would have to assume that everything in common has the target attribute, in case it has it in another program unit.

A further practice that is permitted but which we do not recommend is to mix different storage units in the same common block. When this is done, each position in the storage sequence must always be occupied by a storage unit of the same category.

The total number of storage units must be the same in each occurrence of a named common block, but blank common is allowed to vary in size and the longest definition will apply for the complete program.

Yet another practice to be avoided is to use the full syntax of the common statement:

```
common [ / [ cname ] / vlist [ [ , ] / [ cname ] / vlist ] ...
```

which allows several common blocks to be defined in one statement, and a single common block to be declared in parts. A combined example is

```
common /pts/x,y,z /matrix/a(10,10),b(5,5) /pts/i,j,k
```

which is equivalent to

```
common /pts/ x, y, z, i, j, k
common /matrix/ a(10,10), b(5,5)
```

which is certainly a more understandable declaration of two shared data areas.

The `common` statement may be combined with the `equivalence` statement, as in the example

```
real a(10), b
equivalence (a,b)
common /change/ b
```

In this case, a is regarded as part of the `common` block, and its length is extended appropriately. Such an equivalence must not cause data in two different `common` blocks to become storage associated, it must not cause an extension of the `common` block except at its tail, and two different objects or subobjects in the same `common` block must not become storage associated. It must not cause an object that is not permitted to appear in a `common` block to become associated with an object in a `common` block.

A `common` block may be declared in a module, and its variables accessed by use association. Variable names in a `common` block in a module may be declared to have the `private` attribute, but this does not prevent associated variables being declared elsewhere through other `common` statements.

An individual variable in a `common` block may not be given the `save` attribute, but the whole block may. If a `common` block has the save attribute in any scoping unit other than the main program, it must have the save attribute in all such scoping units. The general form of the `save` statement is

save *[[::] saved-entity-list]*

where *saved-entity* is *variable-name* or *common-block-name*. A simple example is

```
save /change/
```

A blank `common` always has the `save` attribute.

Data in a `common` block without the `save` attribute become undefined on return from a subprogram unless the block is also declared in the main program or in another subprogram that is in execution.

A variable in a `common` block may not be referenced by a simple procedure (Section 23.7.1).

A `common` block (but not an individual variable in a `common` block) may be given the `bind` attribute, perhaps with a separate binding label, for example

```
use iso_c_binding
common /com1/ r, s
common /com2/ t, u
real(c_float) :: r, s, t, u
bind(c) :: /com1/
bind(c, name='CommonBlock') :: /com2/
```

All the variables in a `common` block with the `bind` attribute must be interoperable. The `common` block automatically gets the `save` attribute. It is capable of interoperating with a C variable whose name has external linkage. If a binding label is given, this is the name of the corresponding C variable; otherwise it is the lower-case version of the Fortran name. Either the `common` block has only one member and the C variable is interoperable with it, or the C variable is of a structure type whose components are interoperable with the corresponding members of the `common` block.

If a `common` block is specified in a `bind` statement, it must be specified in a `bind` statement with the same binding label in every scoping unit in which it is declared.

The use of modules (Section 5.5) obviates the need for `common` blocks.

B.1.8   The `block data` program unit

Non-pointer variables in named `common` blocks may be initialized in `data` statements, but such statements must be collected into a special type of program unit, known as a `block data` program unit. It must have the form

```
block data [ block-data-name ]
      [ specification ] ...
end [ block data [ block-data-name ] ]
```

where each *specification* is a derived-type definition or an `asynchronous`, `bind`, `common`, `data`, `dimension`, `equivalence`, `implicit`, `intrinsic`, `parameter`, `pointer`, `save`, `target`, type declaration (including `double precision`), `use`, or `volatile` statement. A type declaration statement must not specify the `allocatable`, `bind`, `external`, `intent`, `optional`, `private`, or `public` attributes. An example is

```
block data
   common /axes/ i,j,k
   data i,j,k /1,2,3/
end block data
```

in which the variables in the `common` block `axes` are defined for use in any other scoping unit which accesses them.

It is possible to collect many `common` blocks and their corresponding `data` statements together in one `block data` program unit. However, it may be a better practice to have several different `block data` program units, each containing `common` blocks which have some logical association with one another. To allow for this, `block data` program units may be named in order to be able to distinguish them. A complete program may contain any number of `block data` program units, but only one of them may be unnamed. A `common` block must not appear in more than one `block data` program unit. It is not possible to initialize blank `common`.

The name of a `block data` program unit may appear in an `external` statement. When a processor is loading program units from a library, it may need such a statement in order to load the `block data` program unit.

The use of modules (Section 5.5) obviates the need for `block data`.

B.1.9   Statement functions

It may be that within a single program unit there are repeated occurrences of a computation which can be represented as a single statement. For instance, to calculate the parabolic function represented by

$$y = a + bx + cx^2$$

for different values of *x*, but with the same coefficients, there may be references to

```
y1 = 1. + x1*(2. + 3.*x1)
   ⋮
y2 = 1. + x2*(2. + 3.*x2)
   ⋮
```

etc. In Fortran 77 it was more convenient to invoke a so-called **statement function** (now better coded as an internal function, Section 5.6), which must appear after any `implicit` and other relevant specification statements and before the executable statements. The example above would become

```
parab(x) = 1. + x*(2. + 3.*x)
   ⋮
y1 = parab(x1)
   ⋮
y2 = parab(x2)
```

Here, x is a dummy argument, which is used in the definition of the statement function. The variables $x1$ and $x2$ are actual arguments to the function.

The general form is

```
function-name ( [ dummy-argument-list ] ) = scalar-expr
```

where the *function-name* and each *dummy-argument* must be specified, explicitly or implicitly, to be scalar data objects. The function must not be of a parameterized derived type (Section 13.2). To make it clear that this is a statement function and not an assignment to a host array element, we recommend declaring the type by placing the *function-name* in a type declaration statement; this is *required* whenever a host entity has the same name.

The *scalar-expr* must be composed of constants, references to scalar variables, references to functions, and intrinsic operations. If there is a reference to a function, the function must not be a transformational intrinsic nor require an explicit interface, the result must be scalar, and any array argument must be a named array. A reference to a non-intrinsic function must not require an explicit interface. A named constant that is referenced or an array of which an element is referenced must be declared earlier in the scoping unit or be accessed by use or host association. A scalar variable referenced may be a dummy argument of the statement function or a variable that is accessible in the scoping unit. A dummy argument of the host procedure must not be referenced unless it is a dummy argument of the main entry or of an `entry` that precedes the statement function. If any entity is implicitly typed, a subsequent type declaration must confirm the type and type parameters. The dummy arguments are scalar and have a scope of the statement function statement only.

A statement function always has an implicit interface and may not be supplied as a procedure argument. It may appear within an internal procedure, and may reference other statement functions appearing before it in the same scoping unit, but not itself nor any appearing after. A function reference in the expression must not redefine a dummy argument. A statement function is pure (Section 7.8) if it references only pure functions.

A statement function statement is not permitted in an interface block.

Note that statement functions are irregular in that use and host association are not available.

B.1.10   Assumed character length of function results

A non-recursive external function whose result is scalar, character, and non-pointer may have assumed character length, as in Figure B.1. Such a function is not permitted to specify a defined operation. In a scoping unit that invokes such a function, the interface must be implicit and there must be a declaration of the length, as in Figure B.2, or such a declaration must be accessible by use or host association.

Figure B.1 A function whose result is of assumed character length.

```
function copy(word)
    character(len=*) copy, word
    copy = word
end function copy
```

Figure B.2 Calling a function whose result is of assumed character length.

```
program main
    external copy               ! Interface block not allowed.
    character(len=10) copy
    write (*, *) copy('This message will be truncated')
end program main
```

This facility is included only for compatibility with Fortran 77 and is completely at variance with the philosophy of modern Fortran that the attributes of a function result depend only on the actual arguments of the invocation and on any data accessible by the function through host or use association.

Such a function may be replaced by a subroutine whose arguments correspond to the function result and the function arguments.

B.1.11   Alternate return

The alternate `return` facility involves labels being specified as dummy arguments in a subroutine call and the called subroutine controlling whether the flow of control on return is normal, that is, to the next statement, or alternate, that is, to a statement for which a label has been specified.

In the called subroutine, the corresponding dummy arguments are asterisks and the alternate `return` is taken by executing a statement of the form

```
return int-expr
```

where *int-expr* is any scalar integer expression. The value of this expression at execution time defines an index to the alternate `return` to be taken, according to its position in the sequence of alternate returns in the argument list. If the value is less than 1, or greater than the number of alternate `returns`, a normal `return` will be taken. Here is a simple example:

```
subroutine deal(nshuff, nplay, nhand, cards, *, *)
   ⋮
if (nplay.le.0) return 1
if (nshuff .lt. nplay*nhand) return 2
```

For each asterisk dummy argument, the call must specify an asterisk followed by a label, for example

```
call deal(nshuff, nplay, nhand, cards, *2, *3)
call play
   ⋮
2  ...           ! Handle no-player case
   ⋮
3  ...           ! Handle insufficient-cards case
   ⋮
```

Each label must be that of an executable statement of the same inclusive scope. Any number of such alternate returns may be specified, and they may appear in any position in the argument list.

This feature is also available for subroutines defined by entry statements. It is not available for functions or elemental subroutines.

This feature may be replaced by use of an integer argument holding a return code that is used in a following case construct.

B.1.12   The entry statement

A subprogram usually defines a single procedure, and the first statement to be executed is the first executable statement after the header statement. In some cases it is useful to be able to define several procedures in one subprogram, particularly when wishing to share access to some saved local variables or to a section of code. This is possible for external and module subprograms (but not for internal subprograms) by means of the entry statement. This is a statement that has the form

   entry *entry-name* [([*dummy-argument-list*]) [result (*result-name*)]]

and may appear anywhere between the header line and contains (or end if it has no contains) statement of a subprogram, except within a construct. The entry statement provides a procedure with an associated dummy argument list, exactly as does the subroutine or function statement, and these arguments may be different from those given on the subroutine or function statement. Execution commences with the first executable statement following the entry statement.

In the case of a function, each entry defines another function, whose characteristics (that is, shape, type, type parameters, and whether a pointer) are given by specifications for the *result-name* (or *entry-name* if there is no result clause). If the characteristics are the same as for the main entry, a single variable is used for both results; otherwise, they must not be allocatable, must not be pointers, must be scalar, and must both be one of the default integer, default real, double precision real (Appendix A.4), or default complex types, and they are storage associated as for `equivalence` (Section B.1.6). The `result` clause plays exactly the same role as for the main entry.

Each entry is regarded as defining another procedure, with its own name. The names of all these procedures and their result variables (if any) must be distinct. The name of an entry has the same scope as the name of the subprogram. It must not be the name of a dummy argument of any of the procedures defined by the subprogram. An `entry` statement is not permitted in an interface block; there must be another body for each entry whose interface is wanted, using a `subroutine` or `function` statement, rather than an `entry` statement.

An `entry` is called in exactly the same manner as a subroutine or function, depending on whether it appears in a subroutine subprogram or a function subprogram. An example is given in Figure B.3, which shows a search function with two entry points. We note that `looku` and `looks` are synonymous within the function, so that it is immaterial which value is set before the return.

Figure B.3 A search function with two entry points.

```
        function looku(list, member)
        integer looku, list(:), member, looks
!
!       To locate member in an array list.
!       If list is unsorted, entry looku is used;
!       if list is sorted, entry looks is used.
!
!       List is unsorted.
        do looku = 1, size(list)
            if (list(looku) == member) return
        end do
!
!       Not found.
        looku = 0
        return
!
!       Entry point for sorted list.
        entry looks(list, member)
        do looks = 1, size(list)
            if (list(looks) == member) return
            if (list(looks) > member) exit
        end do
!
!       Not found.
        looks = 0
        end function
```

None of the procedures defined by a subprogram is permitted to reference itself, unless the keyword `recursive` is present on the `subroutine` or `function` statement. For a function, such a reference must be indirect unless there is a `result` clause on the `function` or `entry` statement. If a procedure may be referenced directly in the subprogram that defines it, the interface is explicit in the subprogram.

The name of an `entry` dummy argument that appears in an executable statement preceding the `entry` statement in the subprogram must also appear in a `function`, `subroutine`, or `entry` statement that precedes the executable statement. Also, if a dummy argument is used to define the array size or character length of an object, the object must not be referenced unless the argument is present in the procedure reference that is active.

During the execution of one of the procedures defined by a subprogram, a reference to a dummy argument is permitted only if it is a dummy argument of the procedure referenced.

A simple procedure (Section 23.7.1) must not contain an `entry` statement.

The `entry` statement is made unnecessary by the use of modules (Section 5.5), with each procedure defined by an entry becoming a module procedure.

B.1.13 The `forall` statement and construct

When elements of an array are assigned values by a `do` construct such as

```
do i = 1, n
   a(i, i) = 2.0 * x(i)      ! a is rank-2 and x rank-1
end do
```

the processor is required to perform each successive iteration in order and one after the other. This represents a potentially severe impediment to optimization on a parallel processor so, for this purpose, Fortran has the `forall` statement. The above loop can be written as

```
forall (i = 1:n) a(i, i) = 2.0 * x(i)
```

which specifies that the set of expressions denoted by the right-hand side of the assignment is first evaluated in any order, and the results are then assigned to their corresponding array elements, again in any order of execution. The `forall` statement may be considered to be an array assignment expressed with the help of indices. In this particular example, we note also that this operation could not otherwise be represented as a simple array assignment. Other examples of the `forall` statement are

```
forall (i = 1:n, j = 1:m)            a(i, j) = i + j
forall (i = 1:n, j = 1:n, y(i, j) /= 0.) x(j, i) = 1.0/y(i, j)
```

where, in the second statement, we note the masking condition – the assignment is not carried out for zero elements of `y`.

The `forall` construct also exists. The `forall` equivalent of the array assignments

```
a(2:n-1, 2:n-1) = a(2:n-1, 1:n-2) + a(2:n-1, 3:n)      &
                 + a(1:n-2, 2:n-1) + a(3:n, 2:n-1)
b(2:n-1, 2:n-1) = a(2:n-1, 2:n-1)
```

is

```
forall (i = 2:n-1, j = 2:n-1)
   a(i, j) = a(i, j-1) + a(i, j+1) + a(i-1, j) + a(i+1, j)
   b(i, j) = a(i, j)
end forall
```

This sets each internal element of a equal to the sum of its four nearest neighbours and copies the result to b. The `forall` version is more readable. Note that each assignment in a `forall` is like an array assignment; the effect is as if all the expressions were evaluated in any order, held in temporary storage, then all the assignments performed in any order. Each statement in a `forall` construct must fully complete before the next can begin.

A `forall` statement or construct may contain pointer assignments. An example is

```
type element
   character(32), pointer :: name
end type element
type(element)           :: chart(200)
character(32), target :: names(200)
   ⋮
forall (i = 1:200)
   chart(i)%name => names(i)
end forall
```

Note that there is no array syntax for performing, as in this example, an array of pointer assignments.

As with all constructs, `forall` constructs may be nested. The sequence

```
forall (i = 1:n-1)
   forall (j = i+1:n)
      a(i, j) = a(j, i)          ! a is a rank-2 array
   end forall
end forall
```

assigns the transpose of the lower triangle of a to the upper triangle of a.

A `forall` construct can include a where statement or construct. Each statement of a where construct is executed in sequence. An example with a where statement is

```
forall (i = 1:n)
   where ( a(i, :) == 0) a(i, :) = i
   b(i, :) = i / a(i, :)
end forall
```

Here, each zero element of a is replaced by the value of the row index and, following this complete operation, the elements of the rows of b are assigned the reciprocals of the corresponding elements of a multiplied by the corresponding row index.

The complete syntax of the `forall` construct is

```
[ name: ] forall ( [ type-spec :: ] index-spec-list [ , scalar-logical-expr ] )
    [ body ]
end forall [ name ]
```

where *index-spec* is

index-variable-name = lower : upper [: stride]

Each *index-variable-name* is a scalar variable local to the loop, so has no effect on any variable with the same name that might exist outside the loop; however, if *type-spec* is omitted, it has the type and kind it would have if it were such a variable. An index variable must not be redefined within the construct. Within a nested construct, each index variable must have a distinct name. The expressions *lower*, *upper*, and *stride* (*stride* is optional but must be nonzero when present) are scalar integer expressions and form a sequence of values as for a section subscript (Section 7.11); they may not reference any index variable of the same statement but may reference an index variable of an outer forall. Once these expressions have been evaluated, the *scalar-logical-expr*, if present, is evaluated for each combination of index values. Those for which it has the value true are active in each statement of the construct. The *name* is the optional construct name; if present, it must appear on both the forall and the end forall statements.

The *body* itself consists of one or more assignment statements, pointer assignment statements, where statements or constructs, and further forall statements or constructs. The subobject on the left-hand side of each assignment in the *body* should reference each index variable of the constructs it is contained in as part of the identification of the subobject, whether it be a non-pointer variable or a pointer object.[2]

In the case of a defined assignment statement, the subroutine that is invoked must not reference any variable that becomes defined by the statement, nor any pointer object that becomes associated.

A forall construct whose body is a single assignment or pointer assignment statement may be written as a single forall statement.

Procedures may be referenced within the scope of a forall, both in the logical scalar expression that forms the optional mask or, directly or indirectly (for instance as a defined operation or assignment), in the body of the construct. However, since there is a possibility that such a reference might have side-effects, presenting a severe impediment to optimization on a parallel processor (as the order of execution of the assignments could affect the results), all such procedures must be pure (see Section 7.8).

As in assignments to array sections (Section 7.11), it is not allowed to make a many-to-one assignment. The construct

```
forall (i = 1:10)
    a(index(i)) = b(i)        ! a, b and index are arrays
end forall
```

is valid if and only if index(1:10) contains no repeated values. Similarly, it is not permitted to associate more than one target with the same pointer.

---

[2]This is not actually a requirement, but any missing index variable would need to be restricted to a single value to satisfy the requirements of the final paragraph of this section. For example, the statement

```
forall (i = i1:i2, j = j1:j2) a(j) = a(j) + b(i, j)
```

is valid only if i1 and i2 have the same value.

As we have seen, in a `forall` statement or construct all the index variables are local to the construct; for example, in

```
idx = 3
forall (idx=100:200) a(idx, idx) = idx**2
print *, idx
```

the value 3 is printed because the `idx` within the `forall` is not the same as the `idx` outside the `forall`. However, the `idx` within the `forall` has the same type and kind as the one outside would have if it existed; and if it does exist, it has to be a scalar integer variable. This is a bit inconvenient, and although there is no other connection between the index variable and any entity outside the `forall`, changing the type or kind of the outer entity would affect the index variable too.

For this reason we recommend specifying the type and kind of the index variables on the `forall` statement itself; for example, in

```
complex :: i(100)
   ⋮
forall (integer(int64) :: i=1:2_int64**32) a(i) = i*2.0**(-32)
```

the outer variable `i` is a complex array, but has no effect on the index variable because of the declaration in the `forall` statement.

The `forall` construct and statement were added to the language in the hope of efficient execution, but this has not happened, and they have become redundant with the introduction of the `do concurrent` construct (Section 7.16) and the use of array assignment following pointer rank remapping (Section 7.14).

B.1.14   The labelled do construct

A further form of the `do` construct (Section 4.4) makes use of a statement label to identify the end of the construct. In this case, the terminating statement may be either a labelled `end do` statement or a labelled `continue` ('do nothing') statement. The label is, in each case, the same as that on the `do` statement itself. The label on the `do` statement may be followed by a comma. Simple examples are

```
      do 10 i = 1, n
         ⋮
   10 end do
```

and

```
      do 10 while (error>1e-5)
         ⋮
   10 continue
```

and

```
    do 20 i = 1, j
       do 10, k = 1, l
          ⋮
10     continue
20 continue
```

As shown in the last example, each loop must have a separate label.

Labelled do statements are now redundant with the use of a name for the construct and the use of the cycle statement.

B.1.15   Specific names of intrinsic procedures

While all of the intrinsic procedures are generic, some of the intrinsic functions also have specific names for specific versions, which are listed in Tables B.1 and B.2. In the tables, 'Character' stands for default character, 'Integer' stands for default integer, 'Real' stands for default real, 'Double' stands for double precision real, and 'Complex' stands for default complex. Those functions in Table B.2 may be passed as actual arguments to a procedure, or be associated with a procedure pointer, provided they are specified in an intrinsic statement (Section 9.1.4).

Specific names for intrinsic functions have been redundant since Fortran 77. They are now obsolescent as they all have generic names.

Table B.1: Specific intrinsic functions not usable as arguments or targets.

Description	Generic form	Specific name	Argument type	Function type
Conversion to integer	`int(a)`	`int`	Real	Integer
		`ifix`	Real	Integer
		`idint`	Double	Integer
Conversion to real	`real(a)`	`real`	Integer	Real
		`float`	Integer	Real
		`sngl`	Double	Real
Maximum value	`max(a1,a2,...)`	`max0`	Integer	Integer
		`amax1`	Real	Real
		`dmax1`	Double	Double
	`int(max(a1,a2,...))`	`amax0`	Integer	Real
	`real(max(a1,a2,...))`	`max1`	Real	Integer
Minimum value	`min(a1,a2,...)`	`min0`	Integer	Integer
		`amin1`	Real	Real
		`dmin1`	Double	Double
	`int(min(a1,a2,...))`	`amin0`	Integer	Real
	`real(min(a1,a2,...))`	`min1`	Real	Integer
`string_a >= string_b`	`lge(string_a,string_b)`	`lge`	Character	Logical
`string_a > string_b`	`lgt(string_a,string_b)`	`lgt`	Character	Logical
`string_a <= string_b`	`lle(string_a,string_b)`	`lle`	Character	Logical
`string_a < string_b`	`llt(string_a,string_b)`	`llt`	Character	Logical

Table B.2: Specific intrinsic functions usable as arguments or targets.

Description	Generic form	Specific name	Argument type	Function type
Absolute value of	`sign(a,b)`	`isign`	Integer	Integer
a times sign of b		`sign`	Real	Real
		`dsign`	Double	Double
max(x-y, 0)	`dim(x,y)`	`idim`	Integer	Integer
		`dim`	Real	Real
		`ddim`	Double	Double
x × y		`dprod(x,y)`	Real	Double
Truncation	`aint(a)`	`aint`	Real	Real
		`dint`	Double	Double
Nearest whole	`anint(a)`	`anint`	Real	Real
number		`dnint`	Double	Double
Nearest integer	`nint(a)`	`nint`	Real	Integer
		`idnint`	Double	Integer
Absolute value	`abs(a)`	`iabs`	Integer	Integer
		`abs`	Real	Real
		`dabs`	Double	Double
		`cabs`	Complex	Real
Remainder	`mod(a,p)`	`mod`	Integer	Integer
modulo p		`amod`	Real	Real
		`dmod`	Double	Double
Square root	`sqrt(x)`	`sqrt`	Real	Real
		`dsqrt`	Double	Double
		`csqrt`	Complex	Complex
Exponential	`exp(x)`	`exp`	Real	Real
		`dexp`	Double	Double
		`cexp`	Complex	Complex
Natural logarithm	`log(x)`	`alog`	Real	Real
		`dlog`	Double	Double
		`clog`	Complex	Complex
Common logarithm	`log10(x)`	`alog10`	Real	Real
		`dlog10`	Double	Double

Continued...

Table B.2 (cont.)

Description	Generic form	Specific name	Argument type	Function type
Sine	`sin(x)`	`sin`	Real	Real
		`dsin`	Double	Double
		`csin`	Complex	Complex
Cosine	`cos(x)`	`cos`	Real	Real
		`dcos`	Double	Double
		`ccos`	Complex	Complex
Tangent	`tan(x)`	`tan`	Real	Real
		`dtan`	Double	Double
Arcsine	`asin(x)`	`asin`	Real	Real
		`dasin`	Double	Double
Arccosine	`acos(x)`	`acos`	Real	Real
		`dacos`	Double	Double
Arctangent	`atan(x)`	`atan`	Real	Real
		`datan`	Double	Double
	`atan2(y,x)`	`atan2`	Real	Real
		`datan2`	Double	Double
Hyperbolic sine	`sinh(x)`	`sinh`	Real	Real
		`dsinh`	Double	Double
Hyperbolic cosine	`cosh(x)`	`cosh`	Real	Real
		`dcosh`	Double	Double
Hyperbolic tangent	`tanh(x)`	`tanh`	Real	Real
		`dtanh`	Double	Double
Imaginary part	`aimag(z)`	`aimag`	Complex	Real
Complex conjugate	`conjg(z)`	`conjg`	Complex	Complex
Character length	`len(s)`	`len`	Character	Integer
Starting position	`index(s,t)`	`index`	Character	Integer

B.2 Deleted features

The features listed in this section were present in early versions of Fortran but have been progressively deleted from modern versions. Although it can be expected that compilers will continue to support these features for some period, their use should be avoided to ensure long-term portability and to avoid unnecessary compiler warning messages. They are fully described in earlier editions of this book.

Non-integer do indices The do variable and the expressions that specify the limits and stride of a do construct or an implied-do in an input/output statement could be of type default real or double precision real.

Assigned go to and assigned formats Another form of branch statement was actually written in two parts, an assign statement and an assigned go to statement. One use of the assign statement is replaced by character expressions to define format specifiers (Section 10.4).

Branching to an end if statement It was permissible to branch to an end if statement from outside the construct that it terminates. A branch to the following statement is a replacement for this practice.

The pause statement At certain points in the execution of a program it was possible to pause, in order to allow some possible external intervention in the running conditions to be made.

H edit descriptor The H (or h) edit descriptor provided an early form of the character string edit descriptor.

Carriage control Fortran's formatted output statements were originally designed for line printers, with their concept of lines and pages of output. On such a device, the first character of each output record had to be of default kind and was not printed but interpreted as a **carriage control character**. The carriage control characters defined by the standard were:

b	to start a new line
+	to remain on the same line (overprint)
0	to skip a line
1	to advance to the beginning of the next page

Arithmetic if statement The arithmetic if provided a three-way branching mechanism, depending on whether an arithmetic expression has a value which is less than, equal to, or greater than zero. It is incompatible with IEEE arithmetic, and thus virtually all modern computers, because a NaN is neither less than, equal to, nor greater than zero. It involved the use of labels, which can hinder optimization and make code hard to read and maintain. It is replaced by the if statement and construct (Section 4.2).

Shared do-loop termination A do loop could be terminated on a labelled statement other than an end do or continue. Nested do loops could also share the same labelled terminal statement, in which case all the usual rules for nested blocks held, but a branch to the label had to be from within the innermost loop. This offered considerable scope for confusion and unexpected errors. Note that labelled do statements are still included as an obsolescent part of the language; they are now limited such that they form properly nested blocks.

C. Significant examples

C.1 Introduction

In this appendix, we describe some significant examples, for which the code is available on the OUP website; a complete URL is given at the end of each section.

A recurring problem in computing is the need to manipulate a dynamic data structure. This might be a simple homogeneous linked list like the one encountered in Section 2.13.2, but often a more complex structure is required. An example is given in Section C.2.

The aim of parallel computing is to reduce the time to solution for computationally intensive algorithms by exploiting multiple hardware execution units. This usually involves making the share of data and work given to the images as equitable as possible. We illustrate this with three matrix examples. In Section C.3 we consider the multiplication of a large square matrix by a vector; here, a good strategy is to divide the matrix into block rows or block columns, each assigned to an image. In Section C.4 we consider the multiplication of a large triangular matrix by a vector; here, a better strategy is to divide the matrix into more block columns and assign several to each image so that each image has a similar amount of data and work. In Section C.5 we consider the Cholesky factorization of a symmetric matrix; here, blocking by rows or columns is not suitable so we subdivide the matrix into square blocks and assign similar numbers of blocks to each image. The example programs include timing measurements (based on the system_clock intrinsic) and their conversions to floating point performance to give an impression of the performance gain obtained through using multiple images. However, no effort was made to assure highly reproducible timing results; significant variations were observed that seem to be largely due to the communication that is included.

C.2 Object-oriented list example

The example oo.f90 consists of a module that provides two types – a list type anylist and an item type anyitem – for building heterogeneous doubly linked linear lists, plus a simple item constructor function newitem. Operations on the list or on items are provided by type-bound procedures. Each list item has a scalar value which may be of any type; when creating a new list item, the required value is copied into the item. A list item can be in at most one list at a time.

List operations include inserting a new item at the beginning or end of the list, returning the item at the beginning or end of the list, and counting, printing, or deleting the whole list. Printing is to the unit np which by default is set to the standard output unit.

Operations on an item include removing it from a list, returning the next or previous item on the list, changing the value of the item, and printing or deleting the item. When traversing the list backwards (via the prev function), the list is circular; that is, the last item on the list is previous to the first. When traversing the list forwards (via the next function), a null pointer is returned after the last item.

Internally, the module uses private pointer components (firstptr, nextptr, prevptr, and upptr) to maintain the structure of the lists.

The item print operation may be overridden in an extension to anyitem to provide printing capability for user-defined types; this is demonstrated by the type myitem. All the other procedures are non-overridable, so that extending the list type cannot break the list structure.

The example source code can be downloaded as the file *oo.f90* accessible through the bullet for this book on http://www.oup.com/fortran2023

C.3   Matrix–vector multiplication

Here we consider the multiplication of a square matrix with a vector v_j, $j = 1, 2, \ldots, n$:

$$b_i = \sum_{j=1}^{n} A_{ij} v_j, \quad i = 1, 2, \ldots, n.$$

The simplest way to divide up the work among images is by matrix column blocks A^i, $i = 1, 2, \ldots,$ num_images(), of equal or nearly equal sizes, as shown in Figure C.1. The input vector v is distributed among images as the blocks v^i, $i = 1, 2, \ldots,$ num_images() with the same block sizes as the matrix columns, and the result array b must have size n. This is known as a **block distribution**.

We have chosen to hold the details of the distribution in a variable of a derived type that is defined in a module and has private components. Procedures are supplied for constructing an object of the type, for returning the local block size, and for the mappings between the global index (matrix column index) and the combination of image index and local column index. There is a built-in choice of distribution, but this can be changed by altering the module and no change to the code that uses the module is needed. Our module is called mod_block and it chooses a threshold t and a block size b. It uses the block size b for blocks $1, 2, \ldots, t$ and the block size $b - 1$ for the remaining blocks.

The module defines the type block_desc and the following procedures:

block_desc (global_size, coextent) returns an object of type block_desc, given the global problem size and the number of coindex values across which data distribution will be performed.

get_size(desc [, coindex]) returns the local block size for the specified coindex. If the optional second argument is omitted, it returns the global problem size. The value zero is returned in the case of an invalid argument value.

Figure C.1 Block column distribution for matrix–vector multiplication.

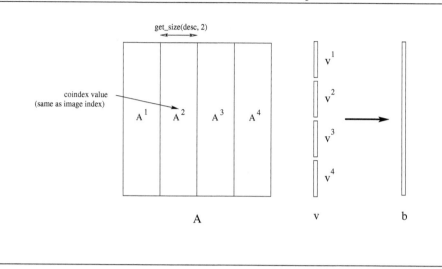

global_index (desc, local_index, coindex) calculates the value of the global
   index from the local index and the coindex. The value zero is returned in the case of
   invalid argument values.

coextent (desc) returns the coextent associated with the descriptor.

coindex (desc, global_index) returns the value of the coindex from the global
   index. The value zero is returned in the case of invalid argument values.

local_index (desc, global_index) returns the value of the local index from the
   global index. The value zero is returned in the case of invalid argument values.

On the executing image, the local index lies between 1 and get_size(desc, this_image()).
All calculations are performed locally on the executing image, but the descriptor object must
be consistently initialized on all images that use it.
   The parallelized processing loop for setting the local blocks reads

```
type(block_desc) :: desc
real, allocatable :: a(:,:)  ! Local part of A
desc = block_desc(n, num_images())
mb = get_size(desc, this_image())
allocate(a(n, mb))
do jl = 1, mb
   j = global_index(desc, jl, this_image())
   do i = 1, n
        a(i, jl) = matval(i, j)  ! Function call
   end do
end do
```

Note that one could allocate a to use local indices with the same values as the global indices:

```
real, allocatable :: a(:,:)   ! Local part of A
⋮
mb = get_size(desc, this_image())
j = global_index(desc, 1, this_image())
allocate(a(n, j:j+mb-1))
```

and make no other calls of global_index or local_index. Such remapping of bounds is not possible in all situations, for example if coarrays are used,[1] or if arrays are statically sized.

The parallel matrix–vector multiplication proceeds as follows:

1. Locally form the matrix–vector product $A^i \cdot v^i$, using the matmul intrinsic. After this step, each array element of b on each image will contain a partial sum of the total result.

2. Execute the collective subroutine co_sum on b to put the final summed result on all images.

Because the only data dependency between images can be resolved through a collective subroutine invocation, none of the objects involved needs to be declared a coarray, and no image control statements are needed. The module matrix_vector contains the implementation of this in the subroutine parallel_mvc, which is invoked from the main program matrix_vector_example.

Alternatively, the matrix may be partitioned by rows into blocks, as illustrated in Figure C.2. The input vector v is replicated on each image. The result vector b is distributed

Figure C.2 Block row distribution for matrix–vector multiplication.

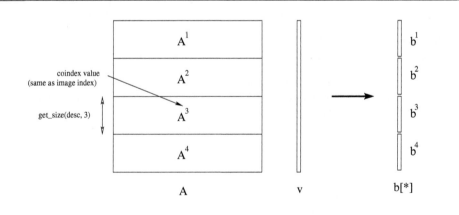

across images in the same way as the rows of A. The alternative parallel matrix–vector multiplication proceeds as follows:

---

[1]An allocatable or pointer component of a coarray of derived type could be assigned image-dependent bounds. It is worth checking the performance impact of such usage for any compiler to be deployed.

1. Locally form the matrix–vector product $A^i \cdot v^i$, using the `matmul` intrinsic. After this step, b^i on image i will hold a section of the result vector.

2. Because matrix–vector operations often arise in the context of an iterative procedure, it is necessary to copy the result data to all other images. To achieve this, b is declared as an array coarray, so the program can address b^i as `b(:)[i]` from any image. The necessary data transfers to v as well as the minimum needed synchronization are included in the procedure.[2]

Note that because b is an array coarray, it must be allocated with the maximum block size on all images. The module `matrix_vector` contains the implementation of this in the subroutine `parallel_mvr`.

The main program `matrix_vector_example` tests and times both versions of matrix–vector multiplication. We use the same array v in both cases, allocating it of different sizes in the two cases. We cannot do the same for b because it is a coarray in only one of the cases. We use the name bb for the replicated whole vector.

The program requests a value for the matrix order from the user, then constructs an example of this size and runs both algorithms, checking the result and writing performance metrics to the default output. This is repeated in a do loop, terminated by the user entering a non-positive matrix order.

The example source code can be downloaded as the file *mv.f90* accessible through the bullet for this book on `http://www.oup.com/fortran2023`

C.4 Triangular matrix by vector multiplication

We now consider multiplying a triangular matrix with a vector $v_j, j = 1, 2, \ldots, n$,

$$b_i = \sum_{j=1}^{i} A_{ij} v_j, \quad i = 1, 2, \ldots, n.$$

A simple block distribution is not suitable for an efficient parallel design because it leads to an imbalance of both the memory resources and the work assigned to each image. Instead, we use a **block-cyclic distribution** that assigns block columns to images in a round-robin manner as illustrated in Figure C.3 for the case of 10 blocks being processed on three images. This leads to better memory and work distribution. We have chosen to give the blocks the same number of columns, say m, except that the final one is smaller if the matrix order is not an integer multiple of m. If the number of images is p, image i hosts blocks $i, i+p, i+2p, \ldots$ For a good load balance $p \times m$ needs to be significantly less than n.

We have again chosen to hold the details of the distribution in a module holding a derived type with private components and procedures for constructing an object of the type and returning information about the distribution. By declaring the components as `private` we allow the distribution to be altered without altering code that uses the module. Our module is called `mod_block_cyclic` and the derived type is called `block_cyclic_desc`. Having

---

[2]Strictly, one could make use of `co_broadcast` on v here, but for illustrative purposes we've decided to keep the coarray-based data exchange.

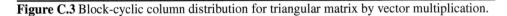

Figure C.3 Block-cyclic column distribution for triangular matrix by vector multiplication.

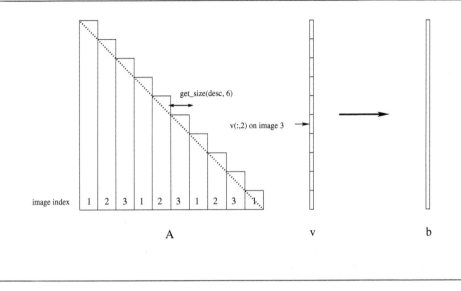

more than one block column on each image makes this more complicated, but there are six procedures very like those of Section C.3. They are:

block_cyclic_desc (global_size, block_size, coextent)   returns an object, given the global problem size, the default block size, and the number of coindex values across which data distribution will be performed.

get_size (desc, block_index)   returns the block size for the specified global block index, or the value zero if the block index is out of range. If the optional second argument is omitted, it returns the global problem size.

global_indices (desc, coindex)   returns the array of global start indices corresponding to the first element of each block associated with the specified coindex value. The array has size zero if the coindex value is out of range.

coextent (desc)   returns the coextent associated with the descriptor.

coindex (desc, block_index)   returns the coindex associated with the specified global block index. The value zero is returned if the global block index is out of range.

local_index (desc, block_index)   returns the local block index associated with the specified global block index. The value zero is returned if the global block index is out of range.

In addition, we need these three functions:

num_blocks (desc, coindex)   returns the number of blocks associated with the specified coindex value, or the value zero if the coindex value is out of range. It returns the total number of blocks if the optional coindex argument is not present.

block_indices (desc, coindex) returns the array of global block indices associated with the specified coindex value. The array has size zero if the coindex value is out of range.

block_sizes (desc, coindex) returns the array of sizes of blocks associated with the specified coindex value. The array has size zero if the coindex value is out of range.

In order to deal with the varying number of rows in each column block, the derived type matrix_block is defined to hold a single block column:

```
integer, parameter :: rk = kind(1.0d0)

type, public :: matrix_block
    real(rk), allocatable :: data(:,:)
end type matrix_block
```

The array component data is allocated to have the same indices as the block of *A* to which it corresponds. The set of blocks on the executing image is held as a rank-one array of this type,

```
type(matrix_block), allocatable :: a_tr(:)
```

and is allocated to have size equal to the number of blocks on the image. For reasonably small block sizes, this wastes only a small amount of memory in the top right corner of each block column, and the per-image aggregate size of all blocks does not significantly decrease with increasing image index, with the overall beneficial effect of reducing the load imbalance.[3] The vector *v* must be distributed in the same way as the columns of *A*; this is achieved by making it a rank-two array, where the second dimension indexes the blocks on the executing image.

The parallelized processing loop for setting up the matrix might read as shown in Figure C.4. The triangular matrix by vector multiplication proceeds in parallel as follows:

1. Locally form the matrix–vector product $A \cdot v(:, ib)$, using the matmul intrinsic for each block *ib*.

2. Add the result vectors for all local blocks. After this step, each array element of *b* on each image will contain a partial sum of the total result.

3. Execute the collective subroutine co_sum on *b* to put together the final result on all images.

The module triangular_blockcolumn_matrix contains the implementation of this in the subroutine parallel_triangular_mvc, which is invoked from the main program triangular_matrix_vector_example. The program requests values for the matrix order and block size from the user, then constructs an example of this order and block size and runs the algorithm, checking the result and writing performance metrics to the default output. This is repeated in a do loop, terminated by the user entering a non-positive matrix order.

The example source code can be downloaded as the file *trmv.f90* accessible through the bullet for this book on http://www.oup.com/fortran2023

---

[3] A lower limit for the block size is usually dictated by the serial processing efficiency for the high-aspect-ratio matrix multiplication.

Figure C.4 A parallelized processing loop for setting up a matrix.

```
type(block_cyclic_desc) :: desc_cy
integer, allocatable :: bsizes(:), start_ind(:)
type(matrix_block), allocatable :: a_tr(:)

desc_cy = block_cyclic_desc(n, default_blocksize, num_images())
bsizes = block_sizes(desc_cy, this_image())
start_ind = global_indices(desc_cy, this_image())

allocate( a_tr(size(bsizes,1)) )
do ib=1, size(a_tr)
   is = start_ind(ib)
   allocate( a_tr(ib)%data(is:n, is:is+bsizes(ib)-1), source=0.0_rk )
   do j = is, is + bsizes(ib) - 1
      do i = is, n
         a_tr(ib)%data(i, j) = matval(i, j) ! Function call
      end do
   end do
end do
```

C.5   Cholesky factorization

We now consider the Cholesky factorization of a real symmetric positive-definite matrix A, so that

$$A = L \cdot L^{\mathrm{T}},$$

where L is a lower triangular matrix. We could subdivide the matrix into block columns as we did in Section C.4, but with subdivision into blocks there are better prospects of balancing the load and limiting the communication between images. We again make all the blocks have the same size, say bs, except that the final block is smaller if the matrix order is not a multiple of the block size, see Figure C.4. Only the lower triangular part need be stored and we eventually overwrite A by L.

We label each block by its block row and block column positions, as illustrated in Figure C.5. We use a pair of coindices to label the images and allocate the blocks to images in a round-robin fashion by both rows and columns. For example, with six images distributed over two rows and three columns, the blocks of Figure C.5 are allocated to images as illustrated in Figure C.6. In general, the images are distributed over nimg_row rows and nimg_col columns, where nimg_row*nimg_col is equal to num_images() with nimg_row and nimg_col as close as possible to each other.

To hold the array A, we use the type

```
integer, parameter :: rk = kind(1.0d0)
type :: matrix_block
   real(rk), allocatable :: data(:,:)
end type matrix_block
```

Figure C.5 Symmetric blocked matrix.

$$
\left[\begin{array}{c|cccc}
A_{11} & A_{21}^T & A_{31}^T & \cdots & A_{k1}^T \\
\hline
A_{21} & A_{22} & A_{32}^T & \cdots & A_{k2}^T \\
A_{31} & A_{32} & A_{33} & \cdots & A_{k3}^T \\
\vdots & \vdots & \vdots & \ddots & \\
A_{k1} & A_{k2} & A_{k3} & & A_{kk}
\end{array}\right]
$$

Figure C.6 The images holding the blocks of Figure C.5.

```
1,1
2,1  2,2
1,1  1,2  1,3
2,1  2,2  2,3  2,1
1,1  1,2  1,3  1,1  1,2
........................
............................
1,1  1,2  1,3  1,1  1,2 ....... 1,2
```

and the array

```
type(matrix_block), allocatable :: tr(:,:)
```

which is allocated on each image thus

```
allocate ( tr(nbr,nbc) )
```

where

- `nbr` is the number of block rows held partially on the image, and
- `nbc` is the number of block columns held partially on the image.

The set of arrays `tr` on all the images is large enough to hold a square matrix of the size of A, but there is little wastage of memory because the `data` components of the upper triangular part remain unallocated. Each `data` component of the lower triangular part is allocated not just to be of the correct size but for its bounds to be exactly those of the corresponding submatrix of A. There is a small amount of wastage for the upper-triangular parts of the blocks on the diagonal.

To calculate, we start as illustrated in Figure C.7. The first stage proceeds as follows:

Figure C.7 First stage of blocked Cholesky factorization.

$$
\begin{bmatrix}
L_{11} & & 0 & & \\
L_{21} & & & & \\
L_{31} & & & & \\
\vdots & & I & & \\
& & \text{(unit matrix)} & & \\
L_{k1} & & & &
\end{bmatrix}
*
\begin{bmatrix}
L_{11}^{\mathrm{T}} & L_{21}^{\mathrm{T}} & L_{31}^{\mathrm{T}} & \cdots & L_{k1}^{\mathrm{T}} \\
 & A_{22}^{(2)} & A_{23}^{(2)} & \cdots & A_{2k}^{(2)} \\
 & A_{32}^{(2)} & A_{33}^{(2)} & & \\
0 & \vdots & & \ddots & \\
 & A_{k2}^{(2)} & & & A_{kk}^{(2)}
\end{bmatrix}
$$

1. Use a serial algorithm to obtain L_{11} such that $A_{11} = L_{11} \cdot L_{11}^{\mathrm{T}}$.
2. For $i = 2,3,\ldots,k$, solve $A_{i1} = L_{i1} \cdot L_{11}^{\mathrm{T}}$ for L_{i1}.
3. For $i = 2,3,\ldots,k$ and $j = 2,3,\ldots,i$, compute $A_{ij}^{(2)} = A_{ij} - L_{i1} \cdot L_{j1}^{\mathrm{T}}$.

In practice, we overwrite A_{11} by L_{11}, overwrite A_{i1} by L_{i1}, $i = 2,3,\ldots,k$, and overwrite A_{ij} by $A_{ij}^{(2)}$, $i = 2,3,\ldots,k$ and $j = 2,3,\ldots,i$. We will refer to these three steps as 'Cholesky factorization', 'pivot-column update', and 'non-pivot update'. Each is performed as a serial calculation on the image on which its block of A resides.

To perform a pivot-column update, we use the coarray

```
real(rk), allocatable :: pivot_block(:,:)[:,:]
```

which is allocated on each image thus

```
allocate( pivot_block(bs, bs)[nimg_row,*] )
```

to hold a copy of the pivot block L_{11} on each image $[i,1]$, $i = 1,2,\ldots,$nimg_row. Immediately after L_{11} has been calculated, the transfer is made and events are posted thus:

```
do ic = 1, nimg_row
   pivot_block(1:bs, 1:bs)[ic, 1] = tr(1,1)%data
   event post( ev_pivot[ic, 1] )
end do
```

Each image $[i,1]$, $i = 1,2,\ldots,$nimg_row, executes an event wait statement for its ev_pivot variable before performing its pivot-column updates. Note that communication of blocks between images occurs only nimg_row-1 times. If we had made tr a coarray and accessed the pivot block L_{11} directly, communication of blocks between images would have occurred $k - k/$nimg_row times.

The non-pivot updates involve all the images. For an image to perform an update, it needs to access both L_{i1} and L_{j1}. Working by columns implies that copies of all blocks L_{i1} need to be accessible until all the block columns have been updated, which means that we need a temporary block for every row and need to retain it for the whole of the update of $A^{(2)}$. We have chosen to use the coarray

```
type(matrix_block), allocatable :: wk(:)[:,:]
```

for this purpose and allocate it thus

```
allocate( wk(nbrmax)[nimg_row,*] ) ! nbrmax is the maximum value
do ib = 1, nbrmax                   ! of nbr across images
   allocate( wk(ib)%data(bs,bs) )
end do
```

before use.[4] We use me_row and me_col to denote the coindices of wk on the current image. Immediately after L_{i1} has been calculated, the transfer is made and events are posted thus:

```
do jc = 1, ucobound(wk, 2)
   wk(ib)[me_row, jc]%data(1:idim, 1:jdim) = tr(ib, jb)%data
   event post ( ev_row(ib)[me_row, jc] )
end do
```

Each of the receiving images executes an event wait statement for its ev_row variables before performing any of its updates. For example, in the case illustrated in Figure C.6 image [1,2] waits for copies of $L_{3,1}, L_{5,1}, \ldots, L_{k,1}$ to be available. Note that communication of blocks between images occurs only nimg_col-1 times. If we had made tr a coarray and accessed the blocks directly, communication of blocks between images would have occurred $k - k/\text{nimg\_col}$ times.

The off-diagonal updates in block column j need access to L_{j1}. Unfortunately, this will have been copied to all the images that hold block row j, many of which will not hold A_{ij}, $i > j$. However, the image that holds A_{jj} accesses L_{j1} via a copy in the coarray wk and will know that this is available before updating A_{jj}. If it posts events thus,

```
do ic = 1, nimg_row
   event post ( ev_col(j)[ic, me_col] )
end do
```

the images that hold block column j can perform a wait,

```
event wait ( ev_col(j) )
```

before making a remote access to the coarray wk to place a copy of L_{j1} in a local array:

```
real(rk), allocatable :: temp(:,:)
```

This can be done just once for all the below-diagonal updates of blocks A_{ij} in column j.

For $\ell = 2, \ldots, k$ we perform the same operations on the submatrix of block rows and columns $i = \ell, \ell + 1, \ldots, k$ and $j = \ell, \ell + 1, \ldots, k$. In each stage, the symmetry of the iterated submatrix $A^{(\ell+1)}$ is preserved. This will result in A being replaced by L. Note that the coarray wk can be reused for each ℓ.

---

[4] An alternative is to use the coarray

```
real(rk), allocatable :: wk(:,:,:)[:,:]
```

and allocate it thus:

```
allocate( wk(bs,bs,nbrmax)[nimg_row,*] )
```

At the time of writing some compilers execute this more efficiently.

For the Cholesky factorization, we use the LAPACK[5] subroutine `dpotrf`, for the pivot-column update we use the Level 3 BLAS[6] subroutine `dtrsm`, for the non-pivot update of a block on the diagonal, we use the Level 3 BLAS subroutine `dsyrk`, and for the non-pivot update of a block not on the diagonal, we use the Level 3 BLAS subroutine `dgemm`. Interfaces with a very brief description of their actions are in the module `la_interfaces`. Optimized versions are provided by most vendors as object code and we strongly recommend that they be accessed. We have written our code in terms of the kind parameter `rk` for the reals, but the BLAS and LAPACK codes were written in Fortran 77 and use `double precision` only. If another precision is desired, the subroutine names will need to be changed.

In our code, each image performs each update of each block that it owns and in the same order as that of the pivot steps from which they originate. Each image does this by executing an outer loop over the pivot steps in which it calculates its part of $A^{(\ell+1)}$. This loop is labelled `pcols` in the code. If the image is involved in the pivot block column, it first performs its actions for this block column (loop labelled `brows`). All images next perform all their non-pivot updates for this block pivot in a loop over block columns (labelled `rcols`) containing a loop over block rows (labelled `rrows`).

The module `triangular_matrix` contains the implementation of this in the subroutine `parallel_cholesky`, which is invoked from the main program `cholesky_example`. The program requests values for the matrix order, block size, and image aspect (1, low, for `nimg_row<=nimg_col`, or 2, high, otherwise), then constructs an example of this order, block size, and image aspect and runs the algorithm. The result is checked for correctness by comparing it with a serial computation, provided the order is no greater than `max_check`. It is timed and a speed in Gflops/s is reported.

The example source code can be downloaded as the file *chol.f90* accessible through the bullet for this book on `http://www.oup.com/fortran2023`

---

[5]*LAPACK Users' Guide*, E. Anderson, Z. Bai, C. Bischof, S. Blackford, J. Demmel, J. Dongarra, J. Du Croz, A. Greenbaum, S. Hammarling, A. McKenney, and D. Sorensen, SIAM, 1999.

[6]'A set of level 3 basic linear algebra subprograms', J. Dongarra, J. Du Croz, S. Hammarling, and I. Duff, *ACM Transactions on Mathematical Software*, 16 (1): 1–17, 1990.

D. Solutions to exercises

Specimen solutions to many of the exercises are in files accessible through the bullet for this book on `http://www.oup.com/fortran2023` For shorthand, we will refer to these files as 'on the OUP site'.

Note: A few exercises have been left to the reader.

Chapter 2

1.

```
b is less than m          true
8 is less than 2          false
* is greater than T       not determined
$ is less than /          not determined
blank is greater than A   false
blank is less than 6      true
```

2.

```
x = y                                correct
a = b+c ! add                        correct, with commentary
word = 'string'                      correct
a = 1.0; b = 2.0                     correct
a = 15. ! initialize a; b = 22. ! and b
                                     incorrect (embedded commentary)
song = "Life is just&                correct, initial line
   & a bowl of cherries"             correct, continuation
chide = 'Waste not,                  incorrect, trailing & missing
     want not!'                      incorrect, leading & missing
c(3:4) = 'up"                        incorrect (invalid form of character constant)
```

3.

-43	integer	'word'	character
4.39	real	1.9-4	not legal
0.0001e+20	real	'stuff & nonsense'	character
4 9	not legal	(0.,1.)	complex
(1.e3,2)	complex	'I can''t'	character
'(4.3e9, 6.2)'	character	.true._1	logical[1]
e5	not legal	'shouldn' 't'	not legal
1_2	integer[1]	"O.K."	character
z10	not legal	z'10'	hexadecimal

---

[1]Legal provided the kind is available.

4.

name	legal	name32	legal
quotient	legal	123	not legal
a182c3	legal	no-go	not legal
stop!	not legal	burn_	legal
no_go	legal	long__name	legal

5.

```
real, dimension(11)      ::  a
real, dimension(0:11)    ::  b
real, dimension(-11:0)   ::  c
real, dimension(10,10)   ::  d
real, dimension(5,9)     ::  e
real, dimension(5,0:1,4) ::  f
```

a(1), a(10), a(11), a(11)
b(0), b(9), b(10), b(11)
c(-11), c(-2), c(-1), c(0)
d(1,1), d(10,1), d(1,2), d(10,10)
e(1,1), e(5,2), e(1,3), e(5,9)
f(1,0,1), f(5,1,1), f(1,0,2), f(5,1,4)

Array constructor: (/ (i, i = 1,11) /)

6.

c(2,3)	legal	c(4:3)(2,1)	not legal
c(6,2)	not legal	c(5,3)(9:9)	legal
c(0,3)	legal	c(2,1)(4:8)	legal
c(4,3)(:)	legal	c(3,2)(0:9)	not legal
c(5)(2:3)	not legal	c(5:6)	not legal
c(5,3)(9)	not legal	c(,)	not legal

7.

```
i) type vehicle_registration
      character(len=2) :: letters1
      integer          :: digits
      character(len=3) :: letters2
   end type vehicle_registration

ii) type circle
       real                 ::   radius
       real, dimension(2) ::   centre
    end type circle

iii) type book
        character(len=20)                  :: title
        character(len=20), dimension(2) :: author
        integer                            :: no_of_pages
     end type book
```

Derived type constants:

```
vehicle_registration('OY', 57, 'ANF')
circle(15.1, [ 0., 0. ])
book("Pilgrim's Progress", [ 'John  ', 'Bunyan' ], 250 )
```

8.

t	array	t(4)%du(1)	scalar
t(10)	scalar	t(5:6)	array
t(1)%du	array	t(5:5)	array (size 1)

9.

  i) `integer, parameter   :: twenty = selected_int_kind(20)`
     `integer(kind=twenty) :: counter`

  ii) `integer, parameter :: high = selected_real_kind(12,100)`
     `real(kind = high)  :: big`

  iii) `character(kind=2)  :: sign`

Chapter 3

1.

`a+b`	valid	`-c`	valid
`a+-c`	invalid	`d+(-f)`	valid
`(a+c)**(p+q)`	valid	`(a+c)(p+q)`	invalid
`-(x+y)**i`	valid	`4.((a-d)-(a+4.*x)+1)`	invalid

2.

```
c+(4.*f)
((4.*g)-a)+(d/2.)
a**(e**(c**d))
((a*e)-((c**d)/a))+e
(i .and. j) .or. k
((.not. l) .or. ((.not. i) .and. m)) .neqv. n
((b(3).and.b(1)).or.b(6)).or.(.not.b(2))
```

3.

$$3+4/2 = 5 \qquad\qquad 6/4/2 = 0$$
$$3.*4**2 = 48. \qquad 3.**3/2 = 13.5$$
$$-1.**2 = -1. \qquad (-1.)**3 = -1.$$

4.

```
ABCDEFGH
ABCD0123
ABCDEFGu          u = unchanged
ABCDbbuu          b = blank
```

5.

`.not.b(1).and.b(2)`	valid	`.or.b(1)`	invalid
`b(1).or..not.b(4)`	valid	`b(2)(.and.b(3).or.b(4))`	invalid

6.

`d <= c`	valid	`p < t > 0`	invalid
`x-1 /= y`	valid	`x+y < 3 .or. > 4.`	invalid
`d < c.and.3.0`	invalid	`q == r .and. s>t`	valid

7.

  i) `4*s`

  ii) `b*h/2.`

  iii) `(4./3.)*pi*r**3`   (assuming pi has value π)

8.
```
integer :: n, one, five, ten, twenty_five
twenty_five = (100-n)/25
ten         = (100-n-25*twenty_five)/10
five        = (100-n-25*twenty_five-10*ten)/5
one         = 100-n-25*twenty_five-10*ten-5*five
```

9.

a = b + c	valid
c = b + 1.0	valid
d = b + 1	invalid
r = b + c	valid
a = r + 2	valid

10.

a = b	valid	c = a(:,2) + b(5,:5)	valid
a = c+1.0	invalid	c = a(2,:) + b(:,5)	invalid
a(:,3) = c	valid	b(2:,3) = c + b(:5,3)	invalid

11.
```
type list
    real :: value
    type(list), pointer :: next, previous
end type
type(list), pointer :: current, new
allocate(new)
new%value = value
new%next => current
new%previous => current%previous
current%previous%next => new
current%previous => new
```

Chapter 4

1. See *ch4.1.f90* on the OUP site.
Note: A simpler method for performing this operation will become apparent in Section 7.11.

2. See *ch4.2.f90* on the OUP site.

6. See *ch4.6.f90* on the OUP site.

7.
```
type(entry), pointer :: first, current, previous
current => first
if (current%index == 10) then
    first => first%next
else
    do
        previous => current
        current => current%next
```

```
        if (current%index == 10) exit
      end do
      previous%next => current%next
  end if
```

Chapter 5

1. See *ch5.1.f90* on the OUP site.
   Note: A simpler method will become apparent in Chapter 9.

2. See *ch5.2.f90* on the OUP site.

3. See *ch5.3.f90* on the OUP site.

4. See *ch5.4.f90* on the OUP site.

5. See *ch5.5.f90* on the OUP site.

7. See *ch5.7.f90* on the OUP site.

Chapter 6

1.
```
  type(real_polynomial) a
  a%coeff = [ 1, 2, 3, 4 ]
  a = real_polynomial([a%coeff, 5., 6.])
```

2.
```
  type(emfield) a, temp
  allocate (a%strength(4, 6))
  a%strength = 1.0
  temp = a                    ! automatic allocation of temp%content
  deallocate (a%strength)
  allocate (a%strength(0:5, 0:8))
  a%strength(1:4, 1:6) = temp%strength
  a%strength(0:5:5, :) = 0
  a%strength(1:4, 0) = 0
  a%strength(1:4, 7:8) = 0
```

3.
```
  type(emfield) a, temp
  allocate (a%strength(4, 6))
  a%strength = 1.0
  temp = a
  a = emfield(reshape( [(a%strength(:,i),0.,0.,i=1,6),      &
                        (0.,0.,0.,0.,0.,0.,0.,0.,i=7,9)], [6,9] ))
```

4.
```
allocate(temp(size(b,1)+2,size(b,2)+2))
temp = 0
temp(2:size(b,1)+1, 2:size(b,2)+1) = b
call move_alloc(temp,b)
```

Chapter 7

1.

i) `a(1, :)`

ii) `a(:, 20)`

iii) `a(2:50:2, 2:20:2)`

iv) `a(50:2:-2, 20:2:-2)`

v) `a(1:0, 1)`

2.
```
where (z.gt.0) z = 2*z
```

3.
```
integer, dimension(16) :: j
```

4.

w	explicit-shape
a, b	assumed-shape
d	deferred-shape, pointer

5.
```
real, pointer :: x(:, :, :)
x => tar(2:10:2, 2:20:2, 2:30:2)%du(3)
```

6.
```
ll = ll + ll
ll = mm + nn + n(j:k+1, j:k+1)
```

7. See *ch7.7.f90* on the OUP site.

8. See *ch7.8.f90* on the OUP site.

Chapter 8

1.

i) `integer, dimension(100) :: bin`

ii) `real(selected_real_kind(6, 4)), dimension(0:20, 0:20) :: &`
`                                    iron_temperature`

iii) `logical, dimension(20) :: switches`

iv) `character(len=70), dimension(44) :: page`

2.
The value of the first i is 3.1, but may be changed;
the value of the second i is 3.1, but may not be changed.

3.

   i) `integer, dimension(100) :: i = (/ (0, k=1, 100) /)`

   ii) `integer, dimension(100) :: i = (/ (0, 1, k=1, 50) /)`

   iii) `real, dimension(10, 10) :: x = reshape((/ (1.0, k=1, 100) /), &`
                                                `(/10, 10/) )`

   iv) `character(len=10) :: string = '0123456789'`

Note: the reshape function will be met in Section 9.15.3.

4.

   i) `type(person) boss = person('Smith', 48.7, 22)`

   ii)   a) `type(entry) new = entry(0.0, 0, null())`

          b) `type(entry) current`
              `data current%value, current%index /1.0, 1/`

5.
All are constant expressions except for:

   iv) because of the real exponent; and

   viii) because of the pointer component.

Chapter 9

1. See *ch9.1.f90* on the OUP site.
Historical note: A similar problem was set in one of the first books on Fortran programming – *A FOR-TRAN Primer*, E. Organick, Addison-Wesley, 1963. It is interesting to compare Organick's solution, written in FORTRAN II, on p. 122 of that book, with the one above. (It is reproduced in the *Encyclopedia of Physical Science and Technology*, Academic Press, 1987, vol. 5, p. 538.)

2. See *ch9.2.f90* on the OUP site.

3.
F	p1 and p2 are associated with the same array elements, but in reverse order
T	p1 and p2(4:1:-1) are associated with exactly the same array elements, a(3), a(5), a(7), a(9)

4.
5	1	a has bounds 5:10 and a(:) has bounds 1:6
5	1	p1 has bounds 5:10 and p2 has bounds 1:6
1	1	x and y both have bounds 1:6

5. See *ch9.5.f90* on the OUP site.

6. See *ch9.6.f90* on the OUP site.

Chapter 10

1.
```
character, dimension(3,3) :: tic_tac_toe
integer                   :: unit
 ⋮
write (unit, '(t1, 3a2)') tic_tac_toe
```

3. See *ch10.3.f90* on the OUP site.

5. See *ch10.5.f90* on the OUP site.

Chapter 11

1.
  i) `print '(a/ (t1, 10f6.1))', ' grid', grid`
  ii) `print '(a, " ", 25i5)', ' list', (list(i), i = 1, 49, 2)`
  or
  `print '(a, " ", 25i5)', ' list', list(1:49:2)`
  iii) `print '(a/ (" ", 2a12))', ' titles', titles`
  iv) `print '(a/ (t1, 5en15.6))', ' power', power`
  v) `print '(a, 10l2)', ' flags', flags`
  vi) `print '(a, 5(" (", 2f6.1, ")"))', ' plane', plane`

2.
  i) `read (*, *) grid`
  `1.0 2.0 3.0 4.0 5.0 6.0 7.0 8.0 9.0 10.0`
  ii) `read (*, *) list(1:49:2)`
  `25*1`
  iii) `read (*, *) titles`
  `data transfer`
  iv) `read (*, *) power`
  `1.0 1.e-03`
  v) `read (*, *) flags`
  `t f t f t f f t f t`
  vi) `read (*, *) plane`
  `(0.0, 1.0),(2.3, 4)`

3. See *ch11.3.f90* on the OUP site.

4. See *ch11.4.f90* on the OUP site.

The program produces the output
```
 default:   1.00  2.00 +3.00  4.00
suppress:   1.00  2.00 +3.00  4.00
    plus: +1.00  2.00 +3.00  4.00
```

Chapter 13

1.
```
type cmplx (kind)
    integer       :: kind = kind(0.0)
    real, private :: r, theta
end type cmplx
```

2. See *ch13.2.f90* on the OUP site.

Chapter 14

1. See *ch14.1.f90* on the OUP site.

2. See *ch14.2.f90* on the OUP site.

Chapter 15

3. See *ch15.3.f90* on the OUP site.

4. For the complete solution see *ch15.4.f90* on the OUP site. Here is a description of how to proceed:

i) The full definition of the abstract type (appearing in the module mod_handle) is
```
type, abstract :: file_handle
contains
    procedure (manage_file), deferred, pass :: my_open
    procedure (manage_file), deferred, pass :: my_close
    procedure (write_file), deferred, pass :: send_data
    procedure (read_file), deferred, pass :: get_data
end type file_handle
```
It requires two additional abstract interfaces, the first of which is
```
subroutine write_file(handle, array)
    import :: file_handle
    class(file_handle), intent(in) :: handle
    class(*), intent(inout) :: array(:)
end subroutine write_file
```

ii) The following type extension can be used to specifically enable writing a real array to an input/output unit:
```
type, extends(file_handle) :: file_handle_real
    private
    integer :: unit = 0
contains
    procedure, pass :: my_open => open_real_fh
    procedure, pass :: my_close => close_real_fh
    procedure, pass :: send_data => write_real_array
    procedure, pass :: get_data => read_real_array
end type file_handle_real
```

The obligatory implementation of one of the type-bound procedures (see the full solution for the others) might be as follows:

```
subroutine write_real_array(handle, array)
   class(file_handle_real), intent(in) :: handle
   class(*), intent(inout) :: array(:)
   select type (array)
   type is (real)
      write(handle%unit, fmt='(3F12.5)') array
   class default
      stop 'ERROR(write_real_array): &
            &      incorrect type for argument "array".'
   end select
end subroutine write_real_array
```

For the constructor, see the full implementation.

iii) The program using these facilities can declare a polymorphic handle

```
class(file_handle), allocatable :: fh
```

read a string to indicate the desired type, and allocate fh using the constructor:

```
fh = file_handle(data_type)
```

The type-bound procedures can be invoked:

```
call fh%my_open()
call fh%send_data(arr)
call fh%get_data(arr2)
call fh%my_close()
```

iv) The module would need to have code added for further type extensions but no change to the main program would be needed.

Chapter 16

1. In order to avoid the need for recompiling the module with the abstract type defined in Chapter 15, Exercise 4, when additional type extensions are created, all specifications and module procedures associated with such extensions are moved to other modules. The original abstract type and its interfaces are contained in the file *ch16.1_1.f90*.

The extension type and its facilities are contained in the file *ch16.1_2.f90*.

The procedure for creating a polymorphic object must be accessible from mod_handle. Since it needs access to the extension type, its implementation must be moved to a submodule to avoid a circular module dependency. This is contained in the file *ch16.1_3.f90*.

Of the procedure create_handle, only the interface remains in the first of the above files. Finally, the program using the facilities is contained in the file *ch16.1_4.f90*.

The first three files must be compiled in order of their appearance. The fourth file only depends on the first file for compilation, but linking the program requires all four object files. All files are available on the OUP site.

Chapter 17

1. See *ch17.1_1.f90,*
 ch17.1_2.f90,
 ch17.1_3.f90, and
 ch17.1_4.f90 on the OUP site.

2. See *ch17.2.f90* on the OUP site.

3. See *ch17.3.f90* on the OUP site.

4. See *ch17.4.f90* on the OUP site.

5.

Code with bottlenecks	Code without bottlenecks
| ```
k = this_image()
if (k<=nz) then
    do i = 1, nx
        a(i, 1:ny) = b(1:ny, k)[i]
    end do
end if
``` | ```
k = this_image()
if (k<=nz) then
    do ii = 1, nx
        i = 1 + mod(k+ii, nx)
        a(i, 1:ny) = b(1:ny, k)[i]
    end do
end if
``` |

6. See *ch17.6.f90* on the OUP site.

Chapter 18

1. The procedure-internal coarray is established in the initial team, but during execution of the procedure in the team context, image indices are remapped to those of the current team. See *ch18.1.f90* on the OUP site.

2. The dummy argument of the shift procedure is established in the current team, even though the actual argument is established in its parent team. Its association is with $a(:,:)[i, i]$ of the parent team object for each image $i = 1, \ldots, nc$ of the current team, which has the team number nc and $2 \times (nc - 1)$ sibling teams which remain inactive. See *ch18.2.f90* on the OUP site.

Chapter 20

1. See *ch20.1.f90* on the OUP site.

2. See *ch20.2.f90* on the OUP site.

3. See *ch20.3.f90* on the OUP site.

Index